Soil Processes

Geography, Environment and Planning Series
Series editor: Neil Wrigley

Rebuilding Construction
Economic change in the British construction industry
Michael Ball

Regenerating the Inner City
Glasgow's experience
edited by David Donnison and Alan Middleton

The Geographer at Work
Peter Gould

Satellite Remote Sensing
An Introduction
Ray Harris

Health, Disease and Society
An introduction to medical geography
Kelvyn Jones and Graham Moon

Nuclear Power
Siting and safety
Stan Opensaw

Cities and Services
The geography of collective consumption
Steven Pinch

An Introduction to Political Geography
John R. Short

An Introduction to Urban Geography
John R. Short

Class and Space
The making of urban society
Nigel Thrift and Peter Williams

Soil Processes
A Systematic Approach

Sheila Ross

R

Routledge
London and New York

First published in 1989 by
Routledge
a division of Routledge, Chapman and Hall
11 New Fetter Lane, London EC4P 4EE

Published in the USA by
Routledge
a division of Routledge, Chapman and Hall, Inc.
29 West 35th Street, New York NY 10001

Typeset by Columns of Reading
Printed in Great Britain
at the University Press, Cambridge

Library of Congress Cataloging in Publication Data
Ross, Sheila, 1953–
   Soil processes: a systematic approach/Sheila Ross.
   p. cm.
   Bibliography: p.
   Includes index.
   1. Soil formation. 2. Soil science. 3. Soil management.
   I. Title.
   S592.2.R67 1988
   631.4—dc19

British Library Cataloguing in Publication Data
Ross, Sheila, 1953–
   Soil processes.
   1. Soils.
   I. Title.
   631.4

ISBN 0 415 00205 2

Contents

Preface

The idea behind this text is to introduce, in a systematic way, the fundamental processes of soil development and behaviour, together with the processes associated with, or altered by, man's use of soil. The text is intended for advanced undergraduate courses dealing with soil science, environmental science, agriculture or geography, and as a sourcebook for postgraduate researchers. An introductory knowledge of soil science and chemistry is assumed. While many introductory soil courses are typically descriptive, examining soil profiles and the spatial distribution of soil types, research in soil science tends to focus either on understanding the dynamics of *in situ* soil processes, such as podzolisation or cation/anion exchange reactions at particle surfaces, or to examine the applied aspects of man's use of soil, such as the fate of agrochemicals and how they influence existing soil processes. This text is designed to provide a bridge between these extremes of the soil science spectrum, by looking at genetic soil processes (soil formation and profile development), at chemical and biological soil processes (solute dynamics and the biology of the rooting zone), and processes influenced by soil management practices (cultivation, fertilizing or the application of pesticides). These subjects are dealt with as a progressive sequence of increasing complexity, since soils become more complex through time and with profile development. We cannot attempt to predict how soils will 'react' to anthropogenic influences such as agricultural management practices, or to the environmental inputs of acid rain or other forms of pollution, until we understand how they 'react' without these influences. One objective of this text is an attempt to disaggregate soil formative and dynamic processes and then to examine their response to external influences. It is clearly important to consider the timescales over which individual soil processes operate, ranging from 'instantaneous', or near instantaneous reactions such as cation exchange, to annual cycles such as agrochemical solute dynamics, and even the longer timescales associated with profile development processes. It is equally essential to consider the interaction of processes which operate at these different timescales.

No attempt is made here to deal with world soil processes: the approach throughout is almost exclusively temperate, but there is clearly a need for similar treatment of *in situ* processes and the effects of soil management practices in tropical and semi-arid soils. Throughout the text examples are drawn from a wide range of background subject matter, from geochemistry to microbiology, and from water chemistry to agriculture and forestry. An integrated view of these subjects is vital for soil scientists advising on current environmental issues.

No text of this nature could be written without stimulation and support from a number of sources. For the orientation and many of the ideas included in this book, I must thank the stimulating environment provided by undergraduate, postgraduate and academic colleagues at the Department of Forestry and Natural Resources, University of Edinburgh, the School of Biological and Environmental Sciences, University of Ulster, and the Department of Geography, University of Bristol. My particular thanks go to Malcolm Anderson, Neil Wrigley and Bridget Gregory who introduced me to the idiosyncrasies of the publishing world, to Pauline Kneale who patiently read early portions of the manuscript, to Simon Godden who produced the very high quality artwork, and to the librarian staff at Bristol University whose tolerance and kindness made reference searching a less troublesome task.

Sheila M. Ross
Bristol
June 1988

Acknowledgements

The author and publishers wish to thank the following for permission to reproduce copyright material:

Academic Press – Figs. 3.4, 6.5, 6.6, 8.9, 9.7, 9.10, 9.13; Agricultural Institute of Canada – Figs. 4.3, 8.7, 8.16; American Elsevier Publishing Co. – Figs. 1.2, 1.3; American Geophysical Union – Fig. 5.21; American Society of Agricultural Engineers – Fig. 7.15; American Society of Agronomy – Figs. 8.13, 8.19, 9.12, 9.15; Athlone Press – Fig. 6.16; Australian CSIRO – Fig. 7.16; Blackwell Scientific Publishers – Figs. 1.12, 2.3, 2.5, 2.6, 3.12, 4.8, 5.12, 5.20, 6.1, 6.3, 6.8, 6.9, 7.1, 7.7, 8.21; Cambridge University Press – Figs. 6.14, 6.15, 6.17, 6.18; Centre for Agricultural Publishing and Documentation (Wageningen) – Fig. 7.12; Chemical Society – Fig. 9.9; Elsevier Applied Scientific Publishers – Figs. 6.11, 9.14; Elsevier Scientific Publishers – Figs. 4.2, 7.8, 7.9; Harper and Row – Figs. 1.4, 1.5, 1.6; International Soil Tillage Research Organisation – Figs. 7.3, 7.6; John Wiley and Sons – Figs. 3.3, 3.7, 5.4, 5.6, 5.13, 8.14; Kluwer Academic Publishers – Figs. 6.1, 6.2, 6.12, 7.17; Longman Group – Figs. 5.7, 5.9; Louisiana State University – Fig. 3.6; McGraw-Hill – Fig. 6.4; Methuen – Fig. 7.13; National Research Council of Canada – Fig. 6.10; Oxford University Press – Figs. 1.1, 7.5; Pergamon Press – Figs. 1.9, 1.10, 1.11, 2.4, 2.9, 6.7, 9.16; Scientific American Inc. – Fig. 9.17; Soil Science Society of America – Figs. 2.10, 2.11, 3.5, 3.10, 5.5, 7.11, 8.3, 8.11, 8.15, 8.17, 8.18, 9.8, 9.11; Soil Survey of England and Wales – Fig. 7.14; Springer Verlag – Figs. 3.11, 5.8; VCH Verlagsgesellschaft – Fig. 8.10; Williams and Wilkins Co. – Table 2.9, Figs. 5.16, 5.19, 7.4, 8.5, 8.8.

Introductory Overview of the Study of Soil Processes

In England and Wales alone, around 300 different soil profile types have been mapped at the 1:250,000 scale, while in the United States, at least 11,000 different soil types have been classified. At smaller spatial and temporal scales, variation in soil properties has long been recognised, as testified by difficulties in (a) classifying and mapping soil units; (b) offering agronomic advice based on data from single soil profiles or from single growing seasons; or (c) obtaining accurate predictions from simulation models which incorporate soil properties, but no stochastic routine to handle their variability. A solution to these problems has frequently been to divide up large regions into smaller areas which are considered to be more homogeneous. While this approach may be standard soil survey practice, Webster (1985) suggests that it 'requires appreciation of the scale and abruptness change, the degree of correlation among different soil properties and of soil relations in the landscape'. Exhaustive knowledge of soil properties alone is not sufficient to overcome these problems. Future attempts to handle soil variability would benefit from a change in the way that variability is conceptualised. Instead of being treated as 'background noise' in the system, it should be regarded as the result of a number of complexly interlinked processes, operating at different spatial and temporal scales.

The idea of a hierarchy of interlinked soil processes operating at different spatial and temporal scales is not difficult to conceptualise. While soil processes do not operate strictly hierarchically, this text is divided into four parts that introduce soil processes which operate in a sequence of increasing complexity:

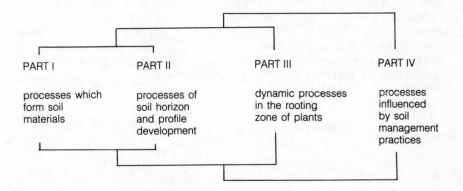

PART I

processes which
form soil
materials

PART II

processes of
soil horizon
and profile
development

PART III

dynamic processes
in the rooting
zone of plants

PART IV

processes
influenced
by soil
management
practices

Within this scheme, the basis of the hierarchy is not just scale, since all four levels of processes could operate at the same scale, in a single soil horizon, for example. The basis could be time, with the suggestion that formative and development processes operate over longest timescales (often tens of years), while dynamic and management-induced processes could be instantaneous, or at least very much faster (growing seasons). The proper hierarchical basis is system complexity. It is possible to visualise a sequence of soil processes, superimposed on each other, starting with the simplest soil forming system and ending with complex geomorphology and a mature ecosystem. At soil initiation, on a bare rock surface for example the formation of new soil material occurs without profile development processes operating. Only once a deep enough regolith has formed, or where soil is forming on a slope, can translocation of soil constituents occur. Even before plants colonise the site, ion exchange and solute transport can occur. Once plants colonise the site, regular inputs of litter provide energy substrates for soil micro-organisms. Roots can influence processes already in operation, such as solute dynamics, through water and nutrient uptake and the generation of moisture potential and solute concentration gradients. Superimposed on all of these are the effects of soil management practices.

To many soil scientists, soil processes tend to mean profile development processes, such as podzolisation and the component processes of eluviation, lessivage and illuviation. Classical studies of these processes have either collected exhaustive, *in situ* soil profile data on the distribution of iron (Fe) and aluminium (Al) hydroxyoxides, or have examined experimental laboratory hardware models, such as simple leaching columns. It is one approach to control inputs and to study outputs in such systems, then to invoke 'process' as a result: the classic black box approach. It is quite another approach to understand and use the component parts (component processes in our case) to construct a jigsaw puzzle of the whole: the grey or even white box approach, depending on our degree of knowledge about process controls. In theory, the main advantage of a component construction approach should be greater flexibility, transportability and wider applicability of the model. These are the main arguments for a systematic approach to the study of soil processes, albeit within the limited confines of a single text. The approach is to build up the picture of a mature soil, comprising very complex, but intimately linked, processes, which operate dynamically through time and space. Although we are still far from realistically simulating soil processes at the field or ecosystem scale, Kirkby (1985) has shown that it is certainly possible to simulate profile development in much simplified soil systems.

PART ONE

Processes which Form Soil Materials

Mineral weathering and organic matter decomposition are the two processes which contribute new inorganic and organic materials to soil profiles. Man mimics these imputs with additions of inorganic fertilisers and organic manures. Since the composition of the soil plays such an important role in determining dynamic processes such as water and solute transport, it is clearly vital to understand both the formative processes themselves and the major controls on direction and rate of operation. Using simple chemical equations to describe formative weathering and decomposition processes, the theory of chemical thermodynamics can be used to indicate the order in which a series of reactions might occur, given a set of specified environmental conditions. The rates at which new soil materials are produced by these processes can then be described using chemical kinetics.

The supply of new mineral materials to soil is limited to (a) *in situ* weathering of parent materials; or (b) imported minerals from other locations, though erosion or mass movements upslope, or through transport by ice, water or wind. Once on site, all inputs to soil through weathering processes are quite slow under temperate humid conditions, spanning years or tens of years. The supply of new organic material to soil is either periodic and annual (deciduous species) or regular and continuous (evergreen species). In stable soil profiles, organic inputs, depending on vegetation type, are frequently larger and more regular than mineral inputs. Over pedogenic timescales, the soil mineral fraction can be considered as finite and non-renewable, while the organic fraction can be considered as renewable and continuously recycling.

Inorganic and organic soil mass balances, together with an understanding of the chemical thermodynamics and kinetics of weathering and decomposition processes, can be used in soil profile modelling, as has been shown by Kirkby (1985). Although this approach simplifies the soil system and makes the important assumption that thermodynamic equilibrium is attained between solutes and the existing soil composition, it provides a useful basic tool for examining the development of different soil horizons.

CHAPTER 1

Production of Soil Material Through Weathering Processes

1.1 INTRODUCTION

Conventionally, processes of weathering and the production of soil fabric from the breakdown of rock minerals has been described by pedologists in terms of physical, chemical and biochemical mechanisms. Like many pedologists, Jackson and Sherman (1953) attempt to differentiate between the chemical weathering of rocks (geochemical processes) and the chemical weathering which takes place in the developing soil (pedochemical processes). This division appears to serve no useful or realistic purpose; the distinction seems to rely simply on the scale of operation of these processes. For this reason, fundamental geochemical principles and processes will be outlined in this chapter as a necessary prerequisite to understanding subsequent soil chemical processes and profile development. Physical weathering, on the other hand, is largely completed during the production of soil parent material and these processes will not be considered in any detail here. In this text, we shall be interested in the physical infrastructure of the soil simply as the mechanical control on the transport dynamics of soil water, solutes, air and colloid particles. The chemical composition of the soil mineral, organic and aqueous phases allows comprehension of mineral instability, solution and ionic exchange, inorganic and organic breakdown and the rates and occurrence of chemical equilibrium reactions. Biological processes act to influence the rates and equilibration of reactions through, for example, the provision of carbon as an energy source and through altering solubilities by chelation.

The geometry of the soil infrastructure of particles and pores is controlled by both physical breakup and also by chemical alteration such as solution (increase in pore/particle ratio) or precipitation (decrease in pore/particle ratio). The translocation of colloidal material from one zone in the soil (for example, eluviation, with resultant increase in pore/particle ratio) to deposition in another (for example, illuviation, with resultant decrease in pore/particle ratio) also causes change in pore geometry.

The mechanical breakup of rock minerals is effected by changes in temperature, pressure and humidity, with expansion and contraction acting primarily between component grains and along planes and joints of weakness. Decriptions of these physical processes are given in Ollier (1969) and Brunsden (1979). The physical breakup of rock minerals exposes a larger surface area to the effects of other weathering

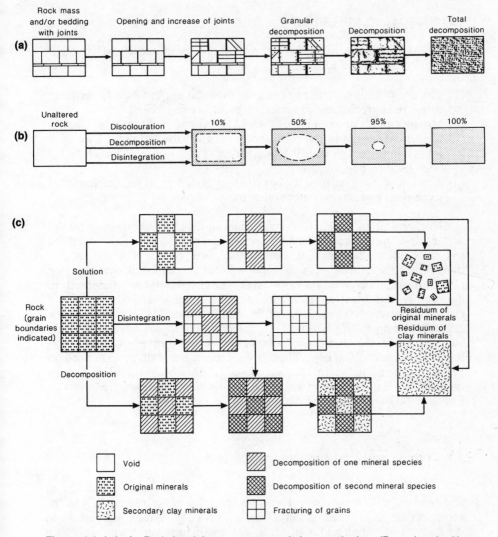

Figure 1.1 (a,b,c) Rock breakdown sequences during weathering. (Reproduced with permission from Selby, 1980, *Hillslope Materials and Processes*, © Oxford University Press

mechanisms, particularly chemical reactions. Fig. 1.1(a), (b) and (c) illustrate the sequence of possible changes occurring as a rock disintegrates. Fig. 1.1(b) represents the progressive alteration from the surface inwards of mineral particles and is a simple visual representation of the surface weathering model proposed by Luce, Bartlett and Parks (1972) to explain the dissolution of magnesium silicates. The model is discussed in detail in Section 1.4.1 as an example of a first order kinetic reaction.

1.2 CHEMICAL WEATHERING

Three major approaches have been adopted in the study of chemical weathering in the landscape:

(1) The use of chemical thermodynamics and, lately, kinetics, to generate theoretical predictions of mineral stability or instability.
(2) Measurements of amounts of soluble (transient) and insoluble (long residence) species in soil compared with unweathered rock. The related approach of devising weathering ratios or indices has also been used by pedologists.
(3) The use of mass balance studies to identify degree of weathering from catchment and streamwater solute budgets.

Geochemical approaches to understanding the chemical composition of surface and stream waters have generally ignored the complexities of the soil zone and have dealt with it on a 'black box' basis in the system. For this reason, the third approach above will not be discussed here. Waylen (1979) notes that only recently have geomorphologists attempted to develop a theoretical approach to rock weathering through the use of chemical thermodynamics, as outlined in method (1) above (see, for example, Curtis, 1976; Kirkby, 1977), while pedologists have exclusively adopted the more traditional composition comparative approach, outlined in method (2) above. These two approaches will be examined in detail. Particularly useful introductions to the background chemistry required for understanding chemical weathering reactions are given in Curtis (1975) and Bohn, McNeal and O'Connor (1985).

1.2.1   Ionic potential and mineral solubility

An initial estimation of the behaviour of mineral ions during weathering is given by their ionic potential which is the ratio of the valency (Z) to the ionic radius (r). Traditionally, ionic potential values are used to differentiate the behaviour of three groups of ions and their likely fate during weathering (Fig. 1.2). Ions can be divided into three groups on the basis of their Z/r ratios. The alkali cations (Na, K, Cs) with Z/r ratios < 3.0 have high ionic potentials and are very soluble and easily weathered. Ions of transition metals (Mn, Fe, Al) with Z/r values of 3–12 have intermediate ionic potentials and are not very soluble. They are weathered from soil minerals, but tend to precipitate as hydroxides. The smallest and most highly charged cations have lowest ionic potentials. They tend to form oxide ligands or oxyanions (Bohn, McNeal and O'Connor, 1985). Phosphates and silicates are least soluble of the group and tend to be retained in soil. A second useful indicator of the fate of mineral species during weathering is their solubility in relation to pH

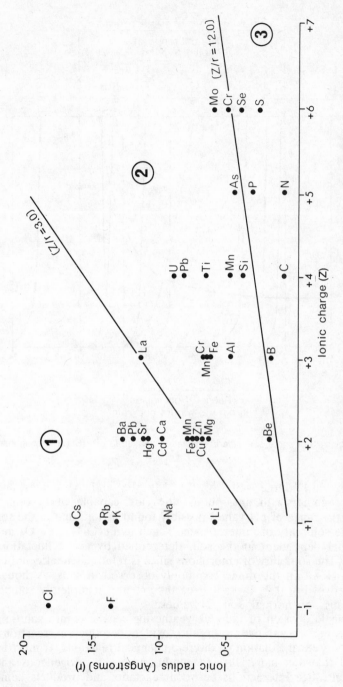

Figure 1.2 Ionic potentials of important ions found in soil. Group ①: alkali cations, Z/r <3; Group ②: transition metals, $Z/r = 3-12$; Group ③: highly charged cations, $Z/r > 12$. (Source: Loughnan, 1969)

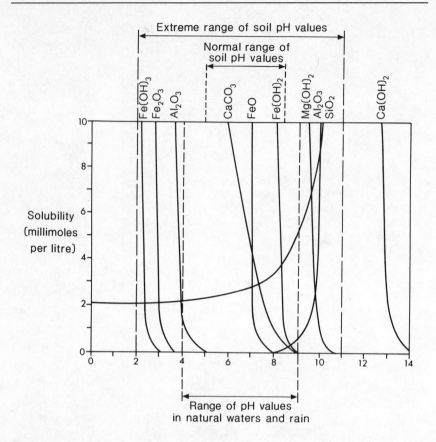

Figure 1.3 Solubility of some soil mineral species in relation to pH. (Source: Loughnan, 1969)

(Fig. 1.3). A fairly large suite of Ca, Mg, Fe and Al oxides and hydroxides, together with amorphous silica are capable of becoming dissolved in the range of pH values possibly found in soils and in the soil solution. The solubility of mineral species such as $Fe(OH)^+$, Fe_2O_3 and Al_2O_3 are pH dependent in the soil, determined by acidic fluctations below pH 4. The solubility of amorphous silica is relatively unaffected by changes in pH within the range commonly observed in soils. Although not illustrated in Fig. 1.3, the alkali earths are also completely soluble within the range of 'normal' soil pH values.

The simplistic division of mineral weathering reactions into solution, carbonation, hydrolysis, hydration and oxidation–reduction reactions, allows the theoretical isolation of discrete chemical reactions. It must be remembered that soil materials and conditions are heterogeneous and complex; multistage interactions between reactants and products being the rule rather than the exception. For this reason, many of the theoretical thermochemical examples examined later may be criticised as being too simplistic to be relevant to real field processes, but the individual components must be understood before comprehensive synthesis is possible.

1.2.2 Weathering processes

The *solution* effects of water on rock minerals depends in part on the dipole nature of water molecules. A positive bias around hydrogen nuclei and negative bias around the oxygen nucleus readily allows the dissolution of ionic solids, such as NaCl, to produce a stable ionic solution (Table 1.1 (a) and (b)). The hydrogen ion activity (pH) and associated anionic species present in the soil solution, such as NO_3^-, SO_4^{2-} and HCO_3^- from, for example, the

$$H_2O \; + CO_2 \rightleftharpoons \; H_2CO_3 \; \rightleftharpoons H^+ \; + HCO_3^- \qquad\qquad (1,1)$$

atmospheric   carbonic   bicarbonate
carbon   acid   ion
dioxide

equilibrium reaction, generate two ionic replacement processes: (a) carbonation, and (b) hydrolysis. During *carbonation* of Ca and Mg carbonates, soluble bicarbonates are formed (see, for example, Table 1.1 (c)). In aluminosilicate minerals, carbonation involves the exchange of H^+ for cations at the crystal surface with the formation of a soluble metal carbonate (Table 1.1 (d)). All soluble carbonate and bicarbonate products are removed by leaching. During *hydration*, minerals such as gypsum absorb water molecules (Table 1.1 (e) and (f)) and often expand in volume. In the case of micaceous clays, water molecules can penetrate between lattice layers, causing expansion and often hastening decomposition of the crystal. Different degrees of hydration of iron oxides are responsible for different soil colours; the more dehydrated versions such as haematite exhibit bright red colours in tropical soils (Table 1.1 (e)). During *hydrolysis*, cations are exchanged by H^+ at the crystal surface. Metallic cations and silica are released into solution. This is the main process by which soluble silica enters the soil solution and hence is transported out of the soil. Released cations and Si may recrystallise to form a new clay mineral such as kaolinite, or may take part in cation exchange (Table 1.1 (g)). Hydrolysis processes are speeded up in the presence of acid precipitation percolating through the soil (Table 1.1 (h)).

In soils where the availability of oxygen varies, such as gleys and peats, the chemical environment equilibrates between *reducing* (under water-logging conditions of low or zero oxygen availability) and *oxidising* (under aerated conditions of optimal oxygen availability) conditions. These *redox reactions* take the general form:

Table 1.1 Examples of common soil mineral weathering reactions

Solution

(a) $NaCl_{(s)}$ + H_2O ⇌ $Na^+_{(aq)}$ + $OH^-_{(aq)}$ + $H^+_{(aq)}$ + $Cl^-_{(aq)}$
   common salt

(b) $KNO_{3(s)}$ + H_2O ⇌ $K^+_{(aq)}$ + $OH^-_{(aq)}$ + $H^+_{(aq)}$ + $NO^-_{3(aq)}$
   saltpetre
   (fertilizer)

Carbonation

(c) $CaCO_{3(s)}$ + H_2O + CO_2 ⇌ $Ca^{2+}_{(aq)}$ + $2HCO^-_{3(aq)}$
   limestone   soluble calcium bicarbonate

(d) $K_2O.Al_2O_3.6\ SiO_{2(s)}$ + $2H_2O$ + CO_2 ⇌ $Al_2O_3.2SiO_2.2H_2O_{(s)}$ + $K_2CO_{3(aq)}$ + $4SiO_{2(s)}$
   orthoclase feldspar   kaolinite   potassium
                                                     carbonate

Hydration

(e) $Fe_2O_{3(s)}$ + H_2O ⇌ $2\ FeO.OH_{(s)}$
   haematite (red)   goethite (brown)

(f) $CaSO_{4(s)}$ + $2H_SO$ ⇌ $CaSO_4.2H_2O_{(s)}$
   gypsum   anhydrite

Hydrolysis

(g) $K_2O.Al_2O_3.6SiO_{2(s)}$ + $11H_2O$ → $Al_2O_3.2SiO_2.2H_2O_{(s)}$ + $4H_4SiO_{4(aq)}$ + $2K^+_{(aq)}$ + $2OH^-_{(aq)}$
   orthoclase feldspar   kaolinite   silicic acid

(h) In the presence of acid rain (eg H_2SO_4; HNO_3; H_2CO_3) generally:
   $aluminosilicate_{(s)}$ + H_2O + H_2SO_4 → clay $mineral_{(s)}$ + $(Ca^{2+}, K^+, Na^+)_{(aq)}$ + $OH^-_{(aq)}$ + $H_4SiO_{4(aq)}$ + $SO^{2-}_{4(aq)}$

(Table 1.1 cont...)

Oxidation/reduction

			Eh (mV) (at pH5)	(at pH7)
(i)	*Inorganic anions*			
	(i)	$NO_3^- + H^+ + 2e^- \rightleftharpoons NO_2^- + H_2O$ nitrate → nitrite	-70	-220
	(ii)	$SO_4^{2-} + 10H^+ + 2e^- \rightleftharpoons H_2S + 4H_2O$ sulphate → hydrogen sulphide	-70	-220
(j)	*Inorganic cations*			
	(i)	$Fe(OH)_3 + 3H^+ + e^- \rightleftharpoons Fe^{2+} + 3H_2O$ ferric → ferrous	170	-180
	(ii)	$MnO_2 + 4H^+ + 2e^- \rightleftharpoons Mn^{2+} + 2H_2O$ manganic → manganous	640	410
(k)	*Organic*			
	(i)	$CH_3COOH + 2H^+ + 2e^- \rightleftharpoons CH_3COH + H_2O$ acetic acid → acetaldehyde		
	(ii)	$C_2H_2(COOH) + 2H^+ + 2e^- \rightleftharpoons C_2H_2(COH) + 2H_2O$ fumaric acid → succinic acid		

$$n\beta A_{(ox)} + n\alpha B_{(red)} \rightleftharpoons n\beta A_{(red)} + n\alpha B_{(ox)} \qquad (1,2)$$

where: α = valency of species A

                β = valency of species B

and $n\alpha, n\beta$ = number moles of A and B respectively.

This overall equation is made up of two half reactions:

$$A_{(ox)} + \alpha e^- \rightleftharpoons A_{(red)} \qquad (1,3)$$
$$B_{(ox)} + \beta e^- \rightleftharpoons B_{(red)} \qquad (1,4)$$

each is called a *redox couple*. Redox reactions can be written either as a transfer of oxygen:

$$4Fe_3O_4 + O_2 \rightleftharpoons 6Fe_2O_3 \qquad (1,5)$$
$$4Fe^{2+} + O_2 + 4H^+ \rightleftharpoons 4Fe^{3+} + 2H_2O \qquad (1,6)$$

or as a transfer of electrons:

$$4Fe_3O_4 + 2H_2O \rightleftharpoons 6Fe_2O_3 + 4H^+ + 4e^- \qquad (1,7)$$
$$4Fe^{2+} \rightleftharpoons 4Fe^{3+} + 4e^- \qquad (1,8)$$

Conventionally, redox couples and their redox potentials (Section 3.2) are expressed as reduction half cells (that is, the 'backward' reaction in Equation (1,7). Although oxidation and reduction of organic soil materials cannot be considered as weathering reactions, two organic examples are given along with inorganic reduction half cells in Table 1.2 (i), (j) and (h).

Two important questions can be applied to the study of chemical weathering reactions:

(i) is a mineral liable to weather under the influence of earth surface conditions and if so, in what direction will the weathering reaction proceed? This question can be answered by the application of *chemical thermodynamics*.

(ii) at what rate will the weathering reaction proceed? This question can be answered by the application of *chemical kinetics*.

1.3   SIMPLE CHEMICAL THERMODYNAMICS OF WEATHERING REACTIONS

Chemical thermodynamics calculate equilibrium reactions and help to identify stable or unstable minerals, given a set of environmental conditions. The most useful tool for predicting whether or not a weathering reaction will occur is to compare the values of Gibbs free energy of formation, ΔG_f° (in kcal mol^{-1}) for reactants and products, to obtain a ΔG_r° for the weathering reaction. For the simplest weathering reaction:

$$A_{(s)} + H_2O_{(l)} \rightleftharpoons B_{(s)} + C_{(aq)} \qquad (1,9)$$

reactant   water   product   soluble
mineral   mineral   product

then:

$$\Delta G_r^\circ = \Sigma\, \Delta G_f^\circ\, [B, C] - \Sigma\, \Delta G_f^\circ\, [A, H_2O] \qquad (1,10)$$

change in Gibbs   sum of Gibbs free   sum of Gibbs free
free energy for   energies for all the   energies for all the
the reaction   *products* in their   *reactants* in their
                   standard states   standard states

When values of ΔG_r° are negative, then ΔG_f° (reactants) $> \Delta G_f^\circ$ (products) and the mineral, A in this case, is unstable and the reaction is likely to take place in the forward direction: 'The driving force of a chemical reaction is the tendency of the free energy of the system to decline until, at equilibrium, the sum of the free energies of the products equals that of the remaining reactants' (Ponamperuma, 1972). ΔG_f° values for solid, liquid and gaseous species can be obtained either directly from empirical values (for example, those listed by Garrels and Christ, 1965) or can be estimated from:

$$\Delta G = \Delta H - T\Delta S \qquad (1,11)$$

where: ΔH = enthalpy (energy input to the reaction)
         ΔS = entropy (unavailable energy or energy lost in
                 rearranging ions and molecules. Entropy diminishes
                 with temperature)
         T = absolute temperature (° Kelvin)

Values of ΔH and ΔS are also available in summarised form (see, for example, Garrels and Christ, 1965). Values of ΔG_f° under set experimental conditions can be determined from:

$$\Delta G = V\Delta P + S\Delta T + F(\text{composition; electrical}, \qquad (1,12)$$
$$\text{gravitational potentials})$$

where: V = volume
         ΔP = change in pressure
         ΔT = change in temperature

By experimentally maintaining P and T constant, $V\Delta T$ and $S\Delta T$ are both zero and ΔG becomes entirely dependent on changes in composition, gravity and electrical potentials. In chemical thermodynamics, changes in composition are of most importance. The gravitational potential, the 'water head' of soil physics, arises due to differences in elevation and has, in the past, been ignored in soil chemical reactions. The electrical potential is an important consideration near charged particle surfaces. To simplify calculations, soil scientists have considered equilibrium reactions to occur in the soil solution phase where ΔG at constant T and P is determined solely by the composition and concentration of the solution (Bohn, McNeal and O'Connor, 1985).

Examples of the use of ΔG calculations in the study of mineral weathering reactions are given in Table 1.2 for the weathering of anorthite (a calcium plagioclase feldspar) and albite (a sodium feldspar) to kaolinite (a secondary aluminosilicate clay mineral). These weathering equations are written very simply as hydration reactions. In a mixed soil mineral substrate containing anorthite and albite, we can use calculated ΔG_f° values for products and reactants to predict which mineral will be the first to weather to kaolinite. These calculations are outlined in Table 1.2. Since the positive ΔG_r° for anorthite weathering > the positive ΔG_r° for albite weathering, we might conclude that under uniform environmental conditions, the production of kaolinite from albite decay will occur preferentially to kaolinite production from the decay of anorthite. Interestingly, both reactions, written as simple hydration equations, appear to produce alkaline soil solutions through the release of OH^- ions. Since soil solutions and drainage waters are more commonly neutral to acid, the relevance of such simple calculations for weathering reactions remains a little doubtful. Curtis (1975) suggests that it would be more realistic to formulate weathering equations to include the influence of acid inputs in rainfall. Carbonic acid, formed by the dissolution of atmospheric carbon dioxide in rainwater, and dilute sulphuric or nitric acids, formed by the dissolution of sulphurous and nitrous oxide pollutants in the atmosphere, are likely contenders for inclusion in weathering equations. ΔG_r° Calculations are given in Table 1.2 for a carbonic acid influence on the hydration of anorthite and albite. With an acid influence, both anorthite and albite weather spontaneously at standard temperature and pressure. Since the [albite $\rightarrow$ kaolinite + quartz] reaction yields a larger negative ΔG_r° value than the [anorthite $\rightarrow$ kaolinite + quartz] reaction, albite will weather preferentially in soil systems, given the more realistic carbonic acid input. Calculations of this type allow us to speculate on the influence of acid precipitation on weathering rates for rocks and soil minerals. An increasing volume of evidence shows that the soil leaching rate of basic cations such as Ca^{2+}, Mg^{2+}, K^+ and Al^{3+} is very much increased with acid precipitation, although rarely have weathering reactions and rates been studied directly. In the River Elbe catchment, Paces (1985) calculated that the rate of chemical weathering of gneisses was doubled with a ten-fold increase in the sulphur dioxide content of rainfall.

1.3.1 Chemical equilibria

In any reaction such as that of Equation **(1,1)**, the driving force to the right or to the left can be related to the concentrations of the reactants and the products. When the rates to the right and the left are equal, the system is said to be in *chemical equilibrium*. The *law of mass action* allows the prediction of the thermodynamic equilibrium constant (K) for any pure, simple, chemical reaction. Equilibrium constants can be calculated either in terms of solute activities $(K°)$ or in terms of solute concentrations (K^c). The activity of a solute can be converted to a concentration by means of an activity coefficient:

$$a = \gamma M \qquad (1,13)$$
where: γ = activity coefficient
M = molarity

The activity coefficient is defined so that:

$$\lim_{M \to 0} \gamma = 1$$

as the solution molarity approaches zero, γ approaches a limit of unity. So, for very low concentration aqueous solutions, the activity of a solute ion can be regarded as the effective solution concentration of that ion. This is the common assumption in soil chemical equilibria. Thus for the hypothetical reaction:

$$aA + bB \rightleftharpoons cC + dD \qquad (1,14)$$

the thermodynamic equilibrium constant (K) is usually written as an activity constant $(K°)$:

$$K° = \frac{[C]^c[D]^d}{[A]^a[B]^b} \qquad (1,15)$$

where square brackets represent activities in very dilute solutions. In words, Equation (1,7) describes the equilibrium reaction between a mol of reactant A plus b mol of reactant B with c mol of product c plus d mol of product D. $K°$ for any reaction of this type varies with temperature, but for any given temperature, is independent of total pressure. In soil solutions of higher ionic strength, an extended form of the Debye–Huckel equation is used to account for the effect of ion charge and concentration on the activity coefficient of an individual ion (Stumm and Morgan, 1970):

$$\log_{10} \gamma_i = - AZ_i^2 \frac{I^{1/2}}{1 + Ba_i I^{1/2}} \qquad (1,16)$$

where: γ_i = the individual ion's activity coefficient
A = constant (0.511 for aqueous solutions at 25 °C)
B = constant (0.33 for aqueous solutions at 25 °C)
Z_i = the ionic charge
I = the ionic strength
a_i = individual ion parameter, determined experimentally ('distance to closest approach' measures, usually given in A and ranging from 2.5 to 11.0)

Table 1.2 Example of chemical weathering thermodynamics (all ΔG^0_f values in kJ mol^{-1} at 25°C and 100 kPa)

(1) Weathering of anorthite

$$\underset{\text{anorthite}}{\text{Ca Al}_2 \text{Si}_2 \text{O}_{8(s)}} + 3\text{H}_2\text{O} \rightarrow \underset{\text{kaolinite}}{\text{Al}_2 \text{Si}_2 \text{O}_5 (\text{OH})_{4(s)}} + \text{Ca}^{2+} + 2\text{OH}^-$$

(a) Free energy

$$\Delta G^\circ_r = \Sigma \Delta G^\circ_f \text{[kaolinite + Ca}^{2+} + 2(\text{OH}^-)] \rightarrow \Sigma \Delta G^\circ_f \text{[anorthite} + 3(\text{H}_2\text{O})]$$
$$= [-905.61^\dagger +(-132.18)^* +2(-37.6)]^* - [-954.3^\dagger + 3(-56.69)]^*$$
$$= (-1112.99) - (-1124.37)$$
$$\underline{\Delta G^\circ_r = + 11.38}$$

(b) Equilibrium constant

$$K^\circ = \frac{\text{[kaolinite] [Ca}^{2+}] \text{[OH]}^2}{\text{[anorthite] [H}_2\text{O]}^3}$$

which reduces to:

$$K^\circ = \text{Ca}^{2+} \text{OH}^2$$

or $\underline{\log K^\circ = \log [\text{Ca}^{2+}] + 2 \log [\text{OH}]}$

since the activities of (a) solid phases (kaolinite and anorthite in this example) and (b) water are assumed to be unity

(c) Combining

Since $\Delta G^\circ_r = -RT \, 1_n K = -1.364 \log K$

$$\log K^\circ = \frac{11.38}{-1.364} = -8.34$$

Thus $\underline{\log [\text{Ca}^{2+}] = -8.34 - 2 \log [\text{OH}]}$

(2) Weathering of albite

$$\underset{\text{albite}}{2 \text{Na Al Si}_3 \text{O}_{8(s)}} + 3\text{H}_2\text{O} \rightarrow \underset{\text{kaolinite}}{\text{Al}_2 \text{Si}_2 \text{O}_5 (\text{OH})_{4(s)}} + 4\text{SiO}_2 + 2\text{Na}^+ + 2\text{OH}^-$$

(a) Free energy

$$\Delta G^\circ_r = [-905.61^\dagger + 4(-192.4)^* + 2(-62.59)^* + 2(-37.6)^*] - [2(-886.31)^\dagger + 3(-56.69)^*]$$
$$= (-1875.59)-(-1886.00)$$
$$\underline{\Delta G^\circ_r = + 10.41}$$

(Table 1.2 cont . . .)

(b) Equilibrium constant

$$K° = \frac{[kaolinite]\,[SiO_2]^4\,[Na^+]^2\,[OH]^2}{[albite]^2\,[H_2O]^3}$$

which reduces to:

$$K° = [SiO_2]^4\,[Na^+]^2\,[OH]^2$$

or $\log K° = 4 \log [SiO_2] + 2 \log [Na^+] + 2 \log [OH^-]$

Since the activities of solid phases and water are assumed to be unity

(c) Combining

Since $\Delta G_r° = -RT \ln K = -1.364 \log K$,

$$\log K° = \frac{10.41}{-1.364} = -7.63$$

Thus $\underline{\log [Na^+] = -7.63 - 2 \log [SiO_2] - \log [OH^-]}$

Rewriting Equations (1) and (2) to account realistically for the generation of anions and their subsequent involvement in the weathering processes[**]:

(3) $CaAl_2Si_2O_{8(s)} + 3H_2O + CO_{2(aq)} \rightarrow Al_2Si_2O_5\,(OH)_{4(s)} + Ca^{2+} + 2HCO_3^-$
anorthite — kaolinite

yields a $\Delta G_r° = -101.73$
and $\log [Ca^{2+}] = 74.58 - 2 \log [OH^-]$

(4) $2Na\,Al\,Si_3\,O_{8(s)} + 3\,H_2O + CO_2 \rightarrow Al_2\,Si_2\,O_5\,(OH)_{(s)} + 4SiO_2 + 2\,Na^+ + 2HCO_3^-$
albite — kaolinite

yields a $\Delta G_r° = -46.01$
and $\log [Na^+] = 33.73 - 2 \log [SiO_2] - \log [OH^-]$

[*] Garrels and Christ (1965)
[†] Helgeson, et al. (1978)
[**] Curtis (1975)

Bohn, McNeal and O'Connor (1985) accept that Equation (1,9) covers the solution concentrations likely to occur in soils and waters (ionic strengths up to 10^{-1} M), but for simplicity, these adjustments are rarely made.

Strictly speaking, soils are always non-equilibrium systems. This is because of the complexity of the chemical system and because pore water retention times are often too short for equilibration to occur. Thus, many so-called equilibrium soil weathering reactions should really be described by a reaction coefficient (K_r) rather than an equilibrium constant (K°). Using the notation of Goulding (1983), the generalised equation for the exchange of cations at the surface of a clay colloid would be:

$$\beta \; \boxed{}\!-\!A_\alpha + \alpha B^{\beta+} \rightleftharpoons \alpha \; \boxed{}\!-\!B_\beta + \beta A^{\alpha+} \tag{1,17}$$

and the equilibrium coefficient would be:

$$K_r = \frac{[\boxed{}\!-\!B_\beta]^\alpha [A^{\alpha+}]^\beta}{[\boxed{}\!-\!A_\alpha]^\beta [B^{\beta+}]^\alpha} \tag{1,18}$$

where A is the adsorbed cation of valency α and B is the solution cation of valency β. Using the exchange, at the colloid surface, of ammonium nitrogen and calcium, the equation becomes:

$$2\boxed{}\!-\!NH_4 + Ca^{2+} \rightleftharpoons \boxed{}\!-\!Ca + 2NH_4^+ \tag{1,19}$$

with a reaction coefficient described by:

$$K_r = \frac{[\boxed{}\!-\!Ca]\,[NH_4^+]^2}{[\boxed{}\!-\!NH_4]^2[Ca^{2+}]} \tag{1,20}$$

Van Breemen and Brinkman (1978) usefully introduced the concept of *partial equilibrium* for soil reactions such as those outlined above. In cation exchange reactions at clay surfaces, the longterm effect of slow alteration of the clay mineral, which may be thermodynamically unstable, is not considered. The structure of the clay mineral is assumed to be stable for the duration of cation exchange. Thus, partial equilibria would be used whenever the reaction rates of minerals not in equilibrium with the system are too low to attain overall equilibrium within the timespan considered.

The relationship between K° for a reaction and its ΔG_r° is given by:

$$\Delta G_r^\circ = - RT \ln K^\circ \tag{1,21}$$

where: R = universal gas constant
T = absolute temperature (° Kelvin)

at 25°C, Equation (1,12) reduces to:

Table 1.3   Three formulations for the weathering of kaolinite in soil (Marshall, 1977)

(1) $Al_2 Si_2O_5 OH_{4(s)} + 6H^+ \rightleftharpoons 2Al^{3+} + 2H_4 SiO_4 + H_2O$

$$K_{(1)} = \frac{[Al^{3+}] [H_4 SiO_4]}{[H^+]^3}$$

$\log K_{(1)} = \log [Al^{3+}] + \log [H_4 SiO_4] + 3pH$

(2) $Al_2 Si_2O_5 OH_{4(s)} + 4H^+ + H_2O \rightleftharpoons 2\, Al\, OH^{2+} 2H_4 SiO_4$

$$K_{(2)} = \frac{[Al\, (OH)^{2+}] [H_4 SiO_4]}{[H^+]^2}$$

$\log K_{(2)} = \log [Al\, (OH)^{2+}] + \log [H_4 SiO_4] + 2pH$

(3) $Al_2 Si_2O_5 (OH)_{4(s)} + 2H^+ + 3H_2O \rightleftharpoons 2\, Al\, (OH)_2 + 2H_4 SiO_4$

$$K_{(3)} = \frac{[Al\, (OH)_2]^2 [H_4 SiO_4]^2}{[H^+]^2}$$

$\log K_{(3)} = 2\log [Al\, (OH)_2] + 2\log [H_4 SiO_4] + 2pH$

$$\Delta G_r^\circ = -\,1.364 \log K^\circ \qquad\qquad (1,22)$$

or, more usefully,

$$\log K^\circ = -\,\frac{\Delta G_r^\circ}{1.364} \qquad\qquad (1,23)$$

This equation is particularly useful for calculating equilibrium constants for chemical weathering reactions in which the thermodynamic data are difficult to measure by conventional methods (Lindsay, 1979). In some weathering reactions, the equilibrium constant is expressed as the pK (or $-\log K$).

Apart from including more realistic environmental acidity in weathering reactions, the actual formulation of the weathering equation is of paramount importance in the estimation of both ΔG_r° and the equilibrium constants or coefficients for the reaction. This problem was well illustrated by Marshall (1977), whose three expressions for the equilibrium coefficient of kaolinite weathering are given in Table 1.3, each resulting in a different stability plot between Al^{3+}, $Al(OH)^{2+}$ or $Al(OH)_2^+$ and H_4SiO_4. A formulation of this type (Equation **1,36**) is used on page 00 to calculate the K° and ΔG_r° for the weathering of kaolinite to gibbsite.

1.3.2 Stability fields of soil minerals

The stability of mineral species in soils depends largely on environmental variables such as pH, Eh and the partial pressure of gases such as O_2 and CO_2. These parameters can be used singly or in combination to produce stability diagrams such as those for haematite and magnetite in equilibrium with ferric and ferrous ions depicted in Fig. 1.4 for the P_{O_2}–P_{CO_2} relationship and in Fig 1.5 for the Eh–pH relationship.

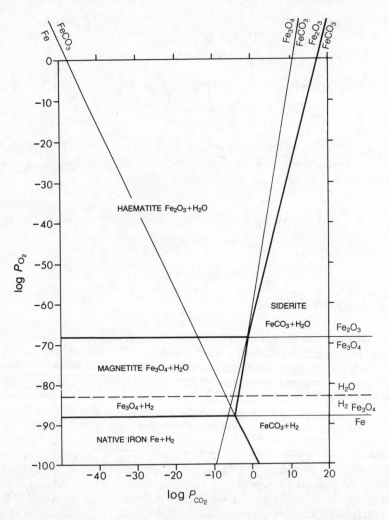

Figure 1.4 Stability relations of some iron compounds as a function of P_{O_2} and P_{CO_2} at 25°C and a few atmospheres of pressure. (Source: Garrels and Christ, 1965)

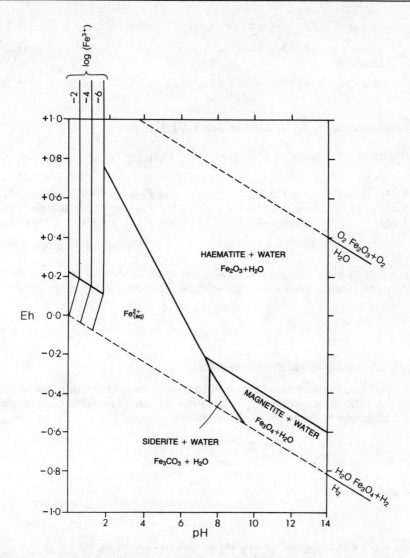

Figure 1.5 Stability of haematite, magnetite and siderite and activity of ferric iron in aqueous solution containing total dissolved carbonate species of 10^{-2} M at 25°C and 1 atmosphere total pressure. (Source: Garrels and Christ, 1965)

$P_{O_2}-P_{CO_2}$ Stability diagram for iron oxides

In Fig. 1.4, the equations required to specify $P_{O_2}-P_{CO_2}$ stability boundaries are fairly simple and are outlined below. Derivation of $P_{O_2}-P_{CO_2}$ stability equations:

$$2Fe_{(c)} + 2CO_{2_{(g)}} + O_2 \rightleftharpoons 2Fe\,CO_{3_{(c)}} \tag{1.24}$$

$$\underset{\text{magnetite}}{2Fe_3O_{4_{(c)}}} + 6CO_{2_{(g)}} \rightleftharpoons \underset{\text{siderite}}{3FeCO_{3_{(c)}}} + O_{2_{(g)}} \tag{1,25}$$

$$\underset{\text{haematite}}{2Fe_2O_{3_{(c)}}} + 4CO_{2_{(g)}} \rightleftharpoons \underset{\text{siderite}}{4Fe\,CO_3} + O_{2_{(g)}} \tag{1,26}$$

using calculations as outlined in Section 1.3 gives:

Equation	Stability boundary	Defining equations in terms of P_{O_2} and P_{CO_2}
(1,24)	Fe–FeCO$_3$	$\log P_{CO_2} = -49.0 - 0.5\log P_{O_2}$
(1,25)	Fe$_3$O$_4$–FeCO$_3$	$\log P_{CO_2} = 10.3 + 0.166\log P_{O_2}$
(1,26)	Fe$_2$O$_3$–FeCO$_3$	$\log P_{CO_2} = 15.9 + 0.25\log P_{O_2}$

(Garrels and Christ, 1965)

In a plot of $\log P_{O_2}$ against $\log P_{CO_2}$, the boundaries between these three pairs are straight lines of slopes -0.5, $+0.166$ and $+0.25$ respectively.

pH–Eh stability diagram for iron oxides

In Fig. 1.5, the equations required to specify Eh–pH stability boundaries are a little more complex: Derivation of Eh–pH stability equations:
(1) Reaction of ferric ion with haematite:

$$\underset{\text{haematite}}{Fe_2O_{3_{(c)}}} + 6H^+_{(aq)} \rightleftharpoons \underset{\text{ferric ion}}{2Fe^{3+}_{(aq)}} + 3H_2O_{(l)} \tag{1,27}$$

calculations give

$$\log [Fe^{3+}] = -\,0.72 - 3pH$$

So log of the activity of $[Fe^{3+}]$ in equilibrium with Fe_2O_3 is a linear function of pH. If pH is stipulated, $[Fe^{3+}]$ is fixed and vice versa. The common method of calculation is to assume a value for $[Fe^{3+}]$ and to solve for pH. Because Eh is not involved, such contours of $[Fe^{3+}]$ will lie parallel to the Eh axis in Fig. 1.5.
(2) Reaction of ferric ion with magnetite:

$$\underset{\text{magnetite}}{Fe_3O_{4_{(c)}}} + 8H^+_{(aq)} \rightleftharpoons \underset{\text{ferric ion}}{3Fe^{3+}_{(aq)}} + 4H_2O_{(l)} + e^- \tag{1,28}$$

since an oxidation is involved, this half cell reaction is calculated using the method outlined in Section 3.2, from

$$Eh = E° + \frac{0.059}{1} \log \frac{[Fe^{3+}]^3}{[H^+]} \tag{1,29}$$

and, since $E° = \Delta G°_r/n$, this becomes:

$$Eh = 0.337 + 0.177 \log[Fe^{3+}] + 0.427pH \tag{1,30}$$

Stability plots can be drawn by substituting values into this equation. For example, for $\log [Fe^{3+}]-6$ and $pH = 1.76$ (which are solutions to Equation (1,30) above, gives:

$$Eh = 0.377 + 0.177 \times -6 + 0.472 \times 1.76 = 0.106 \tag{1,31}$$

Contours in the magnetite field are drawn with a slope of 0.472 volts per pH unit. Other equations required to produce Fig. 1.5 are given in Garrels and Christ (1965, p. 189). The composite plot in Fig. 1.5 can also be drawn by writing equations for reacting ions and solving for the Eh and pH conditions at which they are equal:

$$Fe^{2+}_{(aq)} \rightleftharpoons Fe^{3+}_{(aq)} + e^- \tag{1,32}$$

$$Eh = E° + 0.059 \log \frac{[Fe^{3+}]}{[Fe^{2+}]} \tag{1,33}$$

where $Fe^{2+} = Fe^{3+}$, $Eh = E°$, so the boundary between the ions is at E^0 or 0.771 volt for the half cell.

$$Fe^{2+}_{(aq)} + H_2O_{(l)} \rightleftharpoons Fe(OH)^+_{(aq)} + H^+_{(aq)} \tag{1,34}$$

$$\log \frac{[Fe(OH)^+]}{[Fe^{2+}]} = \log K + pH \tag{1,35}$$

When the ionic activities are equal, $pH = -\log K$. Here, the ions become equal at such a high pH that neither is $> 10^{-6}$ – so $Fe(OH)^+$ does not become a dominant species within the field of 'solubility' designated by Garrels and Christ (1965).

One of the other major controls on mineral stability is the activity of dissolved ions or molecules in solution. Stability diagrams of this type have been widely used in stability studies of the aluminosilicate clay minerals. An example of such a plot, with calculations, is given in Fig. 1.6 for gibbsite stability. If an equilibrium exists between haematite, gibbsite, dissolved silica and water during weathering, then:

$$\underset{\text{kaolinite}}{H_4Al_2Si_2O_{9_{(c)}}} + 5H_2O_{(l)} \rightleftharpoons \underset{\text{gibbsite}}{Al_2O_3.3H_2O_{(c)}} + \underset{\substack{\text{dissolved} \\ \text{silica}}}{2H_4SiO_{4_{(aq)}}} \tag{1,36}$$

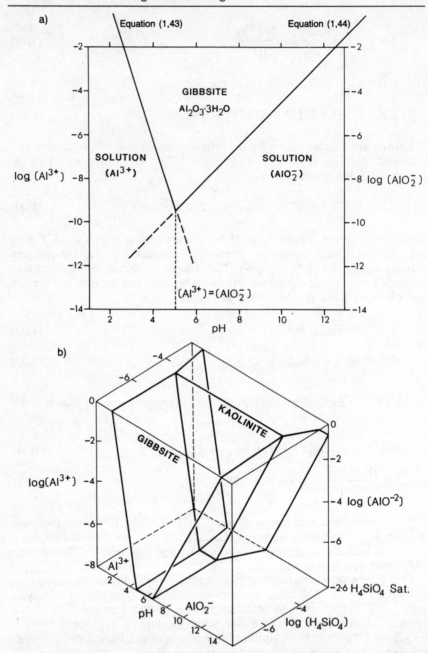

Figure 1.6 (a) Stability of gibbsite expressed in terms of pH and activities of aluminium ion and aluminate ion at 25°C and 1 atmosphere pressure (equation numbers refer to text). **(b)** Gibbsite–kaolinite stability fields expressed in terms of pH and activities of the dissociation products: Al^{3+}, AlO_2^- and H_4SiO_4, at 25°C and 1 atmosphere pressure. (Source: Garrels and Christ, 1965)

$K^0 = [H_4SiO_4]^2$, and, solving ΔG_r° according to: **(1,37)**

$\Delta G_r^\circ = 12.7 \text{ kcal} = -1.364 \log[H_4SiO_4]^2$

$\log[H_4SiO_4]^2 = \log K^\circ = -9.31$ **(1,38)**

$\log[H_4SiO_4] = \log K^{\circ\frac{1}{2}} = -4.65$ **(1,39)**

$K^{\circ\frac{1}{2}} = [H_4SiO_4] = 10^{-4.65}$, so $K^\circ = 10^{-2.325} = 2.1135 \times 10^{-2}$ mol kg^{-1} and, since the molecular weight of H_4SiO_4 in grams is 96, and molality = [ppm(μg ml^{-1})/g molecular weight] $\times 10^{-3}$, then:

$$K^\circ = 2.0289 \ \mu\text{g ml}^{-1} \text{ of } H_4SiO_4 \tag{1,40}$$

so, equilibrium between kaolinite and gibbsite is attained at a fixed value of the activity of dissolved silica of about 2 μg ml^{-1}. Values of dissolved silica < 2 μg ml^{-1} allow the incongruent dissolution (formation of an insoluble product) of kaolinite to leave a residuum of gibbsite, with silica dissolved in solution. If gibbsite is in equilibrium with pore water, it is in equilibrium with various dissociation product:

$$Al_2O_3.3H_2O_{(c)} + 6H^+_{(aq)} \rightleftharpoons 2Al^{3+}_{(aq)} + 6H_2O_{(l)} \tag{1,41}$$

$$Al_2O_3.3H_2O_{(c)} + 2H^+_{(aq)} \rightleftharpoons 2AlO^-_{2(aq)} + 2H_2O_{(l)} \tag{1,42}$$

Calculating K° and ΔG_r°, gives:

$$\log K_1 = 5.7 = \log[Al^{3+}] + 3pH \tag{1,43}$$

and

$$\log K_2 = -14.6 = \log[AlO^-_2] - pH \tag{1,44}$$

Figure 1.6(a) plots Equations **(1,43)** and **(1,44)**. Although not an accurate solubility diagram, the implication is that aluminium content in dilute solutions must be very low to prevent precipitation of aluminium under most pH conditions. Combining these results with equilibration reactions calculated for kaolinite stability, Garrels and Christ (1965) produced Fig. 1.6(b).

Mineral stability plots based on the activities of ions and molecules in bathing solutions have been calculated for an extremely wide range of soil minerals and chemical phases. The most comprehensive range of examples is given by Lindsay (1979).

1.4 SIMPLE CHEMICAL KINETICS OF WEATHERING REACTIONS

Since weathering reactions in soils proceed at different rates, it is the rule rather than the exception to observe chemical reactions which have not reached completion. In any soil medium, weathering will start with unequal rates of removal and a trend towards enrichment of minerals in proportion to their residence times, or their degree of chemical stability, as indicated by thermochemical data (Curtis, 1975). Curtis suggests that, in the soil profile, a steady state will be reached in which the input of fragmented parent mineral to the base of the profile will equal the net removal of soluble breakdown products from the whole profile. The slowest dissolving minerals will then be present in greatest amounts. Chemical kinetics allows the calculation of reaction rates and describes their dependence on reactant concentrations.

In soils, chemical kinetics have been widely used in the study of biochemical reactions such as organic matter decomposition and nutrient, particularly nitrogen, mineralisation, than in studies of chemical weathering. Kinetics of such biochemical reactions is introduced in Chapter 2.

Mercado and Billings (1975) identify five main controls on the overall rate of surface weathering reactions (Fig. 1.7). Rates of processes (a) and (e) are transport controlled while rates of processes (b), (c) and (d) are controlled by chemical kinetics. Any one of processes (a) to (e) may be rate determining for the whole system.

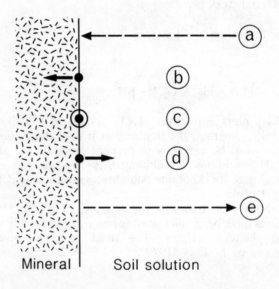

Mineral | Soil solution

Figure 1.7 Processes controlling the overall rate of surface weathering reactions (a) transport of reactant(s) from bulk solution to mineral/solution interface; (b) adsorption of reactant(s) on the surface; (c) chemical reaction on the surface; (d) desorption of product(s) from the surface; (e) transport of the product(s) from the interface to the bulk solution

1.4.1 The kinetic order of chemical reactions

The equation:

$$\alpha A + \beta B \rightarrow \gamma C \tag{1,45}$$

describes the reaction of αmol of A with βmol of B to produce γmol of product C. The reaction is illustrated in Fig. 1.8. The rate of change in concentration of reactants A and B with time is given by: $-\dfrac{d[A]}{dt}$ and $-\dfrac{d[B]}{dt}$ respectively (where square brackets represent concentrations). These values are negative since both A and B diminish through time. The rate of change in concentration of product C is given by $\dfrac{d[C]}{dt}$ since it accumulates through time. For Equation (1,45), α mol of A are used for every β mol of B in the production of γ mol of C. Thus, the rate of change of [A], [B] and [C] are

$$-\frac{1}{\alpha}\frac{d[A]}{dt} = -\frac{1}{\beta}\frac{d[B]}{dt} = +\frac{1}{\gamma}\frac{d[C]}{dt} \tag{1,46}$$

The mathematical expression relating rate of reaction to concentration is called the *differential rate law*. In Equation (1,45), if $\alpha = 2$, $\beta = 3$ and $\gamma = 1$, then the differential rate law may take the form:

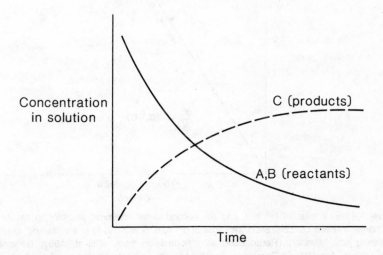

Figure 1.8 Rate of change in concentration of reactants, A and B and product C, during Equation (1, 45).

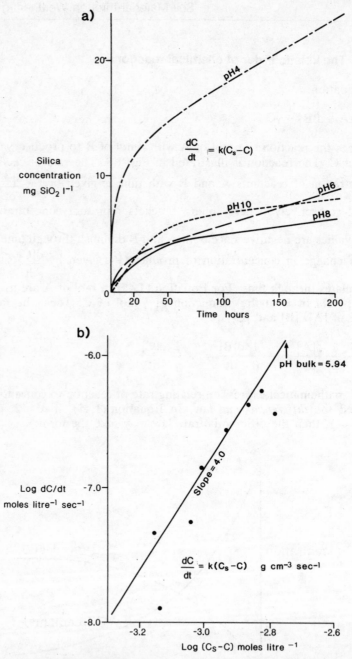

Figure 1.9 Examples of (a) first and (b) second order chemical weathering reactions. **(a)** Change in silica concentration with time at pH 4, 6, 8 and 10, in a 'weathering' solution containing 5% feldspar. (Reproduced with permission from Wollast, 1967, *Geochem. Cosmochem. Acta*, 31, © Pergamon Books Ltd.). **(b)** Log rate of change of concentration versus log saturation deficit in the bulk solution for calcium ions during the dissolution of calcite in CO_2 saturated solution at 25°C and 1 atmosphere pressure. (Reproduced with permission from Plummer and Wigley, 1976, *Geochem. Cosmochem. Acta*, 40, © Pergamon Books Ltd.)

$$-\frac{1}{2}\frac{d[A]}{dt} = -\frac{1}{3}\frac{d[B]}{dt} = \frac{d[C]}{dt} = k[A]^n[B]^m \tag{1,47}$$

where the exponents n and m are integers or half integers; n is called the *order* of the reaction with respect to A and m is the order of the reaction with respect to B. The sum n + m is called the overall order of the reaction. Importantly, the order of any reaction with respect to each reactant or product must be determined experimentally and cannot be predicted or deduced from the equation for the reaction. Figure 1.9 illustrates examples of first and second order weathering reactions. First order reactions are probably the most common in nature, while zero order reactions are quite uncommon.

The alteration of mineral surfaces during weathering, through ion substitution, exchange and leaching, is thought to change the rate of surface reactions. These reactions have been widely studied for feldspar minerals (see, for example, Wollast, 1967; Busenberg and Clemency, 1976) and for magnesium silicates (Luce, Bartlett and Parks, 1972). In an empirical study of the dissolution of serpentine, fosterite and enstatite, Luce *et al.* (1972) found that there was an initial rapid exchange of surface Mg^{2+} ions for H^+ ions, followed by a longer period of hydrogen exchange and extraction of internal Mg and Si. In the experiment, finely ground mineral was leached with solutions whose pH was adjusted in the range 3.2–9.6 using HNO_3 or KOH and rates of release of Mg and SiO_2 were measured over time periods ranging from 16 to 100 h. After an initial period of rapid surface exchange, amounts of dissolved Mg and SiO_2 were proportional to the square root of time such that:

$$Q_{Mg^{2+}} = k_{Mg}t^{\frac{1}{2}} + Q^{\circ}_{Mg^{2+}} \tag{1,48}$$

$$Q_{SiO_2} = k_{SiO_2}t^{\frac{1}{2}} + Q^{\circ}_{SiO_2} \tag{1,49}$$

where: Q = number of moles cm^{-2} in solution at time, t
  k = first order parabolic rate constant
  Q° = number of moles cm^{-2} in solution at time, t = 0

These are examples of the first order parabolic rate law. The dissolution results for serpentine, fosterite and enstatite at an initial pH of 5 are illustrated in Fig. 1.10. Luce *et al.* (1972) proposed three component processes in the magnesium silicate dissolution system:

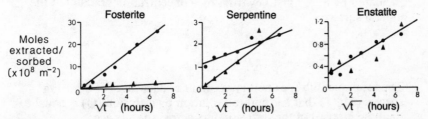

Figure 1.10 Examples of the parabolic extraction/sorption plots for fosterite, serpentine and enstatite at an initial pH of 5.0 •Mg^{2+}; ▲Si^{4+}. (Reproduced with permission from Luce, Bartlett and Parks, 1972, *Geochem. Cosmochem. Acta*, 36, © Pergamon Books Ltd.)

(1) rapid initial ion exchange;
(2) solid state diffusion with $2H^+$–Mg^{2+} exchange occurring faster than $4H^+$–SiO_2 exchange, and
(3) slow, complete dissolution at the mineral surface.

Two models for the parabolic kinetic system were proposed.

In the *first model*, it was assumed that the outer layer of the mineral becomes progressively altered from the surface, the degree of alteration decreasing with depth. Diffusion takes place through the partially altered zone (Fig. 1.11(a)). Fick's second law of diffusion can be integrated to give:

$$C_{(x,t)} = (C_o - C_s) \, \text{erf} \, \frac{x}{2(Dt)^{\frac{1}{2}}} \tag{1,50}$$

where: $C_{(x,t)}$ = concentration of diffusing species in the solid as a function of distance, x, from the surface and time, t
C_o = initial concentration of diffusing species in the solid
C_s = concentration of diffusing species at the aqueous interface
D = diffusion coefficient
erf = error function

The concentration gradient is obtained by differentiating Equation (1,50) with respect to x:

$$\frac{\partial C}{\partial x} = \frac{C_o - C_s}{(\pi Dt)^{1/2}} \exp \left[-\frac{x}{4Dt} \right] \tag{1,51}$$

At the surface of the solid, x=0, so:

$$\left[\frac{\partial C}{\partial x} \right]_{x=0} = \frac{C_o - C_s}{(\pi Dt)^{\frac{1}{2}}} \tag{1,52}$$

The flux of diffusing species (Mg^{2+}) entering solution ($-J$) can be calculated by Fick's first law from the concentration gradient at the mineral surface:

$$-J = D\frac{\partial C}{\partial x} \tag{1,53}$$

The total quantity (per unit area of mineral surface) of diffusing species (Mg^{2+}) that has entered solution by diffusion (Q) is equal to the time integral of the diffusion flux across the surface:

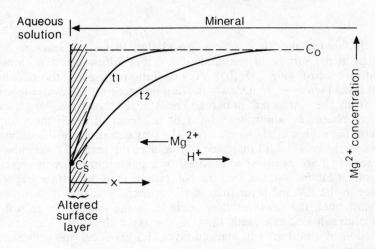

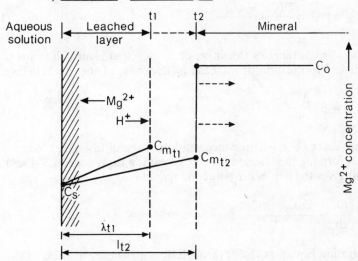

Figure 1.11   (a) Magnesium concentration in the mineral solid for Luce, Bartlett and Parks' (1972) model 1: partially altered primary material; **(b)** Mg concentration in the mineral solid for Luce, Bartlett and Parks' (1972) model 2: leached layer at mineral surface

$$Q = \int_o^t D \frac{\partial C}{\partial x} \, dt \tag{1,54}$$

$$Q = 2(C_o - C_s)\left[\frac{Dt}{\pi}\right]^{1/2} \tag{1,55}$$

$$Q = kt^{\frac{1}{2}} \tag{1,56}$$

In the *second model*, it was assumed that a discrete leached layer develops at the surface of the mineral and that diffusion occurs through this surface 'crust' (Fig. 1.11(b)). As dissolution proceeds, the thickness of this layer increases, so a slowly moving interface separates the leached layer from the unreacted mineral. The leached layer may be a new chemical phase, an amorphous layer or a region of much more rapid diffusion than the underlying mineral. The concentrations of the diffusing species at the solid/liquid interface (C_s) and at the leached layer/mineral interface (C_m) are assumed to be fixed by equilibrium between adjacent phases. If it can be assumed that the concentration of the diffusing species is lower in the leached layer than in the mineral, and that dissolution is not rapid, then the concentration gradient in the leached layer will be linear and will decrease with time as the layer thickens. The rate of production of a unit area of leached layer (R_A) is equal and opposite to the flux of the diffusing species ($-J$):

$$R_A = -J = \frac{D(C_m - C_s)}{\lambda} \tag{1,57}$$

where λ = instantaneous thickness of the leached layer. The rate is also proportional to the rate of increase in thickness of the leached layer:

$$R_A = \rho \frac{d\lambda}{dt} \tag{1,58}$$

where ρ is the equivalent concentration of the diffusing ion (Mg^{2+}) lost in forming the leached layer. Combining Equations (**1,57**) and (**1,58**) gives the *first order parabolic rate law*:

$$\frac{d\lambda}{dt} = \frac{D(C_m - C_s)}{\rho\lambda} \tag{1,59}$$

Integrating Equation (**1,59**) through time gives the thickness of the leached layer at time t as:

$$\lambda^2 = \frac{2D(C_m - C_s)t}{\rho} \tag{1,60}$$

where $Q = \lambda\rho$ (that is, the amount of diffusing species in solution (Q) equals the amount lost in forming the leached layer) then:

$$Q = [2D(C_m - C_s)\rho]^{1/2} t^{1/2} \qquad\qquad (1,61)$$

$$Q = k t^{1/2} \qquad\qquad (1,62)$$

where k is the experimental constant. So, the second model, of a leached layer at the mineral surface, also conforms to the first order parabolic rate law.

1.5 WEATHERING SEQUENCES, RATIOS AND INDICES

Many lists of the relative stabilities of soil minerals have been compiled. Some of these are given in Table 1.4. Perhaps of more interest to pedologists concerned with soil processes are the sequences of clay minerals formed progressively during weathering, since these may help to explain the degree of weathering represented by a particular soil mineralogy. Clay mineral ratios are rarely used directly, but instead, the ratios of major component ions or oxides are used either singly or in combination. From examination of simplified weathering reactions and species solubility diagrams, it is clear that two major trends occur over time: (1) the mobilisation and leaching of soluble species, and (2) the immobilisation and retention of insoluble species. These diverse trends are usually combined in the form of molar ratios to produce weathering indices for regolith and soils. In the simplest molar ratios such as:

$$\frac{SiO_2}{Al_2O_3} \qquad \frac{SiO_2}{Fe_2O_3} \qquad \frac{SiO_2}{Al_2O_3 + Fe_2O_3} \qquad (1,63)$$

the relative rates of loss of the different components during weathering can be assessed. These three ratios in particular have been very widely used as weathering/leaching indices for the E horizon of podsols and will be dealt with in more detail in Chapter 4.

Reiche (1943, 1950) proposed a more complex system to assess degree of weathering. He plotted the weathering potential (W_p) against the product index (P_i) where:

$$W_p = \frac{CaO + Na_2O + MgO + K_2O - H_2O}{SiO_2 + TiO_2 + Al_2O_3 + Fe_2O_3 + Cr_2O_3 + CaO + Na_2O + MgO + K_2O} \times 100 \qquad (1,64)$$

$$P_i = \frac{SiO_2}{SiO_2 + TiO_2 + R_2O_3} \times 100 \qquad (1,65)$$

(all values in moles). Kronberg and Nesbitt (1981) also used the graphical representation of two weathering ratios to illustrate the global scale of chemical weathering. They represented the alteration of feldspars (W_1) by:

Table 1.4 Stability and weathering of soil minerals

(a) Increasing rock stability* →
limestone, dolomite; siltstone, sandstone; basalt, granite; chert, quartzite

(b) Increasing mineral stability[†] →
olivine, augite, labradorite, hornblende, biotite, magnetite, haematite, orthoclase, microcline, albite, muscovite, garnet, quartz

(b) Mineral weathering sequences**

  (i) Potassium feldspar → kaolinite → gibbsite

  (ii) Biotite → chlorite → montmorillonite

    vermiculite → aluminium vermiculite → kaolinite

  (iii) Plagioclase feldspar → sericite → vermiculite → montmorillonite → aluminium montmorillonite → kaolinite → gibbsite

* Birkeland (1974)

[†] Pettijohn (1941) and Smith (1962)

** Tardy et al. (1973)

$$W_1 = \frac{[CaO]+[Na_2O]+[K_2O]}{[Al_2O_3]+[CaO]+[Na_2O]+[K_2O]} \qquad (1,66)$$

and the dominance of oxides of Al and Si (W_2) by:

$$W_2 = \frac{[SiO_2]+[CaO]+[Na_2O]+[K_sO]}{[Al_2O_3]+[SiO_2]+[CaO]+[Na_2O]+[K_2O]} \qquad (1,67)$$

Using a scheme where (W_1) is plotted on the ordinate and (W_2) on the abscissa, the secondary clay minerals, illite, kaolinite and montmorillonite plot below the field of the more abundant primary minerals (Fig. 1.12(a)). Using the same technique, they suggest theoretical weathering curves for regolith materials ranging from bauxite to weathered tropical Amazonian soil and granite (Fig. 1.12(b)).

These very complex indices have more commonly been to assess the weathered status of rocks *in situ* than of whole soil mineralogies. Hodder (1984) has used two complicated ratios, the Miura index and the Parker index, to assess the engineering properties of rocks. The Miura index is the ratio of geochemically mobile to geochemically immobile elements. It is logarithmically related to the free energy (ΔG_r°) of the weathering reaction. Hodder (1984) suggests that it is possible to combine the Miura index with the Parker index (which is believed to show not only the state of weathering, but also the susceptibility to further weathering) to estimate the activation energy for the incongruent dissolution of soil minerals.

Thus, the ratios themselves, or the theoretical chemical thermo-dynamics of weathering processes, are potentially useful as indicators of the degree of rock decomposition and hence rock strength. It is in this geomorphological context that Kirkby (1977, 1985) uses both mineral composition and thermodynamics to predict the rate of loss of rock substrate and changing regolith strength on slopes, through time. A second possible approach in slope and soil profile modelling is to examine the effect of pedogenesis in altering the soil moisture regime, particularly in soils showing progressive illuviation of the soil B horizon. The release of chemical elements during weathering and the subsequent involvement in nutrient cycling has been widely studied in soil science, although rarely with the aid of thermochemical predictions. Garrels (1976) calculated the ionic composition of waters resulting from the weathering of a variety of silicate minerals to kaolinite (Fig. 1.13). Many would argue that such exercises are purely academic for two main reasons. First, it could be argued that the relatively simple chemical equations required for thermochemical calculations can never properly describe the complexity of natural systems, particularly soils. Second, all thermochemical calculations must assume that the system attains equilibrium. With continually changing soil chemistry and hydrology even at the scale of individual soil pores, far less the whole soil profile, this assumption is frequently rejected by empirical field scientists.

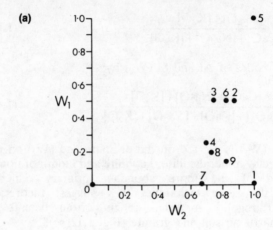

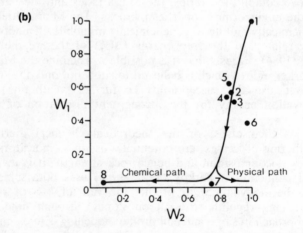

$$W_1 = \frac{CaO + Na_2O + K_2O}{Al_2O_3 + CaO + Na_2O + K_2O} = \text{alteration of feldspars}$$

$$W_2 = \frac{SiO_2 + CaO + Na_2O + K_2O}{Al_2O_3 + SiO_2 + CaO + Na_2O + K_2O} = \frac{\text{dominance of oxides}}{\text{of Al and Si}}$$

Figure 1.12 **(a)** Plot of feldspar alteration (W_1) against oxide retention (W_2), showing mineral distributions; 1: quartz; 2: alkali feldspar, biotite, 3: anorthite; 4: muscovite; 5: diopside, aqueous solutions; 6: jadeite; 7: kaolin minerals; 8: illite; 9: montmorillonite; 10: gibbsite. **(b)** Feldspar alteration–oxide retention curves for theoretical weathering processes 1: river water; 2: crust; 3: granite; 4: andesite; 5: basalt; 6: fertile soil (US); 7: Amazon soil; 8: bauxite. (Source: Kronberg and Nesbitt, 1981)

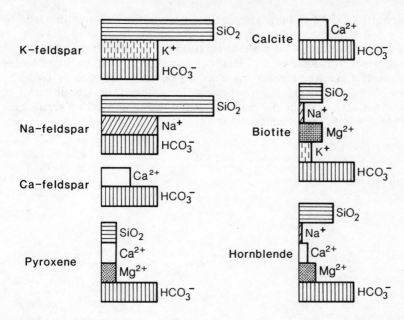

Figure 1.13 Ionic composition of waters, calculated for the weathering of a variety of silicate minerals to kaolinite (compositions shown as mol ratios of dissolved species to HCO_3^-. (Source: Garrels, 1967)

1.6 SUMMARY

An understanding of geochemistry is needed to predict the sequence and rate in which a mixture of soil minerals will weather under the influence of a given set of environmental controls. Basic requirements are a knowledge of component ion valencies and their hydrated radii, since these indicate ionic potential and their relative solubilities and suscept-ibilities to leaching (see Fig. 1.2), and a knowledge of the solubilities of minerals in relation to solution pH (see Fig. 1.3). There are five main chemical weathering processes which occur in soils (see Table 1.1). If balanced chemical equations can be written for these reactions, simple *chemical thermodynamics* can then be used to calculate the relative stabilities of soil minerals. Gibbs free energy values (ΔG_r°) are calculated for weathering process equations, written as equilibrium reactions. ΔG_f° values indicate the direction of any reaction and the order in which a sequence of reactions will occur. Two major assumptions of thermodynamic calculations in soil chemistry are

(1) that the system is simple and that simple reaction equations can be written which are representative of weathering processes occurring in the field; and
(2) that equilibrium is attained for these reaction equations.

Despite the fact that both assumptions are considered oversimplistic by many soil scientists, thermodynamic calculations frequently allow us to evaluate the effect of environmental influences, such as changes in temperature, or increased acid precipitation, on chemical weathering processes. *Chemical kinetics* are used to describe the rate or changing rates of chemical weathering processes over time. The kinetic order of reaction describes the mathematical form of the concentration versus time plot of a weathering reaction.

CHAPTER 2

Production of Soil Material through Organic Matter Decomposition

2.1 INTRODUCTION

In most sites, a combination of mineral inputs from rock weathering processes and organic inputs from decomposition processes produces a recognisable soil material which is in a continual state of evolution. In many conceptual models of soil formation, such as that proposed by Jenny (1940), parent material is the state factor responsible for mineral inputs while vegetation is the state factor responsible for organic inputs. Apart from being a source of litter and soil organic matter, vegetation has several further influences on soil development and these are discussed elsewhere in the text (pp. 178–229). The particularly important effects of roots are discussed in Chapter 6, while chemical and biological processes of the rooting zone, including the rhizosphere, are examined throughout Part III. Plants cycle nutrients from the soil and the economic implications of nutrient uptake and leaching processes are discussed in Chapter 8. Plant canopies are particularly influential on ground surface, and hence soil, microclimates and thus influence rates of chemical and biochemical processes generally. These temperature and moisture influences on rates of decomposition processes are included in this chapter.

2.1.1 Biochemical background to organic matter decomposition

Organic matter decomposition is a general term for a whole sequence of very detailed processes whereby soil organisms use soil organic compounds as a food source. All living organisms are classified on the basis of their main energy source into *phototrophic* (using solar radiation) or *chemotrophic* (using energy released from chemical oxidations). Organisms can be further subdivided on the basis of their principal carbon source into *autotrophs* which use inorganic carbon (CO_2) and *heterotrophs* which use organic carbon compounds (such as carbohydrates). Four groups result from combining these classifications (Table 2.1). The organisms mainly responsible for the decomposition of soil organic matter are chemoheterotrophs, breaking down complex organic molecules to obtain both energy and the simple nutrients that they require to build their own body tissues. Simple nutrients excreted after chemoheterotrophic digestion (Table 2.2) may be taken up by photoautotrophs (Fig. 2.1). Ghilarov (1970) divides the chemoheterotrophic soil organisms

Table 2.1 Energy and carbon nutrition classification of living organisms

Category	Energy source	Carbon source	Examples
Photoautotrophs	Solar radiation	CO_2	Higher plants, algae, blue-green algae, photo-synthetic bacteria
Photoheterotrophs	Solar radiation	Organic compounds	Non-sulphur purple bacteria
Chemoautotrophs	Oxidation of organic compounds	CO_2	Specialised bacteria such as nitrogen and sulphur oxidising bacteria
Chemoheterotrophs	Oxidation of organic compounds	Organic compounds	All animals (vertebrates and invertebrates), most bacteria, most fungi

Table 2.2 Simple inorganic forms of plant nutrients produced during organic matter decomposition

Element	Ionic species (soil solution)			Gaseous species		
Carbon	CO_3^{2-} carbonate	HCO_3^- bicarbonate		CO_2 carbon dioxide	CH_4 methane	
Nitrogen	NO_2^- nitrite	NO_3^- nitrate	NH_4^+ ammonium	NH_3 ammonia	N_2O nitrous oxide	N_2 nitrogen gas
Phosphorus	$H_2 PO_4^{2-}$ primary orthophosphate	$H PO_4^{3-}$ secondary orthophosphate				
Sulphur	SO_4^{2-} sulphate	$S_2 O_3^{2-}$ thiosulphate		H_2S hydrogen sulphide		
Metal cations	K^+ Na^+ Ca^{2+} Mg^{2+} Fe^{3+} Fe^{2+} Mn^{4+} Mn^{3+}					
Micronutrients	$H_2BO_3^-$ borate	MoO_4^{2-} molybdate				
Heavy metals	$Cr_2O_7^{2-}$ chromate	$H A_5O_4^{2-}$ arsenate				

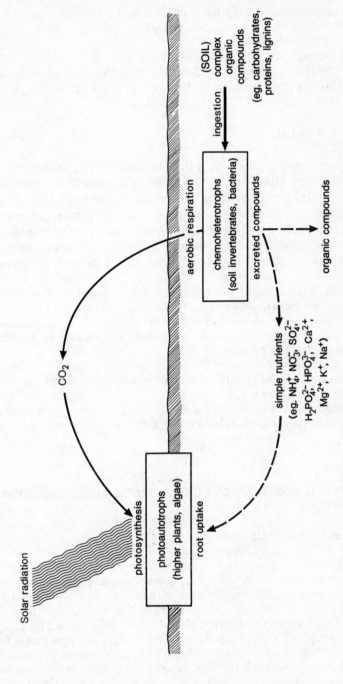

Figure 2.1 Relationships between soil chemoheterotrophs (decomposers) and photoautotrophs (plants) in the soil–plant ecosystem

into three main groups on the basis of their trophic roles:
 (1) phytophagi (ingest plant material),
 (2) zoophagi (carnivores), and
 (3) saprophagi (ingest dead organic matter).

Despite the food substrate used, energy-yielding processes of living organisms involve the transfer of hydrogen according to the general reaction

$$AH_2 + B \rightleftharpoons BH_2 + A \tag{2,1}$$

where substance AH is called the hydrogen donor and substance B is the hydrogen acceptor (Richards, 1974). This is equivalent to the redox reactions given for purely chemical systems in Equations **(1,2)**, **(1,3)** and **(1,4)**, involving the transfer of electrons. The three types of energy-yielding metabolisms found in soil populations are aerobic respiration, anaerobic respiration and fermentation. Examples of these processes, together with the organisms involved and the energy derived, are given in Table 2.3.

Unlike the weathering of soil minerals, organic matter decomposition processes are microbially mediated, so that the factors determining rates of decomposition and nutrient mobilisation are the same as those that determine microbial activity. There are two groups of factors governing rates of decomposition processes:

 (1) substrate quality and quantity – which controls soil nutrient content and influences soil pH;
 (2) environmental factors, primely soil moisture content – which controls soil aeration and influences temperature and leaching.

2.2 SUBSTRATE CONTROLS ON ORGANIC MATTER DECOMPOSITION PROCESSES

2.2.1 Quality of organic matter

A wide diversity of plant and animal materials are added to soil as organic matter. The exact molecular composition of these materials depends not only on the species and part of the plant or animal (for example, foliage or bark of a tree), but also on season and age. The composition of a range of different plant tissues is shown in Fig. 2.2. As plants age, the proportion of cellulose, hemicellulose and lignin rises (and are also higher in woody tissues), while the proportions of proteins and water soluble constituents falls. The 'quality' of organic matter available for microbial decomposition is really an index of its palatability, its energy content and its nutrient value. So quality is determined by

Table 2.3 Energy-yielding reactions of soil chemotrophs

Energy-yielding process	Hydrogen acceptor	Example reaction
Aerobic respiration	Molecular oxygen	$C_6H_{12}O_6 + 6O_2 \rightarrow 6CO_2 + 6H_2O$ glucose many organisms $+ 2800$ kJ mol^{-1}
Anaerobic respiration	Inorganic compound other than oxygen	(i) Denitrification: $NO_3^- + H_2 \rightarrow NO_2^- + H_2O$ $2NO_2^- + 4H_2 \rightarrow N_2 + 4H_2O$ *Thiobacillus denitrificans* (ii) Sulphur reduction: $SO^{2-} + 4H_2 \rightarrow S^{2-} + 4H_2O$ sulphate $SO_3^{2-} + 3H_2 \rightarrow S^{2-} + 3H_2O$ sulphite $S_2O_3^{2-} + 4H_2 \rightarrow 2HS^- + 3H_2O$ thiosulphate *Desulphovibrio desulphuricans*
Fermentation	Organic Compounds	(i) $C_6H_{12}O_6 \rightarrow$ 2 $CH_3CH\text{-}COH$ (with OH O) glucose many organisms lactic acid $+88$ kJ mol^{-1} (ii) $C_6H_{12}O_6 \rightarrow 3CH_4 + 3CO_2$ many bacteria methane $+ 180$ kJ mol^{-1} (iii) $C_6H_{12}O_6 \rightarrow$ 2 $C_2H_5OH + 2CO_2$ alcohol yeast ethanol $+ 75$ kJ mol^{-1}

(1) the amount of fibre or wood,
(2) the content of carbon compounds, which provide the energy source for decomposer organisms, and
(3) the content of nutrients such as nitrogen and phosphorus.

Organic materials containing soluble compounds such as sugars and some amino acids, and easily decomposable substrates such as proteins and carbohydrates, are quickly used as a food source by a wide variety of soil organisms. Since the more resistant compounds tend to have higher calorific values (Table 2.4), an ability to attack these materials should clearly be beneficial to decomposer organisms. Nevertheless, only a few specialist decomposers break down lignin, lipids and chitin and the restricted number of these specialists accounts for the relative abundance

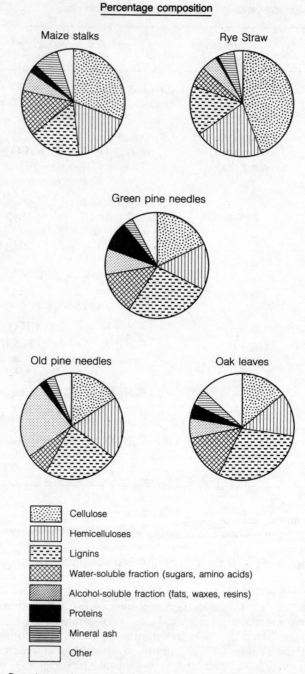

Figure 2.2 Organic constituents of a range of plant materials. (Source: drawn from the data of Waksman, 1936)

and accumulation of these resistant compounds, particularly in organic horizons. As for chemical weathering reactions, it is theoretically possible to calculate organic matter decomposition reactions thermodynamically, but similar assumptions concerning system simplicity and equilibrium must be made (Ross, 1987). The differing abilities of soil micro-organisms to attack the various compounds that makes up plant or animal litter leads to the concept of a succession of decomposers colonising any particular material.

Gray and Williams (1971) liken the colonisation sequence of micro-organisms on soil organic substrates to *autogenic succession*, whereby each wave of colonisation alters the substrate for the next wave, with progressive depletion of the chemical energy sources. Fleshy organic materials decay quickly, rapidly losing water-soluble components, followed by carbohydrates such as cellulose and starch. These losses account for the bulk of the reduction in dry weight that accompanies decay. The decay of woody tissues is very much slower, and the reduction in dry weight is mainly due to the decomposition of cellulose. In the classic work of Minderman (1968), decomposition curves are given for a range of litter and soil organic constituents (Fig. 2.3), measured in field

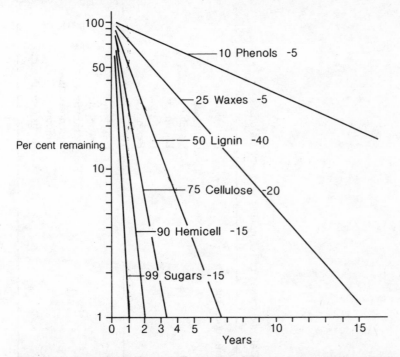

Figure 2.3 Decomposition curves for a range of organic matter constituents measured in the field under *Pinus nigra*, *Pinus sylvestris* and *Quercus robur*. Decomposition represented as a logarithmic function. The number in front of the name constitutes percentage loss after one year. The number after the name constitutes percentage in weight of the original litter. (Source: modified from Minderman, 1968)

Table 2.4 Calorific values of plant, litter and soil materials

(a) Soil organic constituents	Material	Calorific value (kJ g^{-1} ash free dry weight)	
	Cellulose	17.6	(Jenkinson, 1981)
	Starch	17.5	(Jenkinson, 1981)
	Lignin	26.4	(Runge, 1973)
	Plant lipid	38.5	(Jenkinson, 1981)
	Resin, fat	36.8	(Runge, 1973)
(b) Plant litters	Oak leaves	21.0	(Goreham and Sanger, 1967)
	Bryophytes	17–19	(Allen et al., 1974)
	Pteridophytes	20–21	(Allen et al., 1974)
	Gymnosperms	23–25	(Allen et al., 1974)
	Grasses	19–21	(Allen et al., 1974)
	Herbs (aerial parts)	20–21	(Allen et al., 1974)

(Table 2.4 cont . . .)

(c) Woodland ecosystems (Steubing, 1977)

(i) Spruce ecosystem

H-horizon		
litter	15.52	
	19.83	mean = 19.78
bark	23.30	
wood	20.46	

(ii) Beech ecosystem

H-horizon		
litter	11.72	
	19.96	mean = 18.28
bark	21.05	
wood	20.38	

Species (Ovington and Heitkamp, 1961)	Leaves	Calorific value (kJ g^{-1} ovendry weight)			
		Canopy (leaves plus branches)	Bole	litter (L-horizon)	litter (F + H horizons)
Picea abies	20.05	20.46	19.61	20.36	15.39
Pinus nigra	20.88	20.63	19.92	21.32	18.49
Pinus sylvestris	20.30	20.17	19.77	20.72	18.92
Larix decidua	20.54	20.75	20.20	18.47	19.06
Fagus sylvatica	19.63	19.74	19.06	18.89	16.19
Quercus robur	20.08	19.52	19.30	19.69	17.75

studies on litter of *Pinus nigra* (Corsican pine), *Pinus sylvestris* (Scots pine) and *Quercus robur* (sessile oak). These plots indicate the relative resistance to decomposition of phenols, waxes and lignins and the relative ease of decomposition of sugars and hemicellulose.

Some consistencies in the pattern of decomposition reported for a range of substrates allow some useful generalisations about the relative activities and roles of decomposers during the breakdown of litter:

Stage I(a) Many primary decomposers comminute and physically break up litter detritus, such as beetle larvae, centipedes, millipedes, termites in tropical soils.

Stage I(b) Many primary decomposers possess a wide range of extracellular enzymes which enable dissolution of outer protective tissues, such as saprophytic fungi, protozoa.

Stage II Mechanisms in stage I make available a wide range of other organic substrates for further attack by secondary decomposers, such as most soil micro-organisms, saprophytic bacteria, faeces decomposers.

Stage III Corpses and faeces from stage I and II organisms provide a further, diverse substrate for another hierarchy of decomposer organisms.

Many microbial decomposers are limited in their ability to attack certain organic molecules by their inability to produce the enzyme which is required to catalyse molecular hydrolysis (Table 2.5).

From these generalisations and from Minderman's (1968) graph (Fig. 2.3), it would be logical to suggest that a knowledge of soil organic matter composition would allow the prediction of its rate and pattern of decomposition. Many authors have attempted to do just that, using a range of organic matter quality parameters such as lignin, fibre, or nitrogen content (the latter usually represented by the C:N (carbon:nitrogen) ratio). Van Cleve (1974) found a negative correlation

Table 2.5 Organic hydrolysis reactions mediated by extra cellular enzymes

Compound	Method of decomposition	Products
Cellulose	Cellulase	Glucose
Hemicellulose	Hemicellulase	Monosaccharides
Starch	Amylase	Maltose, glucose
Pectin	Pectinase	Pectic acid, methanol
Lipids	Lipase	Glycerol, fatty acids
Lignin	Ligninase	Vanillin, syringaldehyde, quinones, peroxy-radicals

between lignin content of tundra litter and rate of decomposition, while Latter and Howson (1977) found decomposer bacteria numbers to be negatively correlated with crude fibre content and positively correlated with the nitrogen content of the organic substrate. Both of these studies used substrate weight loss as the index of decomposition of materials buried in field soil over an experimental period. Using the same field technique, whereby litter in mesh bags is buried in the soil, Fogel and Cromack (1977) found that lignin content of *Pseudotsuga menziesii* (Douglas fir) needle litter was more important than its C:N ratio in determining the rate of decomposition. Working on the decomposition of heathland litter on deep peat in the Pennines, Northern England, Heal, Latter and Howson (1978) found that dry weight losses after one year ranged from 5 to 8 per cent for *Calluna* wood and 22 per cent for *Eriophorum* leaves, to 35–45 per cent for *Rubus* and *Narthecium* leaves – a sequence that paralleled decreasing fibre content, increasing total N, P and K contents and declining C:N, C:P and C:K ratios of the litters. This pattern of increased organic matter decomposition (see, for example, Van Cleve, 1974) and nutrient mobilisation (see, for example, Haque and Walmsley, 1972) with lower carbon:nutrient ratio has long been known, particularly for nitrogen. Broadbent (1962) explains this general relationship as the ratio of available energy to available nutrients. This concept is the crux of nutrient cycling in relation to organic matter decomposition.

Soil organic matter is the major native source of soil nitrogen, unlike all other chemical elements, apart from sulphur, which are also supplied in appreciable amounts from rock weathering, or atmospheric inputs. For this reason, the utilisation of soil organic matter by the decomposer population has particularly important implications for soil nitrogen dynamics. Thus, the relationship between the energy of a 'food' substrate and its nutrient content has primarily been applied to carbon:nitrogen (C:N) relationships, although in a few cases, carbon:phosphorus (C:P), carbon:sulphur (C:S) and carbon:potassium (C:K) relationships have also been reported.

When substrates with high C:N ratios, such as mor organic matter, which may have values of 25–45 (for example, Williams, 1972; Berg and Bosatta, 1976; Popovic, 1977), are utilised by soil microbes as an energy source, CO_2 is evolved and nutrients, including nitrogen, are immobilised, thus increasing the nutrient content per unit volume of peat and lowering the carbon:nutrient ratio. Availability of inorganic N for plant uptake will be restricted until the C:N ratio is reduced to a level where lack of readily decomposable carbon compounds (energy sources) limits microbial demand for nutrients (Martin and Holding, 1978). Substantial reduction in C:N ratio of various litter types in tundra soils (Dowding, 1974) and beech forest litter (Remacle and Vanderhoven, 1972) have been recorded over periods as short as 8–12 months under field conditions. Birch (1960) demonstrated that the C:N ratio of air-dried, remoistened mineral soils could be significantly lowered over a period of just 3 weeks at 25°C. Simultaneous mobilisation (decomposition and excretion) and immobilisation (uptake and tissue building) act in opposite directions on nutrient cycling, with net mobilisation depending on the value of the C:N ratio of the substrate. Critical C:N values of 20 in agricultural soils (Harmsen and Van Schreven, 1955) and 25 in organic soils (Dowding, 1974) have been noted, above which microbial immobilisation of nitrogen predominates and plants may suffer N deficiency if no additional N supplement is given. In this way, Turner (1977) artificially induced nitrogen immobilisation in the forest floor of a Douglas fir stand by applying large quantities of high energy source carbohydrate (sucrose), without nitrogen. Dowding (1974) used carbon ratios with other nutrients (P and K) as indicators of decomposition of plant litters and soil nutrient status in different Irish peatland types, but found them to be more variable than C:N ratios, presumably due to mineral inputs from sources other than organic matter.

2.2.2   Quantity of organic matter

The frequency and amount of decomposable litter additions to the soil every year is of importance for the maintenance of the soil population. Amounts of litter-fall per year in coniferous and deciduous woodlands are about the same, or only slightly higher in conifers (Bray and Goreham, 1964), but Ovington (1962), in an extensive review, notes much smaller amounts of litter on deciduous forest floors than on coniferous forest floors – an indication of the much faster rates of mineralisation and larger soil population in the former. In independent comparisons of microbial activity in spruce and beech forest litters, both Remacle (1977) and Steubing (1977) report a greater ease of attack of beech litter, despite the higher average calorific value of spruce litter (see Table 2.4).

2.2.3   pH of organic matter

The low pH of relatively undecomposed organic horizons is partly self-perpetuating. Evolution of CO_2 from respiring organisms, together with organic acid products of decomposition, maintain acidic conditions. In addition, the litter of many plants characteristic of organic soils and surface horizons, is acidic. Mattson and Karlson (1944) showed that the pH of heathland litter is strongly acidic, with values for *Calluna* and *Sphagnum* species of 4.4 and 2.8–4.4 respectively.

Heal and French (1974) found that organic matter decomposition rates for the litter of tundra species was lower on sites with pH < 4.5 than on sites with higher pH values. Many authors have noted that nitrogen mineralisation rates are greatly reduced below pH 6.0 (see, for example, Zottl, 1960; Gasser, 1969; Haque and Walmsley, 1972). Collins, D'Sylva and Latter (1976) found that 80 per cent of the aerobic bacteria in deep peat could not grow and mobilise nutrients at pH < 5.5. The only groups growing well in this pH range were fungi and the *Thiobacilli* bacteria which oxidise thiosulphate and thus enhance acidic conditions.

Microbial activity is stimulated by liming these acid peats and organic soils because the pH is brought closer to the optimum for soil organisms, particularly those that mineralise nitrogen, whose optimum pH is about 7.5 (Nommik, 1968; Tusneem and Patrick, 1971).

2.3   ENVIRONMENTAL CONTROLS ON DECOMPOSITION PROCESSES

Soil moisture acts as an overall control on the other edaphic factors that affect organic matter decomposition. Microbial activity is governed by soil moisture content both *directly*: by limiting their movement in drought conditions (for example, Dickinson, 1974), transport of nutrients or toxins (Alexander, 1977) in the soil solution and the build up of oxygen deficiency by waterlogging (Turner and Patrick, 1968); or *indirectly* by the controlling effect of moisture content on soil temperatures (Raney, 1965), or the breakup of decomposable substrates by drying-wetting (Cooke and Cunningham, 1958; Birch, 1959) and freezing-thawing (Gasser, 1958; Hinman, 1970).

Microbial decomposition processes operate over a wide range of soil moisture conditions from the wilting point (1500 kPa) to saturation, although optimum conditions, as measured by peak rates of CO_2 evolution (Miller and Johnson, 1964) and nitrogen mineralisation (Stanford and Epstein, 1974) are around field capacity (33 kPa) in the range considered as readily available water for plant uptake.

2.3.1   Waterlogging effects on decomposition processes

The changes which occur in the soil solution after waterlogging are the lowering of the oxygen content and the build up of acidic and toxic conditions due to stagnation. On waterlogging, continuing aerobic respiration rapidly depletes any oxygen present in the water and anaerobic conditions develop very quickly because the diffusion of replacement oxygen through water is 10^4 times slower than through air. In the absence of oxygen, anaerobic fermentation takes place (see Table 2.3), often by facultative anaerobes which can exist either in the presence or absence of oxygen. In terms of both the energy obtained and the number of microbial cells formed per unit of carbon substrate degraded during such reactions, fermentation processes are only about one-third as efficient as aerobic respiration. So where fermentation predominates, overall rates of decomposition are slower due to the smaller active soil organism population. In the presence of adequate, easily decomposable organic compounds, such as carbohydrates, the first products of anaerobic decomposition are CH_4 (methane, or marsh gas) and smaller amounts of H_2. A whole range of fatty, long-chain and other carboxylic acids are also produced, the main ones according to Stevenson (1967) being:

$$\underset{\text{CH}_3\overset{\text{O}}{\overset{\|}{\text{C}}}\text{OH}}{} \quad \text{(acetic acid)} \quad \underset{\text{CH}_3\text{CH}_2\text{CH}_2\overset{\text{O}}{\overset{\|}{\text{C}}}\text{OH}}{} \quad \text{(butyric acid)}$$

$$\underset{\text{CH}_3\text{CH}_2\overset{\text{O}}{\overset{\|}{\text{C}}}\text{OH}}{} \quad \text{(propionic acid)} \quad \underset{\text{CH}_3\overset{\text{OH}}{\overset{|}{\text{CH}}}\text{--}\overset{\text{O}}{\overset{\|}{\text{C}}}\text{OH}}{} \quad \text{(lactic acid)}$$

Many of the gases which may be produced during fermentation, such as H_2S (hydrogen sulphide), C_2H_6 (ethane) and $CH_2{=}CH_2$ (ethylene), C_3H_8 (propane) and $CH_3CH{=}CH_2$ (propylene), can cause disordered growth of plant roots. Ethylene is a plant hormone, and K.A. Smith (1977) suggests that concentrations as low as 0.01 ppm may affect root growth and seed germination. Values of ethylene recorded in field soils, even those known to be generally well aerated and certainly not water-logged, range from 1.5 to 5.0 ppm, with the lower values representing soils with higher organic matter content (Smith and Restall, 1971). Smith, Restall and Robertson (1969) show that root elongation in a wide range of crop plants, from cereals such as barley, rye and rice to tomatoes and tobacco, is reduced by up to 60 per cent when exposed to ethylene concentrations as low as 0.3 ppm. Plants vary in their ability to withstand H_2S toxicity. Webster (1962), for example, reported positive correlations between concentrations of H_2S in the range 7–43 mg l^{-1} and the growth of both *Calluna* and *Erica tetralix*. Negative correlations were reported for the growth of both *Molinia* and *Myrica* over the same H_2S concentration range.

Although few temperate soils are completely waterlogged, low oxygen concentrations can develop in generally well-aerated soils where gaseous diffusion is restricted by, for example, an ironpan or the tortuosity of the pore space in a heavy clay. Currie (1961) and Greenwood (1962) calculated that crumb aggregates in many soils can contain anaerobic centres even if the soil is not saturated (see Part II, Chapter 3). Anaerobic decomposition processes can, thus, take place in extremely localised pockets, as testified by measurable amounts of ethylene in generally well-aerated mineral soil profiles (Dowdell, *et al.*, 1972).

Waterlogging conditions affect nutrient cycling by altering redox conditions. Inorganic nitrogen, in the form of $NH_4^+{-}N$, accumulates under waterlogging conditions (see, for example, Redman and Patrick, 1965; van Schreven and Sieben, 1972) because nitrification, the production of $NO_2^-{-}N$ (nitrite) and $NO_3^-{-}N$ (nitrate) by the aerobic bacteria, *Nitrosomonas* and *Nitrobacter* respectively, is inhibited. There is also less nutritional demand for inorganic nitrogen by the smaller anaerobic decomposer population. The denitrifying bacteria thrive in waterlogged conditions. The reduction of $NO_3^-{-}N$ is carried out by *Thiobacillus denitrificans* to produce N_2O and N_2 as gaseous end products (see Table 2.3). Such volatile forms of nitrogen constitute a very important loss of nitrogen from both native soil sources and from fertilizer inputs. In a similar sequence of reactions to denitrification, *Desulphovibrio desul-*

phuricans reduces oxidised forms of sulphur, such as SO_4^{2-} (sulphate) and $S_2O_3^{2-}$ (thiosulphate) to produce sulphides including gaseous H_2S (hydrogen sulphide) (see Table 2.3).

2.3.2 Intermittent wetting and drying

The effects of alternate wetting and drying cycles on organic matter decomposition are likely to be twofold:

(1) If soil is saturated for long enough on each wetting cycle, the alternation of wetting and drying will be analogous to anaerobic and aerobic conditions.
(2) Since wetting of soil colloids, particularly the smectite clays, causes expansion and drying causes contraction, the alternation of wetting and drying can cause breakup of aggregates and particles.

Wetting and drying periodicities in soils are likely to range from a few days in the case of saturation caused by impedance after heavy rain, to seasonal waterlogging as a result of a general rise of the water table in wetter winter months. Laboratory studies have shown that the periodic drying and rewetting of soil increases organic matter decomposition, measured both by CO_2 evolution (see, for example, Sorensen, 1974) and nitrogen mineralisation (see, for example, Patrick and Wyatt, 1964). On a seasonal scale, Piene and Van Cleve (1976) have shown that peaks in microbial respiration on the forest floor of a white spruce stand in Alaska followed summer rainfall after a dry spring. In intermittent drying and rewetting experiments, flushes of mineralisation have been noted in the first one or two cycles, with a 'tailing off' in subsequent cycles (see, for example, Sorensen, 1974; Ross and Malcolm, 1988). Birch (1958) noted that the magnitude of the initial flush of nutrients increased with increasing length of the drying period prior to waterlogging. Sorensen (1974) demonstrated that addition of straw to periodically dried and rewetted soil curtailed the tailing off in mineralisation (Fig. 2.4) and suggested a lack of decomposable substrate as the reason for reduced nutrient release. Two main hypotheses have been proposed to explain the nutrient flush:

(1) that initial drying and rewetting shatters soil aggregates and exposes previously unavailable organic substrates for decomposition (Cooke and Cunningham, 1958; Haque and Walmsley, 1972);
(2) that micro-organisms, killed by soil drying, are decomposed to release the nutrients immobilised in their tissues (Birch, 1960).

The high nutrient concentrations in tissues of microflora and soil fauna listed in Table 2.6 indicate that this source of nutrients may be of more importance than previously supposed.

Using alternate cycles of oxygen and argon gases introduced into

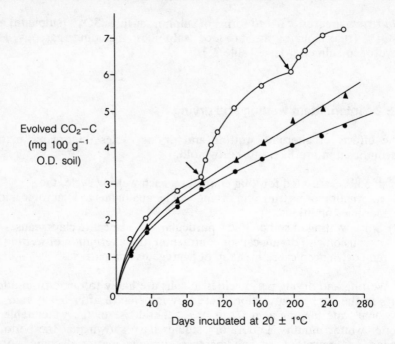

Figure 2.4 Evolution of CO_2 from incubated soil; carbon was originally added as straw. ● Soil kept moist continuously; ▲ soil air-dried every 30th day; ○ soil to which was added straw in amounts corresponding to 250 mg C 100 g^{-1} soil at the start and where indicated by arrows. (Reproduced with permission from Sorensen, 1974, *Soil Biology and Biochemistry*, 6, © Pergamon Books Ltd.)

laboratory incubation vessels to simulate aerobic and anaerobic conditions, Reddy and Patrick (1976; 1977) produced patterns of organic matter decomposition and nutrient mineralisation similar to those of wetting and drying treatments. In field soil, fluctuations in oxygen content in the centres of crumb aggregates and associated with impeding ironpans or subsoil compaction may locally increase mineralisation rates as long as the profile remains generally aerobic.

2.3.3 Temperature effects on organic matter decomposition

Apart from diurnal and season fluctuations in solar radiation receipt at the soil surface, variation in soil moisture content is the main factor determining thermal properties of a soil. The specific heat* of water is 4.18 J and the mean specific heat of soil minerals is around 0.84 J. Thus a soil containing 50 per cent moisture will have a specific heat of 2.47 J, while the same soil at 25 per cent moisture will have a value of 1.57 J.

---

* Specific heat is the energy (in joules) required to raise the temperature of 1 kg of substance by 1 °C.

Table 2.6 Nutrient concentrations in microflora and soil fauna

Organism	Nutrient content (% dry weight)				
	N	P	K	Ca	Mg
Fungal mycelium (mean) (Swift, 1977)	1.74	0.48	0.33	4.28	0.13
Bacteria (Ausmus, Edwards and Witkamp, 1976)	4.00	0.91	1.50	–	–
Actinomycetes (Ausmus, Edwards and Witkamp, 1976)	4.20	1.00	4.00	–	–
Oligochaetes (segmented worms) (Allen et al., 1974)	10.5	1.1	0.5	0.3	0.2
Diplopods (centipedes, millipedes) (Allen et al., 1974)	5.8	1.9	0.5	14.0	0.2
Insects (Allen et al., 1974)	8.5	6.9	0.7	0.3	0.2
Arachnids (spiders, mites) (Allen et al., 1974)	9.0	1.0	0.6	0.4	0.3
Molluscs (snails, slugs) (Allen et al., 1974)	6.5	1.5	0.6	1.9	0.2

Simply, the higher the soil moisture content, the more energy is required to heat up the soil. Martin and Holding (1978) suggest that the lowering of soil temperatures in waterlogged soils may be a major factor in determining the rate of anaerobic microbial activity. In wet soils, some radiant energy is lost in evaporation at the soil surface and this, together with the high specific heat of water compared to air, means that they warm up only slowly. Since the thermal conductivity of water is about 100 times that of air, wet soils retain their warmth longer than dry soils.

Soil temperatures markedly influence soil nutrient mineralisation by controlling the rate of microbial activity. The Q_{10} value is an expression used to compare rates of processes at different temperatures such that:

$$Q_{10} = \frac{\text{rate at t} + 10\,^{\circ}\text{C}}{\text{rate at t}\,^{\circ}\text{C}} \tag{2,2}$$

where t = initial temperature ($^{\circ}$C)

Q_{10} values of around 2 have been reported for aerobic microbial respiration (for example, MacFadyen, 1967) and for nitrogen mineralisation (Stanford, Frere and Schwaninger, 1973) in litter and soil for the temperature range 10–30°C. Different organisms have slightly different Q_{10} values, all changing over different temperature ranges. Seasonal variation in microbial respiration due to soil temperature changes has been widely reported for a range of forest soils and litters (see, for example, Anderson, 1973; Phillipson et al., 1975; Steubing, 1977). Most incubation experiments which have studied the influence of temperature on rates of decomposition processes have adopted a constant, usually elevated temperature over prolonged periods. Understandably, there has been difficulty in relating these artificial estimates of organic matter decomposition and nutrient mobilisation to rates and processes operating in the field. Comparative studies for a range of fluctuating temperatures (Stanford, Frere and Vander Pol, 1975) and for simulated diurnal field conditions (Ross, 1985) indicate that frequent and regular fluctuation of temperatures does not appear to produce significantly different decomposition rates when compared to constant temperature (20°C). The results of many incubation studies suggest that the optimum temperature for organic matter decomposition and nitrogen mineralisation are around 25–35°C. This suggests that few, if any, soil micro-organisms operate optimally in temperate soils.

2.3.4 Freezing and thawing

The temperature at which soil water freezes decreases with decreasing soil moisture content. The expansion and contraction which accompanies freezing and thawing physically disrupts soil aggregates and particles. Freezing is also lethal to the soil microbial biomass. These two factors, as in the case of wetting and drying, are thought to be the reason for increased rates of nitrogen mineralisation (Jager, 1968; Hinman, 1970) and phosphorus mineralisation (Allen and Grimshaw, 1960) after freezing and thawing.

2.3.5 Leaching

Leaching is usually considered to be the process whereby readily soluble soil components, organic and inorganic, are removed in percolating water. Heal (1979) includes in his definition of leaching the removal of fine, particulate organic matter, which, in nutrient-poor upland soils, may constitute a substantial loss of plant nutrients from the soil system. The balance between removal and replacement of nutrient ions in soil depends on the rate of leaching and occasionally volatilisation, compared to inputs from organic matter decomposition, the weathering of soil minerals and atmospheric precipitation.

The main factors influencing the rate of leaching processes are considered by Wiklander (1974) to be

(1) soil structure and texture and hence soil pore space distribution,
(2) the amount and distribution of precipitation, and
(3) topography, surface vegetation and the depth and duration of ground frost.

Leaching of soluble nutrients may initially decrease numbers of soil micro-organisms by reducing the amounts of nutrients readily available for uptake. On the other hand, the removal of particulate organic matter may enhance nutrient mobilisation rates by exposing organic matter surfaces previously unavailable to the soil decomposer population. The chemical transformations and leaching processes associated with podzolisation is dealt with in detail in Chapter 4.

2.4   MODELLING DECOMPOSITION PROCESSES

As in other branches of the environmental sciences, many different modelling approaches have been adopted in the study of decomposition processes. The most widely reported of these fall into four main categories:

(1) iconic, material (hardware) models (small-scale versions of reality),
(2) conceptual models (abstract, theoretical),
(3) mathematical models (theoretical and predictive, often including empirical elements): this group is divided into two types, stochastic and deterministic, depending on whether the model variables are considered to be random, having distributions in probability, or whether the model variables are considered to be free from random variation.

Many of the decomposition experiments discussed earlier in the chapter can be considered to be iconic models. The most widely used of these are laboratory incubation and leaching studies, where field litter, peat or soil material is maintained under controlled environmental conditions, usually of temperature and moisture, for a predetermined experimental period. The aim of such studies is generally to reproduce field decomposition processes using environmental conditions considered to be optimal for micro-organisms. The more sophisticated incubation studies have examined the kinetics, or rate reactions, of biochemical processes such as nitrification and denitrification.

2.4.1   Simple conceptual models of decomposition processes

A widely used concept in soil organic matter studies is the *turnover time*. This can be strictly defined as:

$$t_c = \frac{C_s}{C_p} \qquad\qquad (2,3)$$

where: t_c = turnover time (years)
C_s = soil organic carbon (top 50 cm only)
C_p = net primary production of carbon on land (per year);

t_c is really only calculable if steady state input–output of carbon to and from the soil is assumed. This, of course, is a dramatic oversimplification if variations in climate, vegetation and soil processes are considered. However, this simple calculation allows a useful comparison of organic matter dynamics on a world scale (Table 2.7). As expected, the tropical rainforest has the fastest organic matter turnover time, about three to four times faster than in temperate agriculture systems.

Jenny, Gessel and Bingham (1949) were the first to introduce decomposition into a simple expression describing steady state in an ecosystem:

$$\frac{X}{D} = 1 \qquad (2,4)$$

where: X = total standing crop of dead organic matter
D = total decomposition

The rate of decomposition processes acting in the system is conventionally expressed as:

$$k = \frac{I}{X} \qquad (2,5)$$

where I = total annual input of dead organic matter (litter)
k = annual fractional weight loss

Values of 3/k and 5/k can be calculated to give estimates of the time taken for 95 per cent and 99 per cent respectively of the standing dead organic matter to decompose. Swift, Heal and Anderson (1979) indicate that the main limitation to the use of this simple expression is the problem of accurately measuring below-ground inputs of dead organic matter from root death and exudates. The alternative approach is to calculate organic matter dynamics for the above-ground portion of the system only. Thus:

$$k_L = \frac{L}{X_L} \qquad (2,6)$$

where k_L = annual fractional weight loss of above-ground organic matter
L = annual input of dead plant litter
X_L = standing crop of dead organic matter at soil surface (litter)

k_L values can be used only as a general guide to the decomposition rate in terrestrial ecosystems. This is because the standing crop of plant litter, or soil surface organic matter, is spatially and temporally variable and can rarely be considered to be in a steady state.

A sequence of conceptual decomposition models have been developed

Table 2.7 The turnover of organic carbon in the soil of some contrasting ecosystems (compiled by Jenkinson, 1981)

	Soil sampling depth (cm)	Net primary production[a] (tonnes carbon ha⁻¹y⁻¹)	Carbon entering the soil (tonnes ha⁻¹y⁻¹)	Carbon in the soil (tonnes ha⁻¹)	Turnover time (years)	Reference
Continuous wheat, unmanured (Rothamsted)	0–23	2.6	1.2	26	22	Jenkinson and Rayner (1977)
Continuous ley unmanured (Rothamsted)	0–23	2.7–3.2	20.–2.5	77	31–38	Jenkinson (1969)
Humid savannah	0–30	5.0	1.5[b]	56	37	Greenland and Nye (1959)
Tropical rainforest	0–30	9–10	4.9[c]	44	9	Greenland and Kowal (1960)
Cold temperate beech forest	0–30	7.1	2.4[cd]	72	30	Nihlgard (1972)

a Assuming 40 per cent carbon in plant material where the carbon content is not given.
b Assuming that all above-ground production not harvested is destroyed annually by burning.
c Assuming that the biomass in wood does not enter the soil.
d Assuming that all the litter fall and two-thirds of the roots enter the top 30 cm.

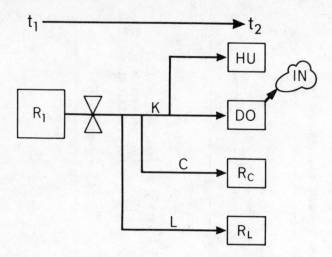

Figure 2.5 Decomposition of resource (R) over a short period of time (t_1–t_2). The three component processes: catabolism (K), comminution (C) and leaching (L) result in chemical changes, such as mineralisation, giving rise to inorganic forms (IN), resynthesis of decomposer tissues (DO) and humus (HU) as well as physical changes, such as reduction in particle size of chemically unchanged litter (R_c) and removal of soluble resource materials in unchanged form to other sites (R_L). (Source: Swift, Heal and Anderson, 1979)

by Swift, Heal and Anderson (1979), based on one simple module (Fig. 2.5) in which three decomposition processes are defined:

(1) *Leaching* (L): the abiotic process whereby percolating water removes soluble mineral and organic materials. Weight loss and chemical alteration may result. Leaching transfers material down-profile where further decomposition may occur.
(2) *Catabolism* (K): the energy-yielding reactions or sequences of reactions described earlier as aerobic respiration, anaerobic respiration and fermentation. Fate of catabolised products: (a) resynthesised into microbial tissues; (b) incorporated into soil humus; (c) lost – due to volatilisation or leaching.
(3) *Comminution* (C): the physical breakup of particles, due to (a) passage through microbial mouthparts and gut, or (b) environmental processes such as wetting/drying and freezing/thawing.

The switch symbol in Fig. 2.5 represents the rate controlling mechanisms of decomposition processes. Swift, Heal and Anderson (1979) give the three major rate controls as

(1) factors of the physico-chemical environment such as moisture, temperature and soil parameters such as pH,

(2) organic matter resource quality, which is an indication of substrate palatability, and

(3) the suite of decomposer organisms present.

The products of the unit decomposition module: HU (humus), DO (decomposer tissue), IN (inorganic nutrients), R_C (chemically unchanged organic matter) and R_L (translocated organic matter), can, themselves, be acted upon by another series of microbial decomposers. If it is assumed that the inorganic component (IN) is not further decomposed, then a *cascade* structure of organic resource model can be formulated (Fig. 2.6). In the Swift, Heal and Anderson (1979) cascade model, organic materials entering the decomposition model directly from primary production are termed the primary resources (R_1), made up of plant litter; organic materials formed by secondary production are termed the secondary resources and comprise the corpses of microbial decomposers. In Fig. 2.6, R_2, R_3 and R_4 represent changing mixtures of plant litter, microbial corpses, faeces and humus, accepting that

(a) rates of operative processes between time periods t_1 and t_4 may vary considerably, and

(b) that inorganic nutrients released during one decomposition stage may be immobilised in microbial tissues at successive stages.

Some useful refinements have been made to these simple conceptual models to relate them to specific field conditions. Jones and Gore (1978), working on accumulation and decomposition in the blanket bog at Moor

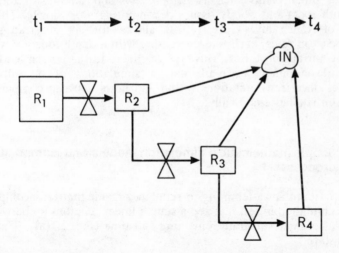

Figure 2.6 Cascade model of decomposition processes. Over t_1–t_4 progressive change of the primary resource (R_1) through states R_2, R_3 and R_4 occurs. For any intermediate stage (for example, t_2–t_3), the products of the preceding module (for example, R_2) is the starting resource. (Source: Swift, Heal and Anderson, 1979)

House National Nature Reserve, in the Pennines, Northern England, imposed a two-tier structure on their decomposition model. Firstly, they decided that the decay rate would vary with depth due to changing aeration status. Instead of adopting the simple two-phase decay scheme of aerobic (above the water table) and anaerobic (waterlogged, below the water table) such as that proposed by Gore and Olson (1967), they divided the peat depth into a series of 1 cm layers or 'boxes' in order to give a more continuous change in decomposition rate with depth. Once input of litter and dead organic matter to the top box exceeded a thickness of 1 cm, overflow and output from decay became the input to the next box where accumulation continued, with decay rate modified by the required depth factor. This sequence continued down the peat profile. They introduced a second tier of decay control into their model by subdividing the peat in each 1 cm layer into six independently decomposing organic substrate:

(1) *Calluna*: shoots, wood and below-ground parts;
(2) *Eriophorum*: leaves and below-ground parts; and
(3) *Sphagnum*.

This choice of substrate division reflected the results of decomposition studies by a range of authors, working within the International Biological Programme, Tundra Biome group (see, for example, Clymo, 1965; Heal and French, 1974; Heal, Latter and Howson, 1978). Such studies report the expected sequence of decomposition, with leaves and shoots which are fleshy and soft, such as *Sphagnum*, decomposing fastest, fibrous leaves of other plants such as *Eriophorum* and *Calluna* decomposing next, with roots and woody tissues decomposing most slowly. Thus, the model of Jones and Gore (1978) allows for six substrate-specific decomposition rates, each rate decreasing with a depth function which is the same for all substrates. This type of theoretical approach lends itself to the development of mathematical simulations of decomposition processes once transformations within and flows between compartments have been studied empirically.

2.4.2 Simple mathematical simulation models of decomposition processes

The earliest studies attempting to simulate organic matter decomposition and accumulation in soil rejected a simple linear relationship between the weight of organic substrate remaining and time (Fig. 2.7(a)). The linear relationship:

$$X_t = X_o - kt \qquad (2,7)$$

where: x_t = amount of organic substrate remaining after time, t
x_o = amount of organic substrate at start (t = 0)

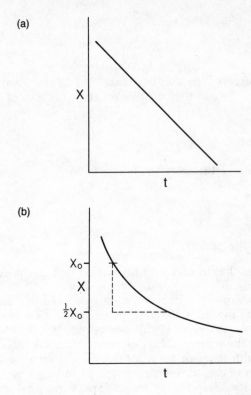

Figure 2.7 Examples of **(a)** linear ($X_t = X_o - kt$), and **(b)** exponential $X_t = X_o e^{-kt}$ decay

$$t = \text{time}$$
$$k = \text{reaction rate constant}$$

appeared to work satisfactorily for fast decomposing tissues such as fleshy and fresh (young) substrates, but for most decomposition studies a negative exponential function, analogous to the first-order kinetic decay of a radioactive substance, is more applicable. This relationship implies the loss of a constant fraction of the weight of substrate remaining rather than the loss of constant increments of weight over successive equal intervals of time. So, from Fig. 2.7(b),

$$\frac{-dX}{dt} = kX \tag{2,8}$$

rearranging,

$$\frac{dX}{X} = -k\,dt \tag{2,9}$$

which integrates to:

$$X_t = X_o e^{-kt} \tag{2,10}$$

Equation **(2,10)** represents exponential decay and is the basis of the concept of a *half-life*. When half of the original amount of organic substrate (X_o) remains,

$$X = \tfrac{1}{2}X_o \tag{2,11}$$

substituting in Equation **(2,10)** gives:

$$\tfrac{1}{2} = e^{-kt_{1/2}} \tag{2,12}$$

$$2 = e^{kt_{1/2}} \tag{2,13}$$

where: $t_{1/2} =$ half-life period

Figure 2.8 illustrates that in each half-life period half of the substrate remaining undergoes decay.

Decay of a pure, organic resource would only depart from the simple first order exponential model if the rate of decomposition changes as decomposition proceeds. As Hunt (1978) points out, this could occur if (a) the size, or (b) the composition of the microbial decomposer population changed with time, or (c) if comminution over time changed the size of the particles of organic resource, exposing a larger surface area for microbial attack. Realistically, we might expect the rate of organic matter decomposition in soil initially to increase after a short lag, during which time the decomposer population becomes established and develops to operate optimally. A complex mixture of soil organic materials, decomposing under constant environmental conditions of temperature

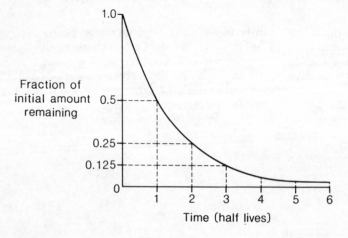

Figure 2.8 The concept of 'half-life'

and moisture, are unlikely to conform to the exponential decay model. Studies such as those by Floate (1970) indicate that actual rates are greater than those predicted by the exponential model in the early stages of decay when the more easily decomposed components, such as carbohydrates, are present. Rates are less than those predicted in the later stages of decomposition when the more resistant components such as lignins and lignin derivatives predominate.

While Minderman's (1968) plot indicates exponential decay curves for the six individual organic components, phenols, waxes, lignins, cellulose, hemicellulose and sugars, neither the summation of individual component curves (see Fig. 2.3, curve S) nor the total litter decomposition for a mor oak forest (see Fig. 2.3, curve M) fit the exponential decay model. So, by using a single, simple exponential model with a rate constant based on the first few years of decay, longterm decomposition would be overestimated. It would be feasible to model decomposition of the total substrate using the sum of the individual components, each decaying according to first order kinetics:

$$\frac{dX}{dt} = \sum_{i=1}^{n} \frac{dX_i}{dt} = - \sum_{i=1}^{n} k_i X_i \qquad (2,14)$$

where X = total organic substrate
$\quad\quad n$ = number of individual organic components
$\quad\quad k_i$ = rate constant for the decay of component X_i

This approach has been adopted with some degree of success in the rather complex model developed by Bunnell et al. (1977b). Hunt (1978) has suggested that there would be several complications connected with the idea of separately modelling each chemical component of a complex resource:

(a) all pure organic substances may not conform to exponential decay;
(b) some chemical components are formed from others during decomposition, so the generation of new substances must be included; and
(c) the presence of certain organic compounds may inhibit the decomposition of others.

In addition, the exact composition of some soil organic matter, such as the lignin derivatives, is not accurately known, so could not be incorporated into a model. Since we do know that some components of soil organic matter decay rapidly and some are more resistant, the two-component approach adopted by Hunt (1978) is a useful simplification. He divides soil organic matter into two fractions: one labile, rapidly decomposing fraction, consisting of sugars, starch and proteins; and one resistant, slowly decomposing fraction, consisting of cellulose, lignin, fats, resins, etc. Each fraction decomposes exponentially. Using the notation of Equation (2,10), the two component model gives:

$$X_t = L\,e^{-kt} + (1 - L)\,e^{-ht} \qquad\qquad (2,15)$$

where L   = initial proportion of labile organic material
   (1−L) = initial amount of non-labile (resistant) organic material
   k   = decomposition rate coefficient for labile organic material
   h   = decomposition rate coefficient for resistant organic material

Hunt found that calculated k and h values showed the same order of magnitude for a range of organic substrates (Table 2.8). This suggested broadly similar proportions of labile and resistant organic constituents in most fresh litter materials and allowed a useful simplification of decomposition dynamics.

Table 2.8   Estimate of parameters of decomposition model (Equation **2,15**)

Material added to soil	L	$k\,10^2$	$h\,10^4$
Oak leaves	0.223	4.09	9.71
Oats	0.602	4.37	5.21
Wheat straw	0.324	2.36	7.35
Soybeans	0.597	4.82	1.93
Corn stover	0.270	4.49	11.07

2.4.3   Some applications of mathematical simulation models

A plethora of mathematical models have been developed which not only allow for the decomposition of a range of differentially resistant organic compounds, but also attempt to simulate decomposition as influenced by environmental controls such as temperature, moisture content and oxygen availability (for example, Bunnell et al., 1979a; Smith, 1979a; 1982). In many cases, these models also deal with soil nitrogen dynamics, since soil organic carbon and organic nitrogen are so closely related. Smith (1979b) uses his model to simulate mobilisation and immobilisation of N as a function of substrate C:N ratio. His model simulations for additions of low C:N ratio substrate (protein) and high C:N ratio substrate (sugar) (Fig. 2.9) mimic closely the thoeretical schemes proposed by Black (1968).

   Jenkinson and Rayner (1977) developed a five-fraction model to fit field data derived from long-term cereal plots at Rothamsted Experimen-

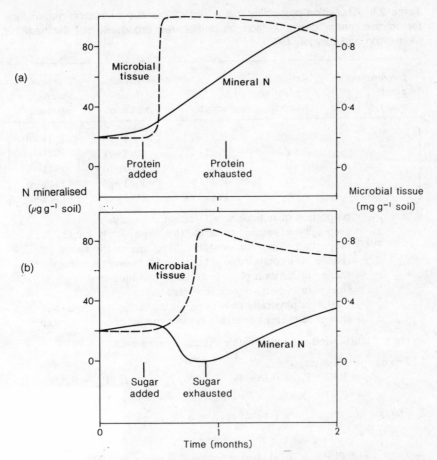

Figure 2.9 Model calculation of changes in content of mineral nitrogen and microbial tissue with time in fallow soil: **(a)** protein added; **(b)** sugar added. (Reproduced with permission from Smith, 1979b, *Soil Biology and Biochemistry*, 11, © Pergamon Books Ltd.)

tal Station, England. Like the decomposition submodel formulated by Kirkby (1977), these fractions are fitted into a transition-probability matrix, in which each loci represents the probability of the transition from one fraction to any other fraction after one time step (one year in their model) (Table 2.9). The transition probabilities after two time steps is obtained by squaring the matrix (Table 2.10), for three time steps by cubing the matrix and so on, until equilibrium (steady-state) conditions are reached. At this stage, matrix values do not change on being raised to a higher power. Each row of the matrix is the same and this stage defines the limit at which probabilities passing from one state to another are independent of the starting state. This condition is called the fixed probability vector, since the matrix can now be described by a single row.

Table 2.9 Transition probability matrix, with input and simulation parameters, for organic matter decomposition on unmanured Broadbalk plot, Rothamsted (Jenkinson and Rayner, 1977)

Decomposable plant material	Resistant plant material	Biomass	Physically stable products	Chemically stable products
d	o	$P_B(1-d)$	$P_P(1-d)$	$P_C(1-d)$
o	r	$P_B(1-r)$	$P_P(1-r)$	$P_C(1-r)$
o	o	$b+P_B(1-b)$	$P_P(1-b)$	$P_C(1-b)$
o	o	$P_B(1-p)$	$p+P_P(1-p)$	$P_C(1-p)$
o	o	$P_B(1-c)$	$P_P(1-c)$	$c+P_C(1-c)$

Where: P_B = proportion of organic substrate that is biomass
P_P = proportion of organic substrate that is physically stable
P_C = proportion of organic substrate that is chemically stable
d = change in decomposable plant material in one time step
r = change in resistant plant material in one time step
b = change in biomass in one time step
p = change in physically stable organic matter in one time step
c = change in chemically stable organic matter in one time step

In the model, the following parameter values were adopted:

1. *Fraction proportions*
   $P_D = 0.837$; $P_R = 0.163$; $P_B = 0.076$; $P_P = 0.125$; $P_C = 0.0035$

2. *Fraction change per time period* $\left(\text{eg: } d = e^{-k_D} = \dfrac{D_1}{D_0} \right)$

   Erratum

   p.68 Table 2.9
   No.2 end of second line
   should read c = 0.99965

   $d = 0.015$; $r = 0.741$; $b = 0.664$; $p = 0.986$; $c =$

3. *Rate constants* (year^{-1}):
   $k_D = 4.2$; $k_R = 0.3$; $k_B = 0.41$; $k_P = 0.14$; $k_C = 0.00035$

4. *Half-lives of fractions* (yrs.):
   $t_D = 0.165$; $t_R = 2.31$; $t_B = 1.69$; $t_P = 49.5$; $t_C = 1980.0$

For the unmanured Broadbalk plot at Rothamsted, an annual input of 1.2 t carbon ha^{-1} is assumed. Using this annual input and the initial proportions of the five fractions given in Fig. 2.10, the Jenkinson and Rayner (1977) model predicts that after 10,000 years (considered to be the time taken for the system to reach steady state), the soil would contain 24.2 t carbon ha^{-1}, of which 0.01 t is decomposable organic matter, 0.47 t is resistant organic matter, 0.28 t is decomposer biomass, 11.3 t is physically stable organic matter, and 12.2 t is chemically stable organic matter. The model thus successfully simulates the production

Table 2.10 Transition probabilities for a simple two-fraction case after two time steps (two years)

$$P^2 = P \times P$$

$$\begin{bmatrix} p_{11}^{(2)} & p_{12}^{(2)} \\ p_{21}^{(2)} & p_{22}^{(2)} \end{bmatrix} = \begin{bmatrix} p_{11} & p_{12} \\ p_{21} & p_{22} \end{bmatrix} \times \begin{bmatrix} p_{11} & p_{12} \\ p_{21} & p_{22} \end{bmatrix}$$

$$= \begin{bmatrix} (p_{11} \times p_{11}) + (p_{12} \times p_{21}) & (p_{12} \times p_{12}) + (p_{12} \times p_{22}) \\ (p_{21} \times p_{11}) + (p_{22} \times p_{21}) & (p_{21} \times p_{12}) + (p_{22} \times p_{22}) \end{bmatrix}$$

And similarly, for three years:

$$P^3 = P^2 \times P$$

$$\begin{bmatrix} p_{11}^{(3)} & p_{12}^{(3)} \\ p_{21}^{(3)} & p_{12}^{(3)} \end{bmatrix} = \begin{bmatrix} p_{11}^{(2)} & p_{12}^{(2)} \\ p_{21}^{(2)} & p_{22}^{(2)} \end{bmatrix} \times \begin{bmatrix} p_{11} & p_{12} \\ p_{21} & p_{22} \end{bmatrix}$$

And, for n years:

$$P^n = P^{(n-1)} \times P$$

through time of predominantly stable organic matter decomposition end products.

Since we know how many transition states the model must complete to arrive at any given matrix configuration, it is thus possible to determine not only the half lives of each fraction (Fig. 2.10), but also the passage times required to obtain a given organic matter composition.

2.4.4 Mathematical modelling of biochemical processes

The close relationship between carbon and nitrogen dynamics during organic matter decomposition was indicated in Section 2.2.1. The immense agricultural importance of understanding soil nitrogen supply has undoubtedly been the reason for the huge investment and research interest in nitrogen dynamics generally (see, for example, Clark and Rosswall, 1981; Frissel and van Veen, 1981; Stevenson, 1982; Stewart and Rosswall, 1982), and more specifically in the study of soil nitrogen mineralisation potentials (see, for example, Stanford and Smith, 1972; Smith, Young and Miller, 1977; Juma, Paul and Mary, 1984).

A whole suite of incubation techniques have been developed for the study of potentially mobilisable, or 'active', soil organic nitrogen. A curvilinear relationship between cumulative inorganic N mobilised in

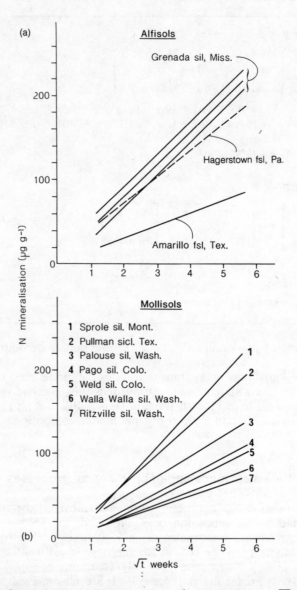

Figure 2.10 Cumulative N mineralisation (μg g^{-1}) in relation to $\sqrt{t}$ (weeks) for **(a)** alfisols (includes gleys and gleyed brown earths); and **(b)** mollisols (includes calcareous gleys, rendzinas and brown forest soils). Fsl = Fine sandy loam; sil = silty loam; sicl = silty clay loam. (Source: adapted from Stanford and Smith, 1972)

incubated soil samples with time has been indicated. All studies aim to define the rate of N mineralisation processes and thus can be classified as studies of N mineralisation kinetics. Studies of this type have generally described the relationship using a first order equation. Stanford and Smith (1972) estimated N mineralisation potentials (N_o) graphically from cumulative amounts of nitrogen mineralised in incubation experiments (Fig. 2.10) for 39 mineral soils in the United States, representing five soil orders in the USDA classification. It is assumed that nitrogen mineralisation reactions follow first order kinetics according to:

$$\frac{dN}{dt} = -kN \qquad (2,16)$$

which integrates over t_o to t as:

$$N_t = N_o \, (1 - e^{-kt}) \qquad (2,17)$$

which Stanford and Smith (1972) alternatively write as:

$$\log(N_o - N_t) = \log N_o - \frac{k}{2.303} \, (t) \qquad (2,18)$$

where: N_o = initial content of inorganic nitrogen
N_t = content of inorganic nitrogen after time t
k = rate constant
t = time

They obtained first approximations of N_t and k from the plot of $\frac{1}{N_t}$ versus $\frac{1}{t}$, which generally gives a linear plot, described by:

$$\frac{1}{N_t} = \frac{1}{N_o} + \frac{b}{t} \qquad (2,19)$$

where: b = slope of regression equation

in which the reciprocal of the intercept is an estimate of N_t (Smith, Schnabel and McVeal, 1980). Stanford and Smith (1972) subsequently evaluated N_t by regressing $\log(N_o - N_t)$ against t, using different successive values of N_o to derive the best-fit regression. Values of k were then obtained from the slope of $\log(N_o - N_t)$ versus t which is $\frac{k}{2.303}$ (Fig. 2.11). Several authors have recently published results of an alternative solution to Equation (2,17) using a non-linear least squares (NLLS) iteration method for the estimation of the parameters in the non-linear regression model (see, for example, Campbell, Jame and Winkleman, 1984; Juma, Paul and Mary, 1984).

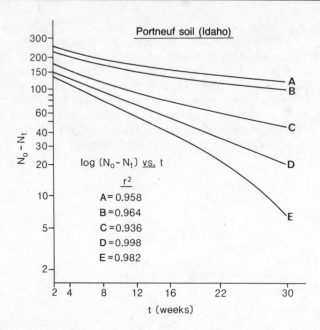

Figure 2.11 Stanford and Smith's method for obtaining N_o values, using 2–30 week incubations and deriving the best linear fit of the log (N_o-N_t) versus t relationship using successive values of N_o for Portneuf soil. (Source: Stanford and Smith, 1972)

Despite which mathematical solution to Equation (**2,17**) has been adopted, a range of values for N mineralisation rate in mineral, agricultural soils has been reported (Table 2.11), this range partly being explained by Campbell, Jame and Winkleman (1984) as a result of moisture and temperature functions. Perhaps the most useful result of these studies is the value for the mean proportion of mineralisable N released per week (column 6 in Table 2.11). For the results of Juma, Paul and Mary (1984), the mineralisable N fraction is released at an average rate of 6.9 per cent per week at 28°C, based on the quantity of mineralisable N remaining after each succeeding week of incubation. Obviously for field agricultural predictions, Q_{10} values must be used to relate N mineralisation rates obtained at elevated incubation temperatures to likely field soil conditions. That Campbell, Jame and Winkleman (1984) found significant correlation between laboratory (35°C) incubation and field (15°C) incubation results is an indication that such predictions are justifiable and potentially useful.

2.5 SUMMARY

Since the exact composition of soil organic matter is unknown, it is impossible to write equations to describe the reactions which occur during

Table 2.11 Range of values for N mineralisation rate in mineral soils

Reference	Soils	Incubation temperature	Estimation technique	k Values	Mean proportion of mineralisable N released per week (%)
Stanford and Smith (1972)	Wide range of agricultural (mineral) soils	35°C	Graphical (double reciprocal plot)	0.044–0.069	5.4
Mary and Remy (1979)		35°C		0.028–0.040	
Campbell et al. (1981)	Queensland mineral soils	35°C	NLLS	mean 0.058	5.8
		25°C	NLLS	mean 0.031	3.1
		15°C	NLLS	mean 0.018	1.8
Campbell, Jame and Winkleman (1984)	Canadian prairie soils	35°C	NLLS	0.050–0.228	10.91
		25°C	NLLS	0.014–0.093	5.17
		15°C		0.009–0.044	3.55
Juma, Paul and Mary (1984)	Canadian chernozem soils	28°C	NLLS	0.036–0.164	6.9

organic matter decomposition. For this reason, it is unusual to predict decomposition sequences using chemical thermodynamics. More commonly, substrate parameter (organic matter quality and quantity) and environmental factors (such as temperature, moisture, oxygen concentration, pH) are used to estimate rates of organic matter decomposition empirically. Organic matter quality is a measure of its palatibility (amount of fibre and lignin present), its energy content (amount of carbon compounds present, which provide an energy source for decomposer organisms) and its nutrient content (particularly N, P, K and Ca). Generally, organic materials or litters containing large amounts of lignin, resins and waxes is decomposed most slowly, while litters containing carbohydrates, such as polysaccharides, proteins and amino acids, are decomposed most readily. The C:N ratio of litter is a useful index of decomposability: the lower the value, the more readily will decomposition occur. Rates of organic matter decomposition are temperature sensitive, with Q_{10} values of around 2. Conceptual and mathematical models of organic matter decomposition are available to predict the rates of decomposition of vegetation litter at the soil surface, predict organic matter loss, particularly in organic soils such as peats and peaty podzols in natural ecosystems, and to anticipate organic matter losses in agricultural soils through overcultivation.

PART TWO

Processes of Profile Development

Unique distributions of meteorological conditions, topography, lithology and vegetation worldwide make for a very large number of different soil profile types. The simplest index for general soil profile development is Thornthwaite's (1931) P-E index, which is the annual sum of the precipitation:evaporation ratios. P-E values range from 128 and above in tropical rainforest, to < 16 in arid desert conditions. These values indicate *moisture effectiveness*; high values representing a net downward movement of drainage water and predominant leaching.

The wide range of translocation and transformation processes in operation worldwide would be impossible to review adequately in a single textbook. Discussion here is restricted to temperate humid conditions and to extremes in profile development: soils with impeded drainage, represented here by gleys, and soils with excessive drainage, represented here by podzols. In theory, gleying and podzolisation processes represent extreme development conditions, but in practice, they operate to different degrees in different soils and in different locations in the same soil. Thus, we may identify podzolised brown earths, gleyed brown earths, and even gleyed podzols, where rather different developmental processes are operating in different parts of the same soil profile. This does not mean to imply that rates of developmental processes are similar. The presence of ochreous mottles along root channels or on the faces of structural peds represent the initiation of gleying and can occur whenever oxygen concentrations become sufficiently depleted to allow iron reduction to occur. A seasonal timescale for this would not be unreasonable. By definition, the initiation of podzolisation means the deposition of iron and aluminium hydroxyoxides in the B horizon. Even in very coarse textured sands under conifer plantation, initiation of podzol development requires 20–24 years. It is clearly very important to appreciate the relative timescales over which profile development processes operate and to remember that individual horizons within the same soil profile may have been formed, or may be forming, at quite different rates.

CHAPTER 3

Processes in Soils with Impeded Drainage

3.1 INTRODUCTION

Impeded drainage or very slow rates of soil water movement produce waterlogged conditions. These conditions may be due to a whole host of contributory controls such as clay texture, claypan, ironpan, compaction or smearing, presence of a high local water table, surface ponding due to crusting, cementation or the presence of duricrusts. The overriding influence of waterlogging on soil processes is due to anaerobic conditions. Since the rate of diffusion of oxygen through water $(0.226 \times 10^{-4} \text{ cm s}^{-1})$ is ten thousand times slower than the rate of oxygen diffusion through air $(0.209 \text{ cm s}^{-1})$, the relative proportions of air and water in the soil porespace control the oxygen concentration. Once soil oxygen has been utilised by soil micro-organisms during aerobic respiration, their future activity is determined by the rate at which it can be replaced from the surrounding soil or from the outside atmosphere. Thus, under waterlogged conditions, in clay soils and in compacted horizons with a fine, tortuous porespace, the rate of oxygen replacement is slow and reducing conditions develop. It is under these conditions that anaerobes and facultative anaerobes carry out biochemical redox reactions. So the oxygen concentration at any location in the soil thus has a very important influence on biochemical transformations, organic matter accumulation, mineral solubilities and the build-up of anoxic compounds. Anaerobic conditions are not restricted to waterlogged soils and have been shown by Smith and Dowdell (1974) to occur within structural peds in cultivated but generally well-aerated soils.

3.2 SOIL REDOX CHEMISTRY

Both organic and inorganic substrates can be used as electron acceptors by facultative anaerobes during anaerobic metabolism. An example of the general form of the reaction is given as:

$$\text{Oxidised state} + \text{electrons} \rightleftharpoons \text{Reduced state} \qquad (3,1)$$

This equation represents one redox couple which is half of a fully balanced oxidation–reduction reaction. Table 3.1 summarises a series of such reactions for (a) some inorganic soil substrates, and (b) some organic soil substrates.

Table 3.1 Redox half cells for some (a) inorganic and (b) organic soil substrates

(a) Inorganic substrates

Oxidised state		Reduced state	Eh at pH=7 (mV)
$O_2 + 4H^+ 4e^-$	$\rightleftharpoons$	$2\,H_2O$ water	820
$NO_3^- + 2H^+ + 2e^-$ nitrate	$\rightleftharpoons$	$NO_2^- + H_2O$ nitrite	420
$Fe^{3+} + 2H^+ + 2e^-$ ferric	$\rightleftharpoons$	$Fe^{2+} + 3H_2O$ ferrous	−180
$SO_4^{2-} + 10H^+ + 8e^-$ sulphate	$\rightleftharpoons$	$H_2S + 4H_2O$ hydrogen sulphide	−220
$CO_2 + 8H^+ + 8e^-$ carbon dioxide	$\rightleftharpoons$	$CH_4 + 2H_2O$ methane	−240

(b) Organic substrates

$$CH_3\overset{O}{\overset{\|}{C}}OH + 2H^+ + 2e^- \rightleftharpoons CH_3\overset{O}{\overset{\|}{C}}H$$

acetic acid   acetaldehyde

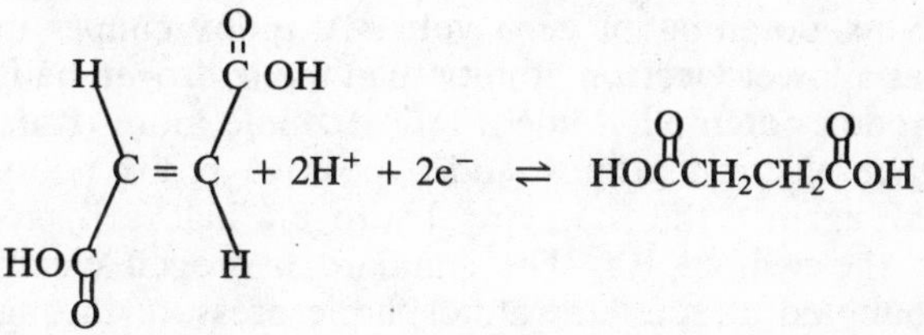

$$\underset{\substack{HO\overset{\|}{\underset{O}{C}} \quad\quad H}}{\overset{\substack{H \quad\quad \overset{O}{\overset{\|}{C}}OH}}{C=C}} + 2H^+ + 2e^- \rightleftharpoons HO\overset{O}{\overset{\|}{C}}CH_2CH_2\overset{O}{\overset{\|}{C}}OH$$

fumaric acid   succinic acid

In any redox reaction, the couple with the greater affinity for electrons assumes the oxidising role while it itself is reduced. The second couple which has a smaller affinity for electrons is oxidised. A laboratory electrical cell is a useful example to illustrate this principle (Fig. 3.1). At the negative cathode, the oxidant gains electrons and it itself is reduced while at the positive anode, the reductant loses electrons and is oxidised. The cell operates according to Faraday's (1834) Second Law of Electrolysis: 'if one faraday of electricity passes between two electrodes in an electrolyte, 1 g equivalent of substance will be liberated at each electrode'. So when two redox couples operate in the soil, an electrical potential is generated.

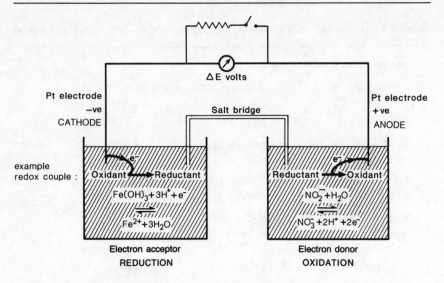

Figure 3.1   Electrical cell made up of two redox half cells or redox couples

The accepted reference point for redox potentials is the standard hydrogen half-cell:

$$2H^+ + 2e^- \rightleftharpoons H_2 \text{ gas} \tag{3,2}$$

which is given a redox potential of zero volts. A redox couple more reducing than this has a lower electron affinity than the hydrogen half cell and has a negative redox potential, while a redox couple more oxidising than this has a higher electron affinity and a positive redox potential (Fig. 3.2). Measured against the standard hydrogen half cell, redox potentials are given the symbol Eh. The standard hydrogen electrode consists of H_2 gas bubbled at standard atmospheric pressure through a solution having unit H activity in which a platinum electrode is placed. This technique is impractical for general purposes and, instead, the slightly more oxidising saturated calomel half cell is used:

$$Hg_2Cl_2 + 2e^- \rightleftharpoons 2Hg + 2Cl^- \tag{3,3}$$

whose Eh at 20°C is +0.248 V
so   Eh = Ecal + 0.248 V

Since redox reactions are thermodynamically reversible, we can calculate the free energy of the reaction in the same manner as we did in Chapter 1. The change in free energy (ΔG) for the reduction couple: $Ox + ne^- \rightleftharpoons Red$ can be calculated by:

$$\Delta G_r^\circ = \Delta G + RT \ln \frac{(Red)}{(Ox)} \tag{3,4}$$

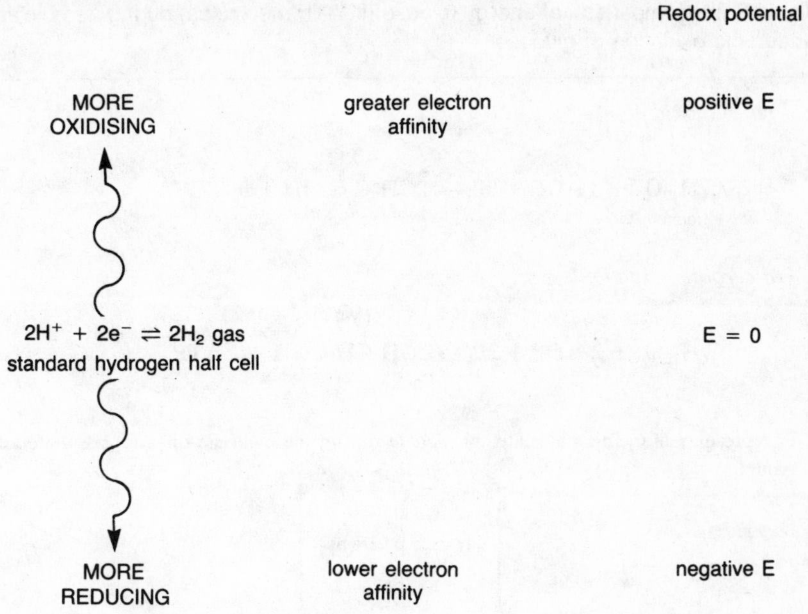

Redox potential

MORE
OXIDISING

greater electron
affinity

positive E

$2H^+ + 2e^- \rightleftharpoons 2H_2$ gas
standard hydrogen half cell

E = 0

MORE
REDUCING

lower electron
affinity

negative E

Figure 3.2 Generalised soil redox potentials

where (Red) and (Ox) represent the activities of the reduced and oxidised species and ΔG is the change in free energy when the activities are unity (Ponamperuma, 1972). Using the relationship:

$$\Delta G^\circ = - nFE^\circ \qquad (3,5)$$

to convert calories into volts, we obtain:

$$Eh = E^\circ + \frac{RT}{nF}\ln \frac{(Red)}{(Ox)} \qquad (3,6)$$

where: Eh = voltage of the reaction measured against the standard hydrogen electrode
E° = voltage when (Ox) and (Red) are each unity
F = Faraday constant in heat units
n = number of faradays (or moles) of electrons in the reaction
R = Universal gas constant
T = absolute temperature (°K)

E° is the *standard redox potential* and is the Eh value the cell would have if both oxidised and reduced states had unit activity. As in Chapter 1, it is usual to use concentrations of oxidised and reduced species instead of their activities. Thus the E° represents the Eh the cell would have if

Table 3.2 Comparison of energy release (in ATP molecules) during aerobic and anaerobic oxidation of glucose

1. *Aerobic oxidation**

$$C_6H_{12}O_6 + 8\ ADP + 8P_i \rightarrow 2CH_5\overset{O}{\overset{\|}{C}}\text{-}\overset{O}{\overset{\|}{C}}\text{-OH} + 8\ ATP$$
glucose pyruvic acid

2. *Anaerobic oxidation**

$$C_6H_{12}O_6 + 2\ ADP + 2P_i \rightarrow 2CH_3\overset{OH}{\overset{\|}{C}H}\text{-}\overset{O}{\overset{\|}{C}}OH + 2\ ATP$$
lactic acid

* P_i represents inorganic phospate, phosphate during the cleavage of one phosphate acid group:

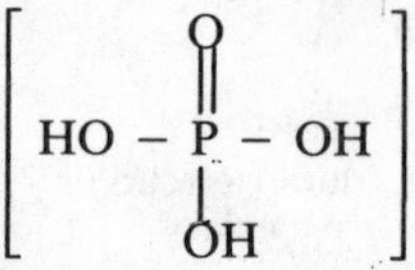

$$\left[\ HO - \overset{\overset{O}{\|}}{\underset{\underset{OH}{|}}{P}} - OH \ \right]$$

concentrations of oxidised and reduced states were equal. If redox couples are written as reduction potentials and arranged in descending order (Table 3.2) then, theoretically, a given system can oxidise any system below it in the table.

Since the H^+ ion itself contributes to the redox system both through the $2H^+ + 2e^- \rightleftharpoons H_2$ gas reaction and by its often implied involvement in every other redox system, for example:

$$NO_3^- + 2H^+ + 2e^- \rightleftharpoons NO_2^- + H_2O \tag{3,7}$$

the solution pH is an important control on redox reactions and on redox potentials. Solution pH may also affect Eh through its influence on the dissociation of participating oxidants and reductants. Thus, Eh values are usually quoted at a standard pH of either pH 5 or pH 7. Equation (3,6) can be modified to take account of the H^+ ion activity:

$$Eh = E° + \frac{RT}{nF} \ln \frac{(Ox)}{(Red)} (H^+)^m \tag{3,8}$$

where m = number of moles of H ions reacted

so

$$Eh = E° + \frac{RT}{nF} \ln \frac{(Ox)}{(Red)} + 2.303 \frac{RT\ m}{nF} \log(H^+) \tag{3,9}$$

$$Eh = E° + \frac{RT}{nF} \ln \frac{(Ox)}{(Red)} - 0.059\frac{m}{n} pH \qquad (3,10)$$

thus:

$$E°_{pH7} = E° - 0.413\,\frac{m}{n} \qquad (3,11)$$

where $E°_{pH7}$ is the standard redox potential ($E°$) at $pH = 7$

Eh values are also quoted at standard temperatures of either 20°C or 25°C, since temperature also has an important effect on dissociation reactions.

Sillen (1967) suggested that it would be useful to express the electrons in any redox reaction in the same way as the participating oxidants and reductants. The term pE was introduced to mean the −log of the electron activity in very much the same way that $pH = -\log H^+$ ion activity. So:

$$pE = -\log(e) = \frac{Eh}{2.303\ RTF^{-1}} \qquad (3,12)$$

or

$$pE = \frac{Eh}{0.0591} \qquad (3,13)$$

The resistance to change in the redox potential when a small amount of oxidant (removal of electrons) or reductant (addition of electrons) is added to the system is termed the *poise*. Poise can be strictly defined as the number of electrons, in the form of reducing substance, which must be added to give a unit reduction in redox potential. Rowell (1981) points out the similarities between the buffering of pH by a weak acid and its conjugate base and the poise of the redox potential by the oxidant and conjugate reductant. Poise increases with increasing concentration of oxidant and reductant and, for a fixed total concentration of oxidant and reductant, it is maximum when the ratio of oxidant to reductant is 1. In the plots produced by Rowell (1981) showing the composition of the redox couple against Eh (Fig. 3.3), the slopes of the curves depend on the number of electrons involved in the reduction of one mole of oxidant (n). For a bivalent couple, the slope at the mid-point (position of most effective poise) is half that of a univalent couple. So the poise produced by a couple increases (that is, slope decreases) as the value of n increases (Rowell, 1981).

In waterlogged soils containing iron, poise is dominated by the ferric–ferrous redox couple. If this is upset by the introduction of a more dominant couple, for example, the introduction of oxygen during soil aeration, equilibrium is lost and adjustments will occur in the system to eventually make the

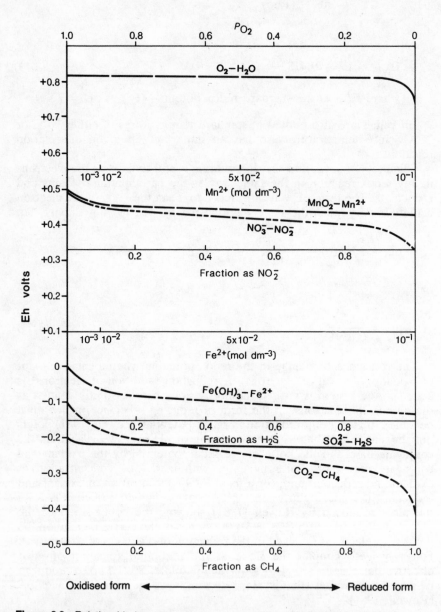

Figure 3.3 Relationship between redox potential and the composition of important redox couples in soils (Reproduced with permission from Rowell, 1981 in: *The Chemistry of Soil Processes*, eds. Greenland, D.J. and Hayes, M.H.B. © John Wiley and Sons, Ltd.)

$$O_2 + 4H^+ + 4e^- \rightleftharpoons 2H_2O \tag{3,14}$$

reaction the main determinant of poise in the newly aerated soil.

3.3 REDOX REACTIONS IN FIELD SOILS

A very large number of studies have examined a wide range of redox effects on the chemistry and biochemistry of field soils. The possible range of effects is large, including:

(1) waterlogging and compaction effects on soil profile development and the production of gleying conditions,
(2) changes in pH accompanying redox processes,
(3) the accumulation of organic matter due to slower anaerobic decomposition or fermentation processes,
(4) the production of toxic byproducts which affect the growth of plants and other organisms, and
(5) the influence of reducing conditions on retention, transformations and losses of native plant nutrients and applied inorganic fertilisers, particularly the denitrification of nitrates.

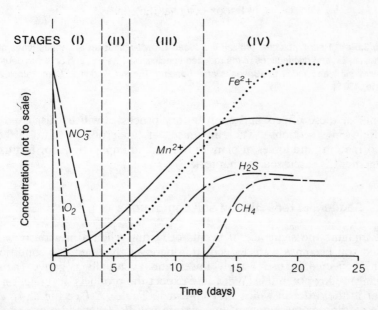

Figure 3.4 Theoretical sequence of loss of oxidised compounds and production of reduced compounds with time after waterlogging a soil. (Source: modified from Patrick, 1978)

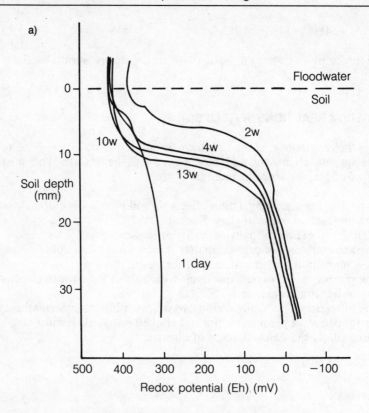

Figure 3.5(a) Redox potential profiles in flooded Crowley silt loam at various times after flooding. **(b)** Eh and distributions of the reduced species: Mn^{2+}, Fe^{2+} and S^{2-} with depth in a Crowley silt loam, 1 day to 13 weeks after flooding. (Source: modified from Patrick and Delaune, 1972)

Some of these effects and contributory processes will be discussed in the following sections. The influence of anaerobic conditions on transformations and losses of plant nutrients and implications for fertiliser management is discussed in chapter 8.

3.3.1 Sequential reduction of soil compounds

If we imagine the sequence of events occurring when a previously well-aerated soil becomes waterlogged, the increasingly anaerobic conditions will be reflected in increasingly lower and eventually negative (hence reducing) redox potentials. We would expect the products of reduction to appear in the order in which they appear in Table 3.2 (see Fig. 3.4). All of the oxidised components of one system will be reduced before any of the components below it in the table can begin to be reduced. Some overlap in the reduction of other redox systems then follows oxygen

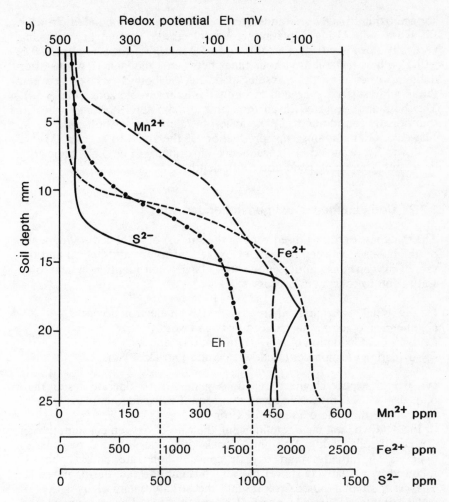

depletion. In Patrick's (1978) general sequence of reduction, four stages are envisaged:

(1) NO_3^- reduction begins before complete removal of O_2. Reduction of $Mn^{4+} \rightarrow Mn^{2+}$ occurs during the reduction of O_2 and NO_3^-.
(2) The next reduction system, $Fe^{3+} \rightarrow Fe^{2+}$ is not operative as long as either O_2 or NO_3^- is present in the system. Reduction of NO_3^-, Mn^{4+} and Fe^{3+} is carried out by facultative anaerobic bacteria.
(3) SO_4^{2-} reduction to S_2 (usually Fe_2S or H_2S) is carried out by true anaerobic bacteria and indicates the complete absence of O_2, NO_2^- and NO_3^-.
(4) CH_4 is not produced in soil until most of the SO_4^{2-} has already been reduced to sulphide.

Patrick and Delaune (1972) found a similar reduction sequence in the surface oxidised layer and underlying reduced layer of a flooded Crowley silt loam soil. Using a platinum microelectrode capable of advancing vertically downwards at a steady rate of 2 mm h^{-1}, they monitored Eh with depth in the soil at various times between 1 day and 13 weeks after flooding. They used these results to define the boundary layer between the aerobic surface zone and the underlying anaerobic zone (Fig. 3.5a). This boundary occurred in all tests at approximately +200 mV, the Eh traditionally considered to be the anaerobic/aerobic transition zone. Then 13 weeks after flooding, the production of the reduced species, Mn^{2+}, Fe^{2+} and S_2, occurred in the sequence illustrated in Fig. 3.5b, indicating increasingly reducing conditions with depth.

3.3.2 Consumption of oxygen in flooded soils

The thickness of the oxidised zone at the surface of a flooded soil depends on the balance between the rate of oxygen diffusion into the soil and the rate of oxygen consumption in the soil. Oxygen consumption in anaerobic soil is due to four main processes:

(1) microbial respiration, where it is used as an electron acceptor,
(2) chemical oxidation of reduced Fe^{2+} and Mn^{2+},
(3) biological oxidation of NH_4^+ (nitrification), and
(4) oxidation of sulphides (Reddy, Rao and Patrick, 1980).

We would expect chemical reactions generally to operate faster than biological ones. This would mean that some Fe^{2+} oxidation is likely to be more rapid than the oxidation of C or NH_4^+ $-N$.

In a study of soil factors influencing the rate of oxygen consumption in a reduced soil, Reddy, Rao and Patrick (1980) suggest that two phases operate: an initially rapid oxygen consumption rate during phase I, followed by a relatively slower oxygen consumption rate during phase II. They suggest that phase I represents the strictly chemical oxidation of water soluble Fe^{2+} while phase II represents the slower chemical and biological oxidation of exchangeable Fe^{2+}

3.4 ORGANIC MATTER ACCUMULATION IN WATERLOGGED SOILS

Anaerobic decomposition of organic matter generates different end products and occurs at a slower rate than does aerobic decomposition. The main reasons for the less rapid decay of carbonaceous compounds in waterlogged soils are:

(1) a lack of electron acceptors for respiration processes,
(2) the production of end products such as hydrogen sulphide and ethylene which are toxic to soil micro-organisms, and
(3) the presence of higher concentrations of fatty acids such as acetic and

butyric acids which inhibit microbial activity, particularly at low pH (Kilham and Alexander, 1984).

Two soil environmental parameters which also control decomposition rates are themselves affected by waterlogging: soil temperature, and soil solution pH.

Soil temperatures in wet and waterlogged soils generally show a lower range of values than do well-aerated soils, since the specific heat of water is four times higher than the specific heat of air. This means that a larger amount of energy is required to heat up a unit of waterlogged soil by unit temperature than is required if the same soil unit is in a dry condition. Second, changes in soil pH accompany many redox reactions. During aerobic respiration in well-aerated soils, protons and electrons are produced and consumed in equal numbers. In redox reactions in waterlogged soils, protons and electrons are not always produced and consumed in equal numbers. Any redox couple which uses more protons than electrons (such as Fe $(OH)_3 \rightarrow Fe^{2+}$) will cause the pH to rise. An additional factor that causes a rise in pH waterlogged soil is the accumulation of NH_4^+ $-N$ due to the inhibition of nitrification (see Section 3.2.4). Since carbon dioxide diffusion through water ($1.63 \times 10^{-9}\,m^2\,s^{-1}$) is 10^4 times slower than through air ($1.53 \times 10^{-5}\,m^2\,s^{-1}$), the accumulation of evolved CO_2 in waterlogged soils counteracts increases in pH and contributes to acidic conditions, particularly in organic soils and peats, where the pH is already low due to the production of organic acids. Studying the effects of submergence on the chemical properties of 26 mineral soils in the United States, Redman and Patrick (1965) found that soils with initial pH values below 7.4 increased in pH after submergence, and soils with initial pH values above 7.4 decreased in pH after submergence (Fig. 3.6). In organic soils and peats, acidification usually accompanies waterlogging. Generally lower temperatures and pH values result in a less varied decomposer population with slower rates of decomposition and the accumulation of organic matter.

As we have seen in Table 3.1, organic compounds as well as inorganic compounds can serve as electron acceptors in anaerobic respiration processes. Yoshida (1975) classifies the metabolic pathways under reducing soil conditions into:

(1) anaerobic respiration processes in which inorganic compounds other than oxygen are used as electron acceptors; and
(2) fermentation processes, in which organic compounds act as electron acceptors.

The example of glucose oxidation is used by Rowell (1981) to illustrate these two pathways and their resultant end products (Fig. 3.7). The main end products of carbohydrate fermentation include acetic, lactic and butyric acids, ethanol, molecular hydrogen and carbon dioxide. During the fermentation of nitrogenous and organic sulphur compounds, additional products such as amines, mercaptans and hydrogen sulphide

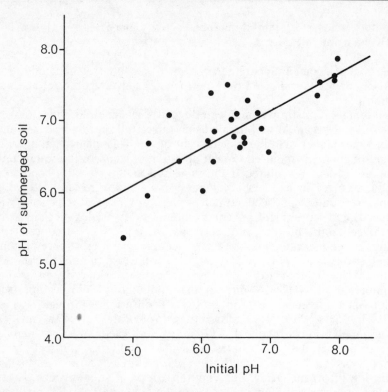

Figure 3.6 Relationship between pH before and after submergence of 26 mineral soils; $y = 3.45 + 0.53 x$; $r = 0.777$. (Source: Redman and Patrick, 1965)

can accumulate. Organic acids and alcohols may themselves be fermented to produce methane.

It is widely accepted that the metabolic degradation of carbohydrate probably proceeds similarly under both aerobic and anaerobic conditions until the formation of pyruvic acid as the final end product of glycolysis. Under aerated conditions, pyruvic acid enters the respiration cycle (Krebs cycle) where oxidation converts a whole range of carboxylic acids to CO_2 and H_2O. Under anaerobic conditions, pyruvic acid is fermented to produce lactic acid and ethanol. The important difference between aerobic and anaerobic oxidation of pyruvic acid is the smaller energy yield during fermentation. During biological respiration, energy is stored in the form of high energy chemical bonds such as in the ATP (adenosine triphosphate) molecule. When energy is required for chemical reactions, the ATP molecule is hydrolysed. This involves the cleavage of the molecule into ADP (adenosine diphosphate) and the production of inorganic phosphate, symbolised as P_i. The comparative energy release during aerobic and anaerobic oxidation of glucose (see Table 3.2) indicates that the anaerobic oxidation of glucose has a low energy yield. With less energy available for soil micro-organisms under anaerobic

(a) *Anaerobic respiration*

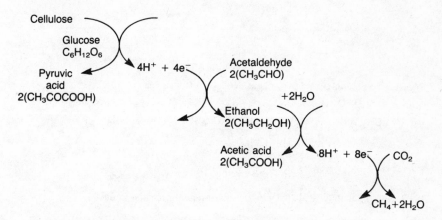

Oxidation of glucose to pyruvic acid linked to the reduction of nitrite to nitrate

(b) *Fermentaton*

(i) Oxidation of glucose to pyruvic acid linked to the reduction of acetaldehyde to ethanol.

(ii) Oxidation ethanol to acetic acid linked to the reduction of CO_2 to methane

Figure 3.7 Oxidation of glucose in **(a)** anaerobic respiration process; and **(b)** fermentation process (Reproduced with permission from Rowell, 1981. In *The Chemistry of Soil Processes*. Greenland, D.J. and Hayes, M.H.B. (eds), © John Wiley and Sons Ltd.)

conditions, decomposition of soil organic matter proceeds at a slower rate than in well-aerated soil.

In anaerobic conditions, such as in a waterlogged soil where oxidation is prevented, NAD (nicotinamide-adenine dinucleotide) acts as an important electron acceptor and pyruvic acid and reduced NADH accumulate. Pyruvic acid can take part in further fermentation reactions such as the production of lactic acid or ethanol (Fig. 3.8). These are 'general' fermentation reactions, carried out by a range of micro-organisms. More specifically, the bacteria *Bacillus fossicularum* and *Clostridium butyricum* are responsible for the production of butyric acid

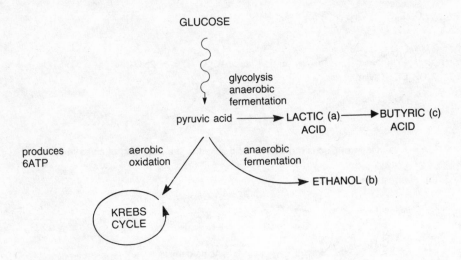

(a) Reduction to lactic acid by reduced NAD

$$CH_3\overset{\overset{O}{\|}}{C}-\overset{\overset{O}{\|}}{C}-OH + NADH + H^+ \rightleftharpoons CH_3\overset{\overset{OH}{|}}{CH}-\overset{\overset{O}{\|}}{C}-OH + NAD^+$$

pyruvic acid   lactic acid

lactic
dehydrogenase

(b) Decarboxylation to acetaldehyde and subsequent reduction to ethanol

$$CH_3\overset{\overset{O}{\|}}{C}-\overset{\overset{O}{\|}}{C}-OH \longrightarrow CH_3\overset{\overset{O}{\|}}{CH} + CO_2$$

pyruvic acid   acetaldehyde

$$CH_3\overset{\overset{O}{\|}}{CH} + NADH + H^+ \rightleftharpoons CH_3CH_2OH + NAD^+$$
ethanol

(c) $$CH_3\overset{\overset{OH}{|}}{C}-\overset{\overset{O}{\|}}{C}-OH \longrightarrow CH_3CH_2CH_2-\overset{\overset{O}{\|}}{C}-OH \quad + 2CO_2 + 2H_2$$

lactic acid   butyric acid

*Clostridium
butyricum*

Figure 3.8 Comparison of aerobic and anaerobic metabolic pathways during the utilisation of glucose

during the reduction of lactic acid and pyruvic acid respectively.

The production of gaseous hydrocarbons under anaerobic soil conditions was reported by Smith and Restall (1971). The most important of these are methane and ethylene. Many of the reactions which reduce organic acids and alcohols to methane are organism-specific. The conversion of hydrogen, ethanol and primary and secondary alcohols to methane is carried out by *Methanobacillus omelianskii*, for example,

$$2CH_3CH_2OH + 2H_2O \rightarrow 4CH_3\overset{O}{\overset{\|}{C}}\text{-}OH + CO_2 + 3CH_4 \qquad (3,15)$$

     ethanol   acetic acid   methane

The conversion of propionic acid to methane is carried out by *Methanobacterium propionicum*:

$$4CH_3CH_2\overset{O}{\overset{\|}{C}}\text{-}OH + 2H_2O \rightarrow 4CH_3\overset{O}{\overset{\|}{C}}\text{-}OH + CO_2 + 3CH_4 \qquad (3,16)$$

    propionic acid   acetic acid   methane

Smith and Restall (1971) found that in anaerobic soil, the concentration of ethylene reached a maximum before the main build-up of methane. They suggest that ethylene production is thus unlikely to be due to the activities of strictly anaerobic bacteria. In their experiments, ethylene was the only hydrocarbon gas that occurred in concentrations high enough to reduce root growth.

3.5 NITROGEN TRANSFORMATIONS IN WATERLOGGED SOIL

During well-aerated soil conditions, litter and soil organic matter are the main source of inorganic nitrogen through the processes of *ammonification* and *nitrification*:

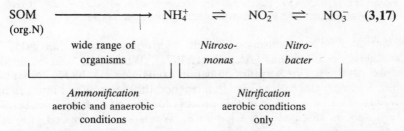

$$SOM\ (org.N) \longrightarrow NH_4^+ \rightleftharpoons NO_2^- \rightleftharpoons NO_3^- \qquad (3,17)$$

where: SOM represents soil organic matter

During the decomposition of soil organic matter, or ammonification, organic forms of nitrogen, such as proteins and amino acids, are mineralised to ammonium (NH_4^+) form. Some of this nitrogen is used in the building of microbial tissues and is thus *immobilised* (see Section 2.2.1). Both NH_4^+ and NO_3^- forms of nitrogen can be taken up by

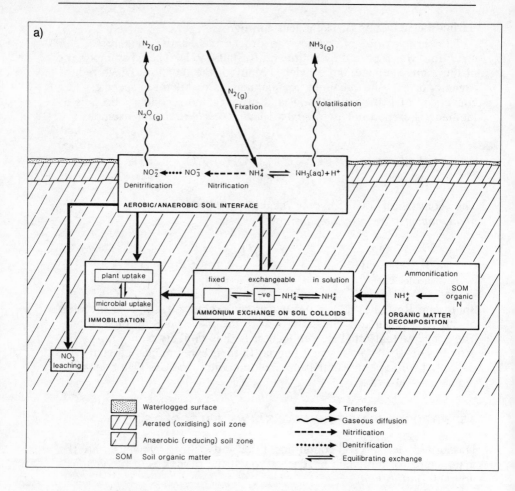

Figure 3.9 (a) Nitrogen transformations occurring in a waterlogged soil. **(b)** Nitrogen transformations occurring in a soil crumb with an anaerobic centre

plant and soil organisms although micro-organisms prefer NH_4^+ for synthesis of cell tissues (Alexander, 1977). While NH_4^+ can be fixed or stored in the soil by sorption on cation exchange sites, $NO_3^- -N$ tends to be particularly mobile and is often leached through soil and into drainage waters. The importance of this process for loss of N from applied NO_3^- fertilisers is discussed in Chapter 8. When soil is flooded, losses of nitrogenous gases occur through both *denitrification* and *ammonia volatilisation*. All of these processes are illustrated in Fig. 3.9.

When soil is waterlogged, there are two predominant effects: (1) the accumulation of $NH_4^+ -N$, and (2) the reduction of $NO_3^- -N$ (denitrification). The fall in oxygen concentrations after soil flooding curtails the activities of the aerobic bacteria *Nitrosomonas* and *Nitrobacter* which are primarily responsible for nitrification processes in soils. Since the

b)

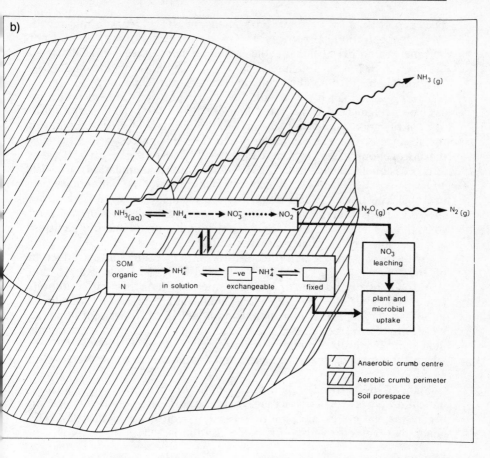

anaerobic decomposer population have a lower nitrogen requirement than aerobic decomposers, a more rapid release of $NH_4^+ - N$ is often observed than would be expected with the much slower rate of anaerobic decomposition (Tusneem and Patrick, 1971). This results in a build up of $NH_4^+ - N$, often leading to a slight rise in pH. Ammonia volatilisation occurs in waterlogged soil by the process:

$$\underset{\substack{\text{ammonium} \\ \text{ion}}}{NH_4^+} \rightleftharpoons \underset{\substack{\text{ammonia} \\ \text{gas}}}{NH_{3(g)}} + H^+ \qquad\qquad \textbf{(3,18)}$$

1 mole of H^+ is released for each mole of NH_4^+ converted to $NH_3(g)$. The pH and buffering capacity of both aerobic and anaerobic soils influence ammonia volatilisation. For nitrogen transformation in floodwater, Savant and DeDatta (1982) write the overall inorganic ammonium nitrogen (equilibrium) system (N_t) as:

$$N_t = NH_{3(aq)} + NH_4^+ \qquad\qquad \textbf{(3,19)}$$

They report a directly proportional change in $NH_{3(aq)}$ concentration with NH_4^+ concentration. Up to pH 9, $NH_{3(aq)}$ increases about ten-fold per unit increase in pH of the aqueous system. The problems associated with high soil $[NH_4^+ -N]$ and ammonia volatilisation from urea and NH_4^+ fertilisers applied to alkaline soils are discussed in Chapter 8.

During waterlogged soil conditions, two types of $NO_3^- -N$ reduction can be distinguished (Buresh and Patrick, 1978). First, NO_3^- respiration, or denitrification, in which NO_3^- acts as the terminal electron acceptor in the absence of O_2. The reduction products of this process are principally the nitrogenous gases: nitrous oxide (N_2O) and nitrogen gas (N_2), but also occasionally nitric oxide (NO). Second, NO_3^- assimilation, in which NO_3^- is reduced to NH_4^+ which can subsequently be immobilised in microbial cell tissues. While denitrification results in nitrogen losses from the soil and is thus an undesirable process, particularly in agricultural soils, the reduction of NO_3^- to NH_4^+ and assimilation by micro-organisms, is a desirable process that conserves soil N.

Of these two processes, rather little work has examined the reduction of NO_3^- to NH_4^+. By using a ^{15}N labelled NO_3^- addition, Buresh and Patrick (1978) found that only under intensely reducing soil conditions (redox potentials of -260 mV), were significant amounts of NO_3^- reduced to NH_4^+ or organic N. Under these conditions, up to 20 per cent of added NO_3^- was recovered as $NH_4^+ -N$ after four days. Their experiments indicated that this was a non-assimilatory conversion of NO_3^- to NH_4^+, in which fermentative anaerobes such as *Clostridium* may use NO_3^- as an electron acceptor.

Since N losses through denitrification are of such great importance to agriculture, a particularly extensive literature exists on denitrification organisms, processes and, more recently, denitrification kinetics. The generally accepted pathway for denitrification is:

$$NO_3^- \rightleftharpoons NO_2^- \rightleftharpoons N_2O \rightleftharpoons N_2 \qquad\qquad (3,20)$$

nitrate nitrite nitrous dinitrogen
oxide gas

Many denitrification studies have examined flooded soils where intensely reducing conditions develop beneath a thin but generally aerated surface soil layer which is in contact with the atmosphere. Since ammonium is the predominant inorganic form of nitrogen in waterlogged and oxygen-deficient soils, denitrification cannot take place until either nitrates are added, for example in the form of fertilisers, or after nitrification of the native ammonium has occurred. Although Firestone (1982) notes that about 23 genera of bacteria are capable of carrying out denitrification in soil, the most commonly cited species is *Thiobacillus denitrificans*. Most denitrifying bacteria are chemoheterotrophs, using chemical energy sources and organic compounds as electron donors and as sources of carbon (see Section 2.1.1). *T. denitrificans* can operate either in this mode, or as a chemoautotroph, in which CO_2 is used as the carbon source.

Apart from an availability of NO_2^- and NO_3^-, five environmental factors control the rate and magnitude of denitrification in soil: (1) soil moisture content, (2) oxygen concentration, (3) temperature, (4) pH, and (5) a source of carbon. Early studies indicated that amount and rate of denitrification was proportional to soil moisture content and that in dryer soil conditions, N_2O was more likely to be evolved than N_2 (see, for example, Nommik, 1956). Focht, Stolzy and Meek (1979) found that N_2 emissions are greater than N_2O when the soil is at field capacity. Increasing emissions of N_2O are frequently reported from soils as they wet up after rainfall (Duxbury et al., 1982). By alternating aerobic and anaerobic conditions in flooded soil during laboratory incubation, Smith and Patrick (1983) found that longer alternating periods increased N_2O dramatically compared to continuously aerated conditions. No N_2O was evolved under continuous anaerobic conditions. It is thought that aerating/drying cycles may enhance the availability to denitrifying bacteria of soil organic matter.

One of the main influences of soil moisture content on denitrification is in controlling the rate of O_2 diffusion (see p. 76). Even in well-aerated soils, the anaerobic centres of soil aggregates (Currie, 1961) provide suitable microsites for denitrification (see Fig. 3.9a). Patrick and Gotoh (1974) and Patrick and Reddy (1976a) show that an increase in concentration of oxygen in the atmosphere at the surface of flooded soil increases the thickness of a surface soil aerated layer. The processes contributing to denitrification in flooded soils are illustrated in Fig. 3.9b. The amount of oxygen consumed during aerobic respiration and nitrification is also increased when the aerobic layer thickens. NO_3^- produced in this surface layer diffuses downwards into anaerobic soil where denitrification occurs. Increasing the concentration of O_2 in the air above waterlogged soil thus increases gaseous N losses. During a laboratory study of the effect of nitrate addition on denitrification, Cho (1982) found that the rate of O_2 consumption was constant. Only after the system became completely anaerobic was N_2O evolved (Fig. 3.10). The reduction of N_2O and production of N_2 gas occurred after 70 h of incubation at 20°C. Cho (1982) found that doubling the nitrate addition to the soil caused a doubling in N_2O evolution and a delay in onset of N_2 production.

Laboratory incubation studies have indicated that the optimum temperature for denitrification lies between 60 and 70°C (see, for example, Nommik, 1956; Keeney et al., 1979). This helps to explain why denitrification results in such important nitrogen losses from rice paddy fields under tropical conditions. Minimum temperatures for denitrification appear to be around 3–10°C (for example, Cho et al., 1979). Keeney, Fillery and Marx (1979) found that the composition of nitrogenous gases evolved during denitrification changed with increasing incubation temperature (Table 3.3). At higher temperatures around 50–60°C, an increased proportion of N_2 is evolved, with a concurrent decline in N_2O.

Denitrifying bacteria are unable to live in acid conditions and most authors report much reduced rates of $(N_2O+N_2)-N$ evolution at values below pH 4.8, with maximum denitrification at pH 8.0. At pH values

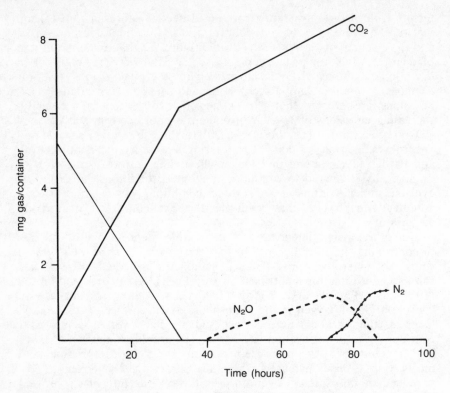

Figure 3.10 Oxygen consumption and the evolution of CO_2, N_2O and N_2 from soil amended with 500 µg NO_3–N/100 g of soil. (Source: Cho, 1982)

< 6.0, the proportion of evolved N_2O relative to N_2 increases, probably because N_2O reduction is inhibited (Nommik, 1956; Koskinen and Keeney, 1982).

The final important factors controlling the rates and gaseous N products of denitrification are an availability of NO_2/NO_3 and decomposable organic matter or carbon supply. At high NO_3 concentrations, Cho and Sakdinan (1978) found that the reduction of N_2O to N_2 was inhibited. They suggested that this may have been due to competition between NO_3^- and N_2O as electron acceptors; the sequence: $NO_3^- \rightarrow NO_2^- \rightarrow N_2O$ operating in preference to: $N_2O \rightarrow N_2$. An adequate supply of decomposable organic matter is vital to maintain the activities of heterotrophic denitrifiers who gain both energy and cellular carbon from this source. Many organic substrates, ranging from glucose to root exudates have been shown to enhance denitrification rates. Not surprisingly, rates of denitrification in incubated organic soils amended with NO_3 are particularly high.

Table 3.3 Amount of N_2 and N_2O evolved from NO_3-N amended soil incubated under helium at different temperatures (from Sahrawat and Keeney, 1986)

Temperature	Form of N	Incubation period (h)		
	(μg N per 50 g soil)	24	48	96
15	N_2	49	84	163
	N_2O	71	127	195
25	N_2	88	337	664
	N_2O	255	361	299
40	N_2	145	295	1117
	N_2O	749	2247	2659
50	N_2	953	7816	7609
	N_2O	4274	158	0
60	N_2	8884	8726	8159
	N_2O	0	0	0
65	N_2	8850	8553	7739
	N_2O	0	0	0

3.6   SULPHUR CHEMISTRY IN WATERLOGGED SOIL

As in the case of iron, the oxidation and reduction of sulphur can occur seasonally or when waterlogged sulphur rich soils are artificially drained. Estuarine and marine derived soils tend to be sulphur-rich due to the presence of sulphates in sea water. Under waterlogged conditions, bacterial reductions are responsible for the production of inorganic sulphides, most notably hydrogen sulphide gas (H_2S) which reacts with reduced ferrous iron to produce black ferrous sulphide (FeS), which subsequently reacts with H_2S to produce the disulphide, usually in the form of pyrite (FeS_2). Two forms of sulphate reduction are generally recognised (see, for example, Anderson, 1978):

(1) *Assimilatory reduction*: carried out by organisms which take in sulphur to build into amino acids and other body tissues. They do not release substantial amounts of sulphide into the soil.
(2) *Dissimilatory reduction*: carried out by bacteria of the genera *Desulfovibrio* and *Desulfotomaculum*. These organisms use oxidised forms of sulphur as electron acceptors during anaerobic respiration (see Section 4.2.3).

During dissimilatory reduction, sulphates (SO_4^{2-}), thiosulphates ($S_2O_3^{2-}$), dithionate ($S_2O_6^{2-}$) and colloidal sulphur are all reduced to sulphide by

Desulfovibrio desulphuricans. Bloomfield and Zahari (1982) summarise this sequence of reduction processes as:

$$SO_4^{2-} \xrightarrow{\text{Anaerobic}} S^{2-}$$

Desulfovibrio desulphuricans
$$S^{2-} + Fe^{2+} \rightarrow FeS$$
$$S^{2-} + Fe^{3+} \text{ or } O_2 \rightarrow S$$
$$FeS + S \rightarrow FeS_2 \tag{3,21}$$

In soil systems lacking iron, the predominant product of sulphur reduction is hydrogen sulphide gas. In concentrations in excess of 10^{-6} M, H_2S is toxic to plant roots, soil organisms and to fish in watercourses draining waterlogged soils.

3.7 PROCESSES IN GLEY SOILS

Gley soils form in temperate regions where intermittent, usually seasonal, waterlogging and drying of mineral soil results in a typical mottled appearance to one or more horizons. The mottles appear orange-brown and grey-blue, indicating the presence of oxidised (Fe^{3+}) compounds and reduced (Fe^{2+}) compounds respectively. The relative proportions of these colours indicates whether soil conditions are predominantly oxidising or reducing. Ochreous orange colours have been ascribed to the presence of the ferric oxyhydroxides: lepidocrocite (γ FeOOH) and goethite (α FeOOH) or to the dehydrated form of maghemite (γ Fe$_2$O$_3$). While both lepidocrocite and goethite are identified by their yellow-brown colour, lepidocrocites tend to be more highly crystallised than goethites. Schwertman (1985) found different crystallinities of lepidocrocite formed in different environments, with most crystalline forms found in mottles within soil aggregates and least crystalline forms found in Fe^{2+} bearing springwater (Fig. 3.11). He suggests that the formation of lepidocrocite requires Fe^{2+} as a necessary precursor. A higher rate of Fe^{2+} supply combined with a fast rate of oxidation appears to produce poor crystallinities, while the much slower oxidation processes operating within soil peds is more conducive to good crystal formation. The blue-grey coloured mottles in gleys have been ascribed to the presence of vivianite (ferrous phosphate), ferrous sulphide (FeS) and to the hypothetical $Fe_3(OH)_8$ which is the hydrated form of Fe_3O_4 (Bloomfield, 1981). Bloomfield (1981) suggests that FeS occurs in waterlogged soil in higher concentrations than vivianite and imparts a black colour to the soil which appears bluish in bright sunlight. Well-crystallised vivianite gives a brighter blue colour which oxidises to a red-brown colour within minutes of exposure. The presence of 'green rusts' in gleys at low per cent oxidation has been cited by Bloomfield (1981). These are ferric/ferrous hydroxy chlorides which can be altered under laboratory conditions to

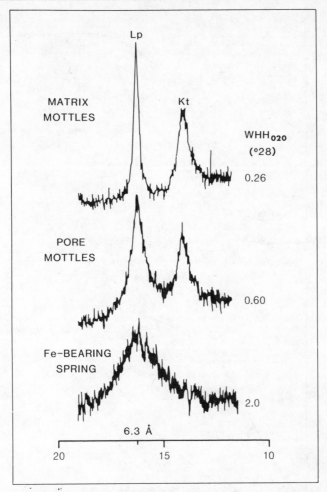

Figure 3.11 X-ray diffraction traces of lepidocrocites from three different environments. WHH_{020} = corrected width at half weight of the (020) line at 6·27 angstroms (Source: Schwertman, 1985)

ferric hydroxyoxides (FeOOH) and may be important contributors to the grey-blue colour of gley mottles.

There seems to be little doubt that the production of oxidised and reduced mottles and chemical species in gleyed soils is due to biochemical processes. Ottow (1971) demonstrated the ability of the freeliving nitrogen fixing bacteria *Clostridium* to reduce ferric to ferrous iron under laboratory conditions. Bloomfield (1950, 1951) had earlier identified the need for an energy input before gleying could be simulated in laboratory incubations. In experiments where glucose was added to anaerobic soil, he found that the rate of glucose loss parallelled the rate of iron mobilisation (Fig. 3.12a). He also confirmed that vegetation litters could be used as an energy source for iron reducing microbes and that fresh

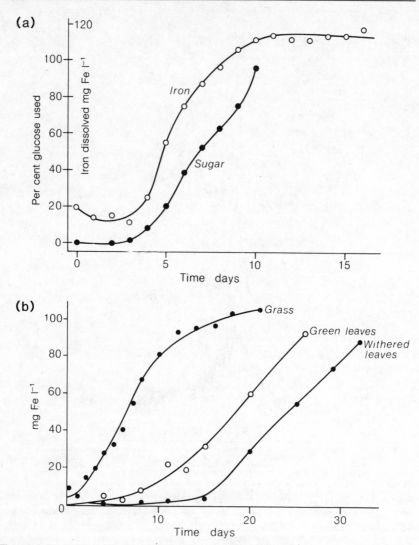

Figure 3.12 Iron mobilisation during gley simulation experiments with added energy sources. **(a)** Under anaerobic conditions, iron mobilisation and glucose consumption are both sigmoidal curves with similar lag times. **(b)** Effect of age and condition of added litter on iron mobilisation in anaerobic soil. (Source: Bloomfield, 1951)

litter mobilised more soil iron than did old, withered litter (Fig. 3.12b). In a short review of microbial processes in gleys, Bloomfield (1981) suggests that as well as the *Clostridia*, other bacteria, such as *Bacillus polymyxa* and various species of *Escherichia* and *Aerobacter* were also capable of reducing iron under anaerobic conditions. Ottow (1971) simplifies the energy involvement in the iron reduction stages of gleying as follows:

(1) Inorganic fermentation
   energy source $\rightarrow$ e$^-$ + H$^+$ + ATP + end product   (3,22)
(an example of this mechanism is illustrated in Fig. 4.7b)

(2) Iron as a H$^+$ acceptor
   \
   FeOH + OH$^-$ + H$^+$ + e$^-$ $\rightleftharpoons$ Fe^{2+} + H$_2$O   (3,23)
   / ferric   ferrous

(3) Gley formation
   $2Fe(OH)_3 + Fe^{2+} + 2OH^- \rightleftharpoons Fe_2(OH)_8$ (3,24)
   brown   green-grey

A wider literature indicates the involvement of soil bacteria in the oxidation of mobilised ferrous iron when aerobic soil conditions are encountered. These reactions are invariably closely linked to the redox reactions of soil sulphur (see Section 3.2.5). It is generally accepted that the major organisms responsible for the oxidation of both Fe^{2+} and S$_2$ in soils is *Thiobacillus ferrooxidans*. This organism is also thought to have a catalytic effect on the oxidation of pyrite. The involvement of *T. ferrooxidans* in the oxidation of pyrite has been studied most intensively in the reclamation of colliery spoil. Gemmell (1977) suggests that the oxidation of pyrite involves both chemical and biochemical reactions:

$$2 FeS_2 + 7 O_2 + 2 H_2O \rightleftharpoons 2 FeSO_4 + 2 H_2SO_4 \qquad (3,25)$$
   pyrite   ferrous   sulphuric
                                       sulphate   acid

$$*4 FeSO_4 + O_2 + 2 H_2SO_4 \rightleftharpoons 2 Fe_2(SO_4)_3 + 2 H_2O \qquad (3,26)$$
   ferrous   ferric
   sulphate   sulphate

$$* \ FeS_2 + Fe_2(SO_4)_3 \rightleftharpoons 3 FeSO_4 + \quad 2S \qquad (3,27)$$
   pyrite   ferric   ferrous   elemental
             sulphate   sulphate   sulphur

$$2S + 3 O_2 + 2 H_2O \rightleftharpoons 2 H_2SO_4 \qquad (3,28)$$
                          sulphuric
                          acid

* Bacterially aided oxidation reactions (*Thiobacillus ferrooxidans*)

Gemmell (1977) suggests that the production of Fe^{2+} in Equations (3,25) and (3,27) above acts to slow down the rate of pyrite oxidation until iron has been converted to Fe^{3+} by reaction 2. Apart from processes in colliery spoil, the above reactions are particularly important in newly installed field drainage schemes where mobilised Fe^{2+} migrates to aerated drains and is oxidised to form an iron ochre deposit of ferric oxide.

Since oxidation and reduction in gleys are microbially controlled, the seasonal variations in soil moisture content and temperature control the rate and occurrence of redox processes. Russell (1973) envisages a five-stage sequence of processes in gley development under temperate conditions:

(1) Reducing conditions start along old root channels in the autumn when soil becomes waterlogged but is still warm.
(2) Fe^{2+} is produced in channels and diffuses into soil peds, forming the channel wall.
(3) The following summer, this ferrous iron is oxidised to ferric hydroxide, giving an ochreous orange colour inside the ped.
(4) Some Fe^{2+} produced during reducing conditions will be leached in slow drainage, moving ferrous iron completely out of some soil horizons.
(5) Fe^{2+} may be moved into new soil horizons and deposited as ferric hydroxide when oxidising conditions arise.

3.8 SUMMARY

In soils with impeded drainage, the overriding influence on soil chemistry is the development of oxygen deficiency and anaerobic conditions. The degree of anaerobism depends on (1) the rate of O_2 consumption in both biological processes such as microbial respiration, and chemical processes, and (2) the rate of O_2 diffusion through soil to the site of consumption. As well as the magnitude of the oxygen partial pressure gradient, the proportions of soil porespace occupied by water or air controls the rate of O_2 diffusion, since oxygen diffuses through water 10,000 times slower than through air. When all oxygen has been reduced, other substrates, such as nitrate, sulphate, or ferric iron, are used as electron acceptors during reduction. Oxidation–reduction reactions (or redox reactions) are characterised by writing equations for the reduction half cells only. Each reduction reaction has an associated electrode potential, called a redox potential, which can be used to characterise the state of anaerobism of the soil system. A second expression used to characterise redox reactions is the pE (analogous to pH) which is the $-\log$ of electron activity. Redox reactions in field soil occur sequentially as the soil becomes more anaerobic, with oxides (NO_3^- and SO_4^{2-}) being reduced first, then Fe^{3+}, Mn^{4+} and other oxidised states of soil transition minerals. The suppression of aerobic decomposer micro-organisms in waterlogged soils leads to the accumulation of organic matter and frequently the formation of peat. The grey and orange mottling which is characteristic of waterlogged mineral soils (gleys) represents zones of reduced ferrous (Fe^{2+}) iron and oxidised ferric (Fe^{3+}) iron respectively. The reduction and oxidation of most soil substrates, including iron, is biologically mediated and requires a readily available source of soil organic matter to provide metabolic energy. The transformation of nitrogen under anaerobic conditions has important implications for fertiliser efficiency and is discussed in detail in Chapter 8.

CHAPTER 4

Processes in Freely Draining Profiles

4.1 INTRODUCTION

This chapter aims to introduce and coordinate the vast literature which describes the development of leached soil profiles and discusses experiments designed to elucidate podzolisation processes. Since the early work on iron oxide translocation by Deb (1949) and Bloomfield (1952), a whole range of studies have been published, representing work world-wide. The aim here is to discuss current theories on the translocation and redeposition chemistry of soil colloids, organic matter and, in particular, soil iron and aluminium in the profile. Current theories on the evolution of podzol profiles have been developed to account for the presence, in B horizons, of imgolite substances (Farmer, 1982) and realistically to include modern ideas in colloid chemistry (De Cornink, 1980). Some recent theories concerning the development of podzolic horizons and hence podzol profiles will be examined.

Despite the wide range of podzol profile types which have been described in humid temperate regions, their formative processes, albeit in varying degrees of intensity, are usually divided rather simply into translocation processes and deposition processes. In podzolisation, these two groups of processes are termed *eluviation* and *illuviation* respectively, where each term summarises a range of component processes that will be discussed in some detail in the following sections.

Improved understanding of podzolisation processes has cast some doubt on the methods used to define podzolic (spodic) horizons for classification purposes (see, for example, Soil Survey Staff, 1975). Therefore, an attempt is made here to examine the conflicts between soil profile morphology and classification based on laboratory chemical criteria.

4.2 THE ROLE OF WEATHERING IN PODZOLISATION

Many authors have reported the distinctive clay mineral distribution that is observed in weathered podzol profiles. In the selection of represent-ative profiles given by Ross (1980) (Table 4.1), two clear mineralogical trends are seen in the clay fraction:

(1) the predominance of smectite clays with some mica-vermiculite in A/E horizons with chlorite absent; and

Table 4.1 Proportions of layer silicates in the clay fraction of representative spodosols (from Ross, 1980)

Site		Parent material	Horizon	Smectite	Vermiculite	Chlorite	Mica	Kaolinite	Chlorite-vermiculite	Mica-vermiculite	Mica-smectite	References
Uplands	Canada	Alluvial sand	AE	++++	−	−	tr	+	−	−	+[a]	McKeague (1965)
			Bfh	−	+	+	+	+	++[b]	−	−	
			C	−	+	++	++	+	+	+	−	
Big Bald Mountain	Canada	Saprolitic granite	AE	++	++	−	+	tr	−	+	tr	Wang et al. (1980)
			Bf	tr	++	++	+	tr	++	+	tr	
			C	−	+	++	++	tr	+	+	tr	
Longlake	Canada	Till from chlorite schists	AE	−	++	−	+	−	−	++[a]	−	Ross (1980a)
			Bf	−	++	+	+	−	+[a]	++[a]	−	
			C₂	−	+	++	++	−	+[a]	+[a]	−	
Eredine	Scotland	Till from chlorite schists	A₂	−	tr	−	+	++	−	+++[a]	−	Bain (1977)
			B₃	−	+	+++	++	+	−	−	−	
			C	−	+	+++	++	+	−	−	−	
Oyer	Norway	Sandstone	A₂	−	tr	−	++	−	−	+++[a]	tr	Kapoor (1972)
			B₂	−	tr	tr	+++	−	−	++[b]	−	
			C₂	−	tr	+	++++	−	−	tr	−	

tr < 10 per cent, trace; + 10–25 per cent, minor; ++ 25–50 per cent, moderate; +++ 50–75–100 per cent, dominant

[a] Regularly interstratified

[b] Randomly interstratified with mica or aluminum hydroxide interlayers

(2) the predominance of chlorite, mica and vermiculite in B and C horizons, with smectites absent.

While it is recognised that quartz and feldspars tend to persist throughout podzol profiles, microscopic study of particle surfaces has indicated much more intensive etching of the surfaces of plagioclase feldspars in Eg and Bg horizons compared to Bs and Cs horizons (Farmer, McHardy and Robertson, 1985). Although the nature of the sand and silt fractions depend more on the type of parent material, A/E horizons generally lack chlorite, amphiboles and feldspars in this size range, while B and C horizons commonly contain chlorite, mica and smaller amounts of augite, hornblende and feldspars (see, for example, McKeague and Brydon, 1970; Bain, 1977; Smith, Coen and Pluth, 1981). The main mineralogical change during the development of podzol profiles, the disappearance of chlorite from the A/E horizon, is usually explained by the decomposition of amorphous aluminosilicate in the prevailing acidic conditions (Bain, 1977). Ross (1980b) identifies five possible weathering stages in the development of podzol profiles:

(1) trioctahedral and dioctahedral mica and trioctahedral chlorite in the C horizon are hydrated;
(2) they undergo additional structural change in the B horizon to form trioctahedral and dioctahedral vermiculite, interstratified with chlorite and mica;
(3) this vermiculite product is further 'cleaned' of chloritic and sesquioxide material in the A/E horizon;
(4) further structural alteration of the vermiculite, including reduction of charge density with loss of K^+ from interlayers and of cations from octahedral layers, resulting in the formation of a beidellitic smectite; and
(5) final destruction of the smectite by further weathering.

Farmer, McHardy and Robertson (1985) also report maximum intensity of mineral weathering, as indicated by (quartz + feldspar):biotite ratios in the podzolic Eg horizon of a peaty podzol in northern Scotland, with minimal weathering in Bs and C horizons. They interpret their results to indicate that weathering in Bs and C horizons changes biotite into vermiculite, during which process interlayer K^+ ions are replaced by more hydratable ions such as Ca^{2+} and Mg^{2+}. In the Eg horizon, they suggest that weathering is much more severe, resulting in the total decomposition of the mineral structure and giving a (quartz + feldspar): biotite ratio in the Eg horizon which is more than ten times higher than in Bs and C horizons.

Prior to translocation, *in situ* weathering reactions are responsible for releasing and mobilising iron and aluminium oxides from primary soil minerals such as the ferromagnesian minerals (for example, olivine, hornblende) and feldspars. In Section 1.2.2 and Table 1.3, the thermodynamic principles of the weathering of kaolinite to various aluminium

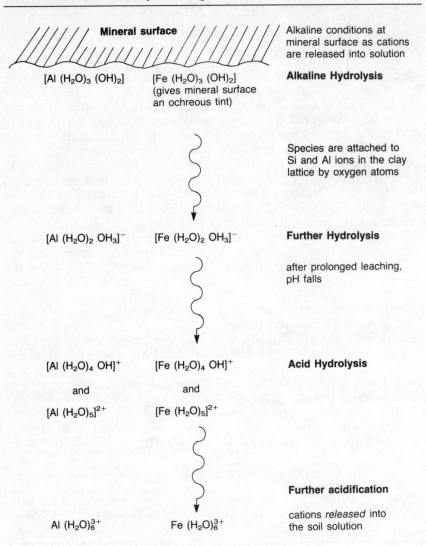

Figure 4.1 Hydrolysis of Al and Fe at a weathered mineral surface. Square brackets indicate that species is still attached to the mineral surface; only at the very last stage are free cations released. (Source: Van Schuylenborgh, 1965)

oxides are outlined (Marshall, 1977). Van Schuylenborgh (1965) writes similar, if rather simplified, reactions for the production of ferric oxides under aerated conditions. These established thermodynamic reactions led Van Schuylenborgh (1965) to suggest the sequence of hydrolysis reactions at the weathered mineral surface shown diagrammatically in Fig. 4.1.

Very similar hydrolysis reactions can occur once organometallic complexes (chelates) form in surface soil horizons. According to Van

Schuylenborgh (1965), chelates of Al and Fe are particularly susceptible to hydrolysis of the form:

$$MR^{n-m} + nOH^- = M(OH)_n + R^{-m} \tag{4,1}$$

where: M   = the metal, Al or Fe
        n   = metal valency (3 for Al; 2 or 3 for Fe)
        R   = organic acid
        m   = basicity of organic acid (for example, compared with HCl which is monobasic and H_2SO_4 which is dibasic)
     MR^{n-m} = the organometallic complex (ligand)

For Equation (4,1), the equilibrium constant will be:

$$K = \frac{[M(OH)_n]\,[R^{-m}]}{[MR^{n-m}]\,[OH^-]} \tag{4,2}$$

as soon as solid $M(OH)n$ exists in the system, $[M(OH)_n]$ becomes constant and the equilibrium (hydrolysis) constant (K_h) becomes:

$$K_h = \frac{[R^{-m}]}{[MR^{n-m}]\,[OH^-]^n} \tag{4,3}$$

where

$$\frac{K}{[M(OH)_n]} = K_h \tag{4,4}$$

This indicates that an increase in the hydroxyl ion activity shifts Equation (4,1) to the right, causing precipitation of the hydroxide. Precipitation can partly be countered by an excess of metal chelate (MR^{n-m}). If, in Equation (4,2), K > 1, the hydrolysis reaction will also proceed to the right. By transforming Equation (4,1) using the stability constant (K_{MR}) and the solubility product of the metal hydroxide (K_{SO}):

$$K_{MR} = \frac{[MR^{n-m}]}{[M^{+n}]\,[R^{-m}]} \text{ and } K_{SO} = \frac{1}{K_{MR}\,K_{SO}} \tag{4,5}$$

then K_h becomes:

$$K_h = \frac{1}{K_{MR}\,K_{SO}} \tag{4,6}$$

This indicates that the magnitude of K_{MR} and K_{SO} control the ease of hydrolysis of any organometallic complex.

As suggested in Chapter 1, silica:sesquioxide ratios, such as the three ratios given in Equation (1,63), have been commonly used to indicate both the degree of weathering occurring in a soil horizon and the relative translocation or deposition of Fe and Al oxides. High values of all three ratios tend to occur in surface soil horizons, particularly the E horizons of podzols, and indicate relative enrichment of silica, with the loss of iron and aluminium oxides. All three ratios have minimum values in Bs horizons, indicating the deposition and accumulation of iron and aluminium oxides (Glentworth and Muir, 1963). These trends are depicted in a podzol sequence studied by Mackney (1961) at Sutton Park, Warwickshire, England (Table 4.2).

Table 4.2 Clay fraction sesquioxide ratios for a sequence of podzol types in Sutton Park, Warwickshire, Central England (Mackney, 1961)

| Soil | Horizon | Sesquioxide ratios | | |
		SiO_2/R_2O_3	SiO_2/Fe_2O_3	SiO_2/Al_2O_3
Sandy brown earth	Ae	2.6	9.7	3.8
	B_1	2.5	9.9	3.3
	B_2	2.3	9.3	3.0
Podzol intergrade	Ae/Bh	3.4	8.9	5.4
	Bs	2.1	7.0	3.1
	C	2.3	10.0	2.9
Iron podzol	Ae	5.9	53.2	6.7
	Ae/Bh	3.1	15.3	4.1
	Bs_1	2.3	5.8	3.7
	Bs_2	2.4	8.0	3.8
Iron–humus podzol	Ae	4.4	29.7	5.2
	Bh	2.5	7.7	3.5
	Bh/s	1.7	3.9	2.6
	Bs	2.2	8.3	2.7

The rather more complex calculation of eluviation/illuviation coefficients was first suggested by Rode (1935). Muir and Logan (1982) summarise the assumptions involved in this technique as follows:

(1) the parent material from which the profile has developed was originally uniform;
(2) the C horizon is closest in composition to that of the original parent material; and
(3) the substance chosen as an *internal index*, from which intensities of eluviation and illuviation are assessed, is relatively immobile.

With these assumptions in mind, calculation of eluviation/illuviation coefficients (τ) according to Rode (1935) can be outlined as:

$$\tau = \frac{S_{hi} \, Q_{hi}}{S_{oi}} - 1 \text{ where } Q_{hi} = \frac{X_{oi}}{X_{hi}} \tag{4,7}$$

where: S_{hi} = per cent constituent S in ignited horizon, h
$\quad\quad\; S_{oi}$ = per cent constituent S in ignited horizon, C
$\quad\quad\; X_{hi}$ = per cent internal index in ignited horizon, h
$\quad\quad\; X_{oi}$ = per cent index in ignited horizon, C
thus, Q_{hi} = ignited parent material quotient for horizon, h

Negative values of τ indicate a percentage loss, such as occurs in podzolic E horizons, while positive values of τ indicate a percentage gain, such as occurs in podzolic B horizons.

Using these calculations, Muir and Logan (1982) assessed the degree of eluviation and illuviation in three Scottish podzols. Their results indicate intense leaching of aluminium in all the profiles studied. The values calculated for the thin ironpan podzol at Hill of Auchlee, Aberdeen (Table 4.3), indicate intensive leaching of aluminium, iron and manganese from the A horizon, with deposition of all three species in the B horizon. The particularly high τ value for $Fe_2 O_3$ in the B horizon indicates an iron pan, with substantial MnO accumulation. Concentrations of sesquioxides were paralleled by those of phosphates, indicating anionic adsorption in all B horizons.

Table 4.3 Per cent composition and τ values (eluvial/illuvial coefficients) for Al_2O_s, Fe_2O_3 and MnO for a thin ironpan podzol in Scotland (Muir and Logan, 1982)

Site	Horizon	Al_2O_3 per cent content	τ	Fe_2O_3 per cent content	τ	MnO per cent content	τ
Hill of	H	10.8	−35	2.5	−52	0.06	−23
Auchlee	A_{2g}	12.1	−29	1.5	−72	0.03	−62
	Bsgh	14.7	−1	5.6	+20	0.05	−28
	B_1	13.8	+17	23.3	+530	0.09	+63
	Bs_1	13.2	−15	3.8	−22	0.08	+10
	Bs_2	14.4	−6	4.4	−8	0.07	−2
	C	15.0	−	4.7	−	0.07	−

Guillet, Rouiller and Souchier (1975) used quartz as the internal index when looking at weathering in podzols in the Vosges. They calculated the *isoquartz balance* for feldspars and for what they called 'weathering complexes' which included silicate clay minerals and amorphous material (sesquioxides of Si, Fe and Al). The isoquartz balance for a component of a horizon in comparison with the parent material is calculated according to:

$$\Delta X_i = d_i \, s_i \, \frac{Q_i}{Q_o} \, (X_i - X_o) \qquad\qquad (4,8)$$

where: ΔX_i = gain (positive) or loss (negative)
d_i = depth of a given soil horizon
s_i = bulk of density of a given soil horizon
X_i = isoquartz content of a specified component in the given soil horizon
X_o = isoquartz content of the same component in the parent material
Q_i = quartz content of the soil horizon
Q_o = quartz content of the parent material

The isoquartz balances for two podzol types in the Vosges (Fig. 4.2) indicate that amounts of primary minerals decrease upwards in the profile due to weathering. Both secondary clay minerals and sesquioxides are almost completely absent from A horizon, indicating relative eluviation, while marked accumulation in B horizons indicates a high degree of illuviation. There appears to be good evidence here that weathering mobilises materials in A horizons, allowing their translocation and deposition in B horizons. The ratio between the amorphous fraction and the clay minerals indicates selective accumulation of Al_2O_3, SiO_4 and Fe_2O_3 sesquioxides in the B horizons.

4.3 PROCESSES OF TRANSLOCATION

Eluviation, or translocation in podzol profiles, consists of three component processes:

(1) *leaching*: the translocation of soluble salts;
(2) *cheluviation*: the translocation of organometallic complexes (chelates); and
(3) *lessivage*: the translocation of colloidal clay particles.

Although it is clear that all three types of migration play important roles in the development of all soil profiles, in podzol development, much emphasis has been placed on the role of soluble organic molecules in forming metal chelates with hydrolysed Fe and Al oxides and altering their mobility. The magnitude of all three processes is ultimately controlled by the intensity of water percolation through the soil profile. Clay mineral colloids for translocation by lessivage are provided by the weathering processes and in the sequences outlined in Chapter 1. The main controls on the occurrence of clay migration are soil physical properties such as the tortuosity of the porespace and hence the magnitude of saturated and unsaturated hydraulic conductivity. Soluble salts, organic molecules and chelates for translocation by leaching and cheluviation are provided by both weathering and the decomposition of soil organic matter. As in the case of lessivage, soil physical properties

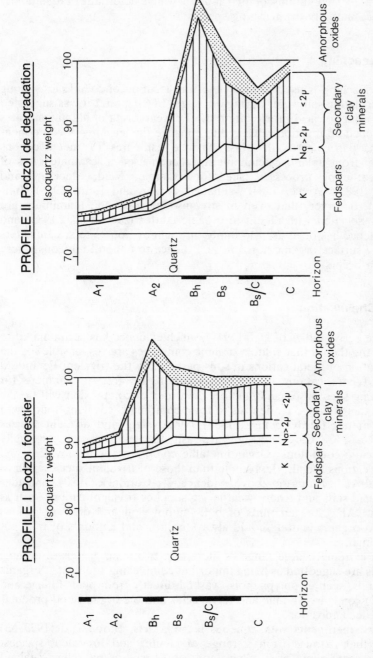

Figure 4.2 Weathering balance, expressed as isoquartz weight, for two podzol types in the Vosges (Reproduced with permission from Guillet, Rouiller and Souchier, 1975, *Geoderma*, 14, © Elsevier Scientific Publications, B.V., Amsterdam)

control leaching and cheluviation but also of importance are the soil environmental conditions of pH and Eh which determine the solubility and hence mobility of migrating species.

4.3.1   Leaching

Strictly, leaching is used to mean the translocation of soluble ions along with percolating soil water during drainage. In Chapter 1 it was suggested that the effect of acid precipitation on the weathering of aluminosilicates was to speed up reaction rates and to enhance the leaching of component minerals, particularly Al. Mass balance estimates of metal cations 'leaching' from upland catchments have been used to estimate rates of rock weathering processes (see, for example, Reid, MacLeod and Cresser, 1981; Grieve, 1984; Williams, Ternan and Kent, 1986). It is apparent, however, that even in streamwater, iron and aluminium are closely associated with dissolved organic matter (see Greive, 1985) and this evidence has been used to imply throughflow routes to streamwaters via peaty surface organic horizons and thence to mineral horizons where chelation occurs.

4.3.2   Cheluviation

Since the early studies of Deb (1949), much evidence has accumulated to support the theory that soluble organic complexes are responsible for the translocation of metal cations in soils particularly the trivalent Fe and Al oxides. Where cations in the soil solution are free to compete for complexing sites on organic matter, the order of complexion will be the same as their relative tendencies to form coordinate–covalent bonds with organic matter (De Corninck, 1980). Monovalent and divalent species such as Na^+, K^+, Ca^{2+} and Mg^{2+} are easily leached from soils, especially under acidic conditions. Organometallic complexes of monovalent and divalent cations are also less stable than those of trivalent species such as Fe^{3+} and Al^{3+}. From the above evidence, De Corninck (1980) concludes that in acid soils and where weathering supplies trivalent cations such as Fe^{3+} and Al^{3+}, the amounts of both monovalent and divalent cations bound to organic matter should always be low and amounts of trivalent cations high.

In an extremely large range of literature, two main groups of organic materials are suggested as being important complexing agents: (1) organic compounds, such as polyphenols, washed directly from plant foliage and litter; (2) condensed humic and fulvic acids and the organic end-products of their decomposition.

Using experiments with aqueous leaf extracts, Bloomfield (1953–55) showed that extracts from a range of conifer and broadleaf species, including scots pine, larch, aspen and ash as well as two New Zealand species: the rimu (*Dacrydium cupressium*) and the kauri pine (*Agathis australis*), were capable of forming water-soluble organometallic complexes with Fe and Al. Malcolm and McCracken (1968) later studied the iron and aluminium mobilising capacities of canopy drip from southern red oak, live oak and longleaf pine in North Carolina. They estimated that about 20 kg ha^{-1} y^{-1} of organic matter could be contributed to the soil from canopy wash alone. Their studies indicated that this amount of soluble organic matter could mobilise up to 1.50 kg ha^{-1} y^{-1} of soil Fe and up to 0.70 kg ha^{-1} y^{-1} of soil Al. Malcolm and McCracken (1968) found that canopy drip from the oaks was more effective in mobilising soil Fe, while canopy drip from longleaf pine more effectively mobilised soil Al. Soils under species other than trees are known to show podzolisation. According to Fisher and Yan (1984), aqueous extracts from the foliage and shoots of the heathland species *Calluna vulgaris* and *Erica tetralix* are capable of mobilising iron from iron-coated quartz sand. Young, pioneer *Calluna* is significantly more effective in this respect than older, established plants. In the same study, *Betula pendula* extracts taken in October and December mobilised twice as much iron as *Calluna* and *Erica*, despite the fact that birch invasion on heathland is often associated with decreased iron migration in soil (Dimbleby, 1962). In an attempt to explain why aqueous leaf extracts from deciduous tree species not usually associated with podzolised soils were capable of mobilising Fe and Al oxides, Bloomfield (1957) suggested that all tree species exert a podzolising effect but that 'opposing influences' dominate under broadleaves. He did not, however, suggest what these 'opposing influences' might be.

While studies such as those outlined above confirmed that metal oxides could be mobilised by leaf leachates, the active complexing agents were not yet identified. Bloomfield (1957) suggested that, since aqueous leaf extracts reduced much of the ferric iron to ferrous form, the active organic constituents could be organic acids and polyphenols. Since these early studies, fractionation techniques have identified substantial quantities of amino acids, organic acids and polyphenols in aqueous leaf extracts. Malcolm and McCracken (1968) calculated that canopy drip from oak and pine contributed 1 kg ha^{-1} y^{-1} of polyphenols to the soil. In beech leaves, Coulson, Davis and Lewis (1960a) found that polyphenol content declined with age of leaves:

decreasing polyphenol content →
growing leaves > senescent leaves > dead leaves > fallen leaves > leaf litter humus

They also found that leaves from beech trees growing on silicious, base-deficient sites, contained higher amounts of simple polyphenols than those growing on base-rich sites. Coulson, Davis and Lewis (1960a) suggested that polyphenol synthesis may be increased in trees growing on N and P deficient acid soils.

Despite confirmation that a wide range of polyphenols occur in the leaves of many tree species, in both litter and soil, there has been much debate surrounding the role of polyphenols in soil Al and Fe translocation. Using epi-catechins and D-catechins, two pure polyphenols known to occur in fresh beech and other tree foliage, Coulson, Davies and Lewis (1960b) confirmed that ferric iron could be reduced and solubilised by simple polyphenols. Confusion was introduced into the debate when Hingston (1963), looking at polyphenols in aqueous extracts of a range of eucalyptus and other Western Australian trees and shrubs, found no correlation between the polyphenol content of leaf extracts and their capacity to solubilise iron. Further evidence supporting the importance of polyphenols in Al and Fe translocation came from the work of Thomas (1967) who fractionated canopy drip from oak and spruce into (a) amino acids, (b) carboxylic acids, (c) lead-precipitable orthohydroxy phenols and phenolic acids, and (d) residual phenolic substances and some polysaccharides. He compared the capacities of these four fractions in mobilising iron over a pH range of 3.1–8.5 from uniform grains of soil minerals, including olivine, augite and hornblende. For olivine, group (c) compounds were the most effective iron mobilising agents with the relative mobilising capacities of the four compounds following the sequence: c > d > b > a. He also identified highest mobilisation capacities between pH 5.1 and 5.2. McHardy, Thomson and Goodman (1974) liken the reduction of iron by phenolic organic compounds and its subsequent oxidation with increasing pH to the fluctuating redox conditions which operate in gley soils.

These findings led Davies (1971) to suggest that, in brown earths, under less intensive rainfall and higher base status, polyphenols in the soil solution are likely to be rapidly metabolised and hence iron mobilisation is small. With greater rainfall and increasing acidity, he suggested that the polyphenol content of leachate would increase, together with its efficiency to mobilise iron. These conditions would be less favourable for oxidative mineralisation and the water-soluble compounds would be washed downprofile, resulting in an eluviated horizon.

A second school of thought in chelate translocation studies, headed by Schnitzer and his colleagues, emphasised the importance of humic and, more particularly, fulvic acid fractions. Soil humic substances can be simply divided into humic acid, fulvic acid and humin fractions. This division is not based on physicochemical differences or on degree of humification but instead on (1) their abilities to dissolve in electrolyte solution, and (2) their molecular weight. These differences are illustrated in Fig. 4.3. Because of their complexity and heterogeneity, the chemical structure of all three fractions is still a matter of some speculation. Tan (1977) describes fulvic acid as a complex soil humic compound, consisting of combinations of molecules with a high proportion of functional end groups. Although Schnitzer and his collaborators had earlier recognised the importance of fulvic acid in the translocation of metal ions in the soil profile, ony recently have the actual functional groups responsible for metal binding been studied in detail.

Using infrared spectroscopy, Schnitzer and Desjardins (1969) identified the fulvic acid fraction as the main organic component (87%) of the

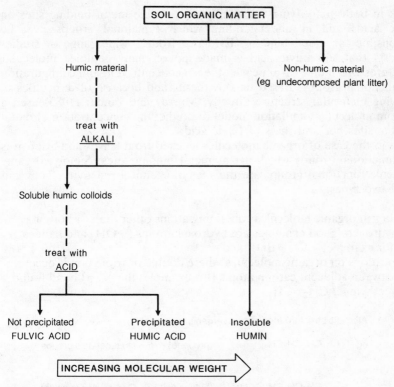

Figure 4.3   Schematic fractionation of soil organic matter

leachate from a humic podzol. This finding was substantiated by Dawson, Ugolini and Hrutfiord (1978) who identified a large mobile fulvic acid fraction in the soil solution passing through the A horizons of a podzol, but detected a sharp decrease in this fraction in the podzol B horizons. Their infrared analysis of the fulvic acid fraction indicated the presence of aromatic hydroxy carboxylic acids. Schnitzer and Skinner (1965) had previously shown, by selectively blocking both acid carboxyl (COOH) groups and phenolic hydroxyl (OH) groups on their fulvic acid fraction, that both COOH and OH groups could be important sites of metal bonding during chelation. In an earlier study, Schnitzer and Skinner (1963) examined the metal-complexing power of the organic fraction extracted from a podzol B horizon. They discovered that 1 mol of organic matter could complex either 6 mol of ferric iron or 1 mol of aluminium, indicating a lower complexing ability for Al than for Fe. They concluded that a range of metal : OM complexes were possible, becoming increasingly water soluble as more metal is complexed. This conclusion has particularly important implications for the redeposition of sesquioxides in podzolic B horizons, as will be seen in Section 4.4.

Despite extensive study, the molecular structure of fulvic acid is still not known in detail. Not surprisingly, this has been the major stumbling block in both quantifying the number of possible metal binding sites on fulvic acids and in understanding which functional groups may be responsible for metal binding. Evidence from a wide range of studies suggests that soil fulvic acid is made up of many complex molecular structures which differ from each other in concentration and configuration of functional groups. Based on this thesis, and because of difficulties in studying molecular structure directly, Murray and Linder (1983) used a random molecular simulation model to predict the concentration of metal binding sites per unit mass of fulvic acid.

As in the case of organic molecules leached from leaves and litter, it is the functional groups which act as metal binding sites. Simply, organic molecule functional groups are the sites of chemical reactivity. They can be of two types:

(1) sites in organic molecules containing atoms other than carbon or hydrogen (good examples are hydroxyl groups ($-OH$) and halides (for example, $-Cl$, $-Br$)); and
(2) locations on organic molecules where double or triple bonds occur between adjacent carbon atoms (for example, the pi (π) bond found in ethylene: $CH_2=CH_2$).

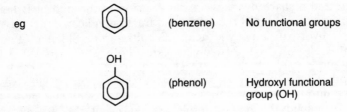

1. Aliphatic structure (non-cyclical organic molecule)

| eg | $CH_3CH_2CH_3$ | (propane) | No functional groups |

$$CH_3CH_2CH_2\overset{\overset{\textstyle O}{\|}}{C}-OH$$

| | $CH_3CH_2CH_2C-OH$ | (butyric acid) | Carboxylic functional group (COOH) |

2. Aromatic structure (cyclical organic molecule with shared bonds between adjacent carbon atoms)

| eg | (benzene) | No functional groups |

| (phenol) | Hydroxyl functional group (OH) |

Figure 4.4 Basic structures of organic molecules

There are two basic organic molecule structures: *aliphatic*, or non-cyclical (chain) molecules and *aromatic*, or cyclical (ring) molecules (Fig. 4.4). Both are likely to occur in fulvic acid. The results of an extremely large number of studies indicates that three configurations of organic structures and functional groups are likely to play a major role in metal bonding (Fig. 4.5).

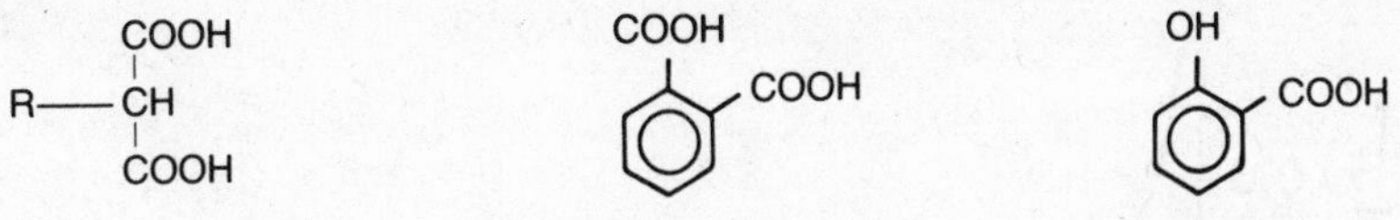

(two adjacent carboxyls on an aliphatic chain)

Phthalate-type
(two adjacent carboxyls on an aromatic ring)

Salicylate-type
(both carboxyl *and* hydroxyl groups on an aromatic ring)

Figure 4.5 Functional group configurations important in metal bonding in fulvic acid

Developing their model to simulate the binding of Mg^{2+}, Ca^{2+}, Mn^{2+}, Zn^{2+}, Cu^{2+} and Fe^{3+}, Murray and Linder (1984) found that phthalate sites were important for all metal ions apart from Fe^{3+} and that salicylate sites were most important for Fe^{3+}. The affinity of the metal ions for fulvic acid as a whole was found to decrease in the order:

$$\text{Fe(III)} > \text{Cu(II)} > \text{Zn(II)} > \text{Mn(II)} > \text{Ca(II)} > \text{Mg(II)}$$

This sequence agrees very well with that reported in an empirical study of fulvic acid with ionic strength 0.1 and pH 3 by Schnitzer and Hansen (1970):

decreasing affinity →

$$Fe^{3+} > Al^{3+} > Cu^{2}+ > Ni^{2+} > Co^{2+} > Pb^{2+} = Ca^{2+} > Mn^{2+} > Mg^{2+}$$

One of the most important controls on the mobility of humic and fulvic acids and their metal complexes is the relative viscosity they impose on the soil solution. Relative viscosity (η_{rel}) can be defined as the ratio of the solution viscosity (η) to that of pure water (η_o). Viscosity is influenced by changes in molecular configuration of the polymer. The addition of an electrolyte to humic substances causes an increase in viscosity. This is due to ionisation occurring on the polymer and resultant expansion of the molecule due to repulsion by similarly charged functional groups along the polymer chain (Hayes and Swift, 1978). Smith and Lorimer (1964) found that the addition of sodium chloride to humic acid compounds extracted from *Sphagnum* peat caused a significant increase in viscosity even at the low concentration of 0.01 M NaCl (Fig. 4.6). The effect of increased pH on humic acid compounds would be to increase dissociation of polymer carboxyls, thus increasing both molecular expansion and solution viscosity (Hayes and Swift, 1978). Thus, increased concentration of Na and K in the soil solution would act to increase the mobility of humic substances, while increased pH would act to decrease their mobility.

The shape and size of humic and fulvic acid complexes also control their ease of movement in the soil solution because frictional forces act on macromolecular surfaces. The frictional ratio (f/f_o), is defined as the ratio

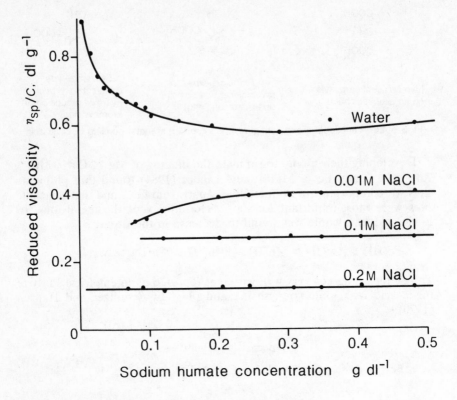

Figure 4.6 Plot of reduced viscosity (η_{sp}/c) against concentration for a sodium humate, showing the effects of added salts. (Source: Smith and Lorimer, 1964)

of the frictional coefficient for the molecule in question (f) to the frictional coefficient for a dehydrated sphere (f_o) occupying the same volume. The frictional ratio is thus a measure of the deviation in shape of the molecule from that of a sphere. Cameron *et al.* (1972) have shown that the frictional ratio is a function of molecular weight and that the precise function depends on the configuration of the macromolecule: spherical or eliptical (rigid structures) or flexible, random coils. As well as size, chain branching, which probably occurs in humic complexes, gives a higher molecular density and results in a molecule that is more compact than a single, linear chain of the same molecular weight. Thus, many-branched polymer complexes are likely to have lower frictional ratios and may be more mobile. The presence on the polymer of functional groups capable of dissociation, can lead to intramolecular repulsion, causing molecular expansion. This could increase the frictional ratio and reduce the mobility of the organic complex.

Studying humic acid–metal complexes, Sipos *et al.* (1978) found that the molecular weight of the complexes increased linearly with divalent metal concentration and exponentially with trivalent metal concentration.

This result was corroborated for Fe and Al by Ritchie and Posner (1982) who also found that these metal additions did not significantly alter the shape or degree of branching of the humic acid fractions. They also showed that the solubility of humic acid–metal complexes increased when more hydrated species of the complexing cations were involved. From Fig. 4.7 it can be seen that the more hydrolysed forms of Fe and Al occur at higher pH values. Again, this result implies greater metal-humate mobility with increasing soil solution pH.

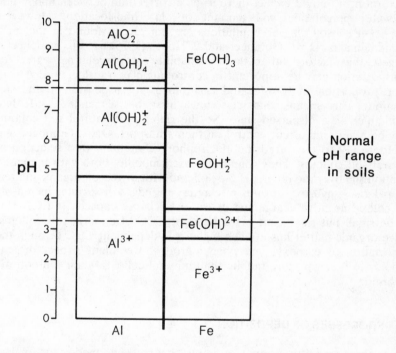

Figure 4.7 Hydration of Fe and Al cationic species in relation to pH

4.3.3 Lessivage

Evidence suggesting that lessivage, or clay translocation, occurs in many freely draining soils, originally came from micromorphological study of soil thin sections from B horizons. These microscopic studies indicated that fine clay particles were deposited in layers or skins around soil particle surfaces and as coatings in voids. These coatings are generally called *cutans*; those composed of fine, translocated clay are called *argillans*. In a whole group of freely drained soils, B horizons characterised by clay accumulation are termed *argillic*. By studying the texture characteristics of whole profiles containing these argillic horizons one can deduce that clay is mobilised in A horizons, perhaps by weathering processes (Guillet, Rouiller and Souchier, 1975), transported downprofile and deposited in B horizons (see Section 4.2 and Fig. 4.2). Wang and McKeague (1982) report substantial losses of clay and of mobile iron and aluminium (as measured by pyrophosphate extraction) from A/E horizons of a series of sandy podzols, with relative enrichment of all of these components in B horizons. Thus, argillic B horizons are cited as evidence for both clay translocation and, in some cases, for illuviation in podzolic soils.

In the search for mechanisms important in clay transport and deposition, we might expect there to be a correlation between magnitude of water percolation and amount of clay translocation. However, significant subsoil clay accumulations are not always found in the most humid climates. Dixit, Gombeer and D'Hoore (1975) use this evidence to suggest that factors other than availability of percolating water for transportation may be important in controlling clay translocation. A very large proportion of translocated clay moves in suspension in the soil solution. This means that clay must first be suspended and then maintained in a dispersed state. So the *stability* of natural soil colloids may be a very important control on their transport. Dixit, Gombeer and D'Hoore (1975) measured the electrophoretic mobility of a selection of natural soil colloids. They found increased mobility both with increased solution pH over the range pH 5.5–8.5 and with increased organic matter. As pH rises, hydroxyl groups at clay crystal edges dissociate, increasing the colloid negative charge and its cation exchange capacity (CEC). This is the small but significant pH-dependent CEC of mineral soil colloids. Since organic matter has a much higher pH-dependent CEC through the dissociation of carboxyl and phenyl groups, we might expect organic colloids to be even more mobile than mineral colloids when solution pH is raised.

4.4 PROCESSES OF DEPOSITION

Over the last fifty years, attempts have been made to explain accumulations of clay minerals, Fe and Al sesquioxides and sometimes organic matter in podzolic B horizons in terms of physical and chemical 'filter' effects. These two groups of theories can simply be described as

(1) a mechanical sieving effect, depending on size of translocated colloidal material compared to B horizon porespace; and
(2) chemical alterations of translocated colloidal and chelated material, resulting in reduced solubility and precipitation in the B horizon.

Soil micromorphological evidence suggests that both groups of processes probably operate at some stage in podzolic B horizon development.

Until recently, much attention in podzolisation studies has been focused on the mobility of iron and iron oxyhydroxides in soil profiles. We have seen how these studies have resulted in thories which propose that organic matter in the soil solution binds with divalent and trivalent metal cations, particularly iron species, produced through mineral weathering in the surface mineral soil horizons, to form soluble chelates which migrate down through the soil profile and are deposited in subsurface horizons. Illuviation of these organometallic complexes has been discussed in terms of their reduced solubility due to (1) reduced acidity, or (2) biodegradation of the organic matter resulting in a higher metal : OM ratio and causing precipitation. Since Farmer and his co-workers (Farmer, Russell and Bellow, 1980; Farmer and Fraser, 1982) demonstrated the presence of imogolite-like allophanes (gel-like amorphous aluminosilicates) in podzolic B horizons, a change of direction has occurred in illuviation studies. Much attention is now concentrated on the study of aluminium migration and deposition. A large suite of studies on B horizon imogolite has necessitated substantial revision of existing podzolisation theories, since Farmer (1982) argues that imogolite could not be deposited from soluble fulvic or humic complexes. An attempt is made here to (1) summarise the chronological development of illuviation theories, and (2) introduce some modern hypotheses for the development of B horizons in podzolic and podzolised soils.

The surface charge on soil organic matter is due to dissociation of phenolic ($-OH$) and carboxylic ($-COOH$) groups as acidity is reduced. Schnitzer and Skinner (1963) were the first to suggest that these functional groups were the sites of metal bonding. When the charge on soil organic matter approaches zero, its character changes from hydrophilic to hydrophobic and it becomes insoluble. The surface charge or acidity of organic matter is thus a measure of both its complexing ability and the solubility of its organometallic complexes (Buurman, 1985). Schnitzer and Skinner (1963) were also the first to show that a whole range of metal complexes are possible, ranging in metal : OM ratio from 1 : 1 to 6 : 1 and becoming increasingly insoluble as more metal is complexed. McHardy, Thomson and Goodman (1974) have subsequently shown that ferric oxides are precipitated from iron-phenolic chelates when the organic matter is oxidised and the iron : OM ratio is increased.

Theoretical calculation of the critical metal content at which organic complexes bcome insoluble is possible if several assumptions are made. For a known organic C content and assuming (1) maximum organic matter acidity, and (2) that trivalent Fe or Al binds only with $-COOH$ functional groups on soil organic matter, Buurman (1985) calculated that organic complexes in podzols would be precipitated at theoretical C : sesquioxide ratios between 4 and 12. This is similar to ratios of 10–14 found in laboratory experiments carried out at pH 4, but rather lower than ratio values up to 30 for precipitated organic complexes in the field (Buurman, 1985). In explaining these findings, Buurman (1985) suggests that the theoretical prediction is too low because laboratory measurements of organic matter total acidity overestimates dissociation at field pH. In podzol eluvial horizons, Mokma and Buurman (1982) found that the titratable acidity to pH 4 was less than half the acidity to pH 7. This implies that at pH 4 in these horizons, less than half the maximum acidity of organic matter is available for binding sesquioxides. Thus at higher pH, higher amounts of metal are necessary to precipitate organic matter. Since the C : sesquioxide ratio at which organic matter in podzol B horizons precipitates is pH-dependent, Buurman (1985) points out that acidification of soil by, for example acid rain, would lead to a lower complexing power of organic matter.

The discovery of imogolite-type allophanes in podzolic B horizons not only cast doubt on existing theories of illuviation, but also seriously questioned whether organic materials played any part in sesquioxide transportation in podzol profiles. Farmer's interpretation of the presence of imogolite in podzolic B horizons (see, for example, Farmer, 1982) is that Al migrates in the form of inorganic, proto-imogolite sols, not in high molecular weight organic matter complexes. Anderson *et al.* (1982) offer two explanations other than the migration of organic matter complexes for the presence of Bh horizons and for the presence of metal humates in Bs horizons: either

(1) that acidic organic matter in drainage water interacts with previously deposited sesquioxide coatings in these horizons to form precipitated organic matter complexes *in situ*; or
(2) that colloidal organic matter is preferentially precipitated at the top of the B horizon with deeper penetration of soluble fulvic acids.

Farmer (1982), however, does not rule out the possiblity that small molecular organic complexes, such as oxalate or tartrate, may be important in Al translocation.

4.5 MODERN INTERPRETATIONS OF PODZOL DEVELOPMENT

Current theories of podzolisation and podzol development resolve themselves into two groups:

(1) those in which translocation and deposition of soil Fe and Al occur as *organic complexes* (here we will call this the 'organic theory', Table 4.4); and
(2) those in which translocation and deposition of soil Fe and Al occur as *inorganic compounds* (here we will call this the 'inorganic theory', Table 4.5).

In the organic theory, the traditional view that humic and fulvic acids are responsible for complexing and mobilising mineral Fe and Al was initially pioneered by Schnitzer and his co-workers. Subsequent modifications of this theory to include (1) complexation by soluble organic materials in leaf leachates (see, for example, Coulson, Davies and Lewis, 1960a; Malcolm and McCracken, 1968); and (2) some modern concepts in colloid organochemistry (De Corninck, 1980) has made it more widely acceptable to soil scientists studying soil processes.

In the inorganic theory, Farmer and his co-workers have identified

Table 4.4   Organic theory of podzol formation (from De Corninck, 1980)

Stage	Development process
(1)	Mobile organic substances are formed during the decomposition of surface litter and soil organic matter.
(2)	If there are sufficient polyvalent cations (Al and Fe) at the top of the mineral soil profile, the mobile organic substances are immobilised immediately and no downward migration occurs.
(3)	If insufficient amounts of Al and/or Fe are available to completely immobilise the mobile organic matter, these cations are complexed by the mobile organic matter and transported downwards.
(4)	Immobilisation of organomineral complexes may occur at depth due to (a) supplementary fixation by cations (b) desiccation (c) on arrival at a level with different ionic concentration

Table 4.5   Inorganic theory of podzol development* (Farmer, 1982)

Stage	Development process
(1)	Early stage of podzol formation, before development of A/E horizon, mobile Al, Fe and Ca are relatively abundant in the A horizon, and any fulvic acid liberated by organic matter decomposition will be precipitated *in situ* as an insoluble salt.
(2)	Once A/E horizon develops, fulvic acid becomes excess of the A horizon capacity and passes down the profile. This fulvic acid can carry only small amounts of complexed Al and Fe, but it will attack imogolite and proto-imogolite already deposited in the B horizon; this liberates silica and forms an insoluble Al fulvate *in situ*. This mechanism can account for the absence of imogolite-type minerals in Bh horizons.
(3)	Thin ironpan (Bf) horizons can provide effective barriers to the passage of fulvic acids – so in profiles with ironpans, imogolites persist up to the ironpan.
(4)	Finally, downward migrating organic matter is sorbed on imogolite-like material of B_2 horizons.

* A very similar sequence of events is proposed for the transport and deposition of Fe (as soluble iron oxides) with localised Fe concentrations in B horizons due to oxidation–reduction processes in the B horizon.

imogolite-like compounds in the B horizon of podzols and used this finding to suggest that translocation of soil Fe and Al also occurs in inorganic compounds such as gel-like amorphous aluminosilicates.

4.6   METHODS FOR ANALYSING SESQUIOXIDES AND RESULTANT PROBLEMS IN SOIL CLASSIFICATION OF PODZOLS

Bascomb (1968) was the first to recognise the need to conceptualise and characterise the mobile (or 'active') and immobile (or 'inactive') forms of iron in soil. His scheme for analysing these forms of iron in soil is outlined in Fig. 4.8. Since then, three extractants have been widely used for this purpose:

(1) 0.1 M sodium or potassium pyrophosphate, at pH 10 to extract *organic plus amorphous 'gel' forms of iron* (Bascomb, 1968; Loveland and Digby, 1984).
(2) 0.2 M acid ammonium oxalate to extract *aged amorphous forms of iron* (Farmer, Russell and Smith, 1983; Chao and Zhou, 1983).
(3) Sodium dithionate (Holmgren, 1967; Pawluk, 1972), or citrate-bicarbonate (Mehra and Jackson, 1960) to extract *crystalline forms of iron*.

The objective of many studies has been to identify a suitable extractant to characterise the forms of iron compounds found in typical horizons of

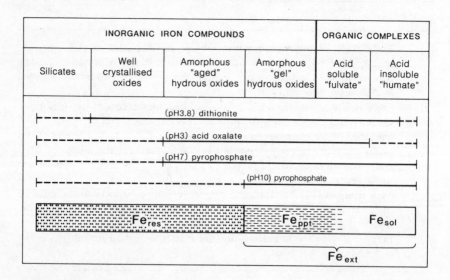

Figure 4.8 Ranges of iron compounds removable by different extractants. (Source: Bascomb, 1968)

podzols, especially the illuviated B horizon (spodic horizon). Apart from the search for a reproducible index for the distribution of different forms of iron in soils for classification purposes, extractants have been sought to aid interpretation of Fe and Al translocation processes. The assumption that fulvate complexes are responsible for the translocation of iron and aluminium has been the justification for using pyrophosphate-soluble Fe and Al as a diagnostic criterion for illuviated podzol B (spodic) horizons (Avery, 1980; Soil Survey Staff, 1975). Since the inorganic theory of Farmer, Russell and Berrow, 1980) for accumulation in podzolic B horizons, it has been suggested that a different extractant may be required to characterise crystalline forms of Fe and Al. Support for a re-examination of Fe extraction techniques for soil classification purposes has come from studies, particularly in North America, which showed that profiles that morphologically appear to be podzols (spodosols) do not meet the chemical requirements for the podzolic B horizon (see, for example, Mokma, 1983; McKeague et al., 1983). In addition to these problems, Loveland and Digby (1984) obtained very inconsistent results during the exhaustive testing of three pyrophosphate extraction solutions, using standard extracting techniques. Despite these problems, new proposals for defining podzolic B horizons use the same extractants, but different criteria (where Al_p, Fe_p and C_p are pyrophosphate extractable Al, Fe and C):

(1) Chemical criteria for podzolic B (spodic) horizon (Soil Survey Staff, 1975):
$$\frac{(Al_p + Fe_p)}{\text{per cent clay}} \geqslant 0.2 \quad \text{or} \quad \frac{(Al_p + C_p)}{\text{per cent clay}} \geqslant 0.2$$

(2) New chemical criteria suggested by Mokma (1983):
$$(C_p + Al_p + Fe_p) > \begin{array}{c} \text{overlying} \\ \text{soil} \\ \text{horizon} \end{array} \quad \text{or} \quad (C_p + Al_p + Fe_p) \geqslant 0.5 \quad \text{or}$$
$$\frac{C_p}{(Al_p + Fe_p)} \text{ is 5.8 to 25.0}$$

(3) New chemical criteria suggested by McKeague et al. (1983):
$$\frac{(Al_p + Fe_p)}{\text{per cent clay}} \geqslant 0.1 \quad \text{or} \quad \frac{(Al_p + C_p)}{\text{per cent clay}} \geqslant 0.1$$

While the above solutions fit classification requirements, they do not help in examination of podzol development, particularly translocation processes. In view of the evidence for the mobilisation of iron and aluminium not only by fulvic and humic acids, but also by leaf leachate, Ross and Smith (1987) compared the amounts of Fe extracted using pine and oak leaf leachates with amounts extracted by 0.1 M sodium pyrophosphate solutions at pH 10. They found that pyrophosphate consistently extracted about twice as much iron as the leaf leachates. They found that an 0.0025 M aqueous solution of commercially available catechol more realistically mimics the effects of tree leaf leachates and recommend the use of a simple organic solvent as a laboratory analogue for field chelation and translocation processes.

4.7　SUMMARY

A large and often confusing literature describes the study of soil podzolisation processes. Podzolisation is the term used to embody three component processes:

(1) *in situ weathering*: the hydrolysis of soil minerals to produce hydroxy ions of Fe and Al;
(2) *translocation*: the eluviation of soluble and chelated hydroxy Fe and Al ions, and fine clay particles from surface mineral horizons; and
(3) *deposition*: the illuviation of hydroxy Fe and Al ions and fine clay particles in subsurface mineral horizons.

Since the pioneering work of Bloomfield in the 1950s, soluble organic compounds derived from live plant leaves, leaf litter and soil organic matter have been implicated in soil translocation processes under forest and heathland vegetation. Current debate surrounds the relative importance in translocation processes of mobilisation by fairly simple organic compounds, such as polyphenols, derived from leaf and litter washings, and the much more complex and high molecular weight organic compounds, called humic and fulvic acids, derived from soil organic matter. There is also some debate over the mechanism(s) of deposition responsible for the development of an illuviated podzolic B (spodic) horizon. The traditional theory for the deposition of chelated Fe and Al cations is that of reduced solubility of the organometallic complex with higher pH at depth and as biodegradation reduced the ratio of organic matter to metal in the complex. The discovery of imogolite-type alophanes in podzolic B horizons has seriously questioned whether organic compounds play any important role in Fe and Al translocation in podzolic profiles.

PART THREE

Processes in the Rooting Zone

Soil in the zone occupied by plant roots is altered in two ways which make it different from the bulk of the soil. First, solute chemistry is altered by plant uptake which creates water and solute potential gradients in soil adjacent to roots. Second, organic enrichment in the rhizosphere immediately adjacent to root surfaces enhances biological activity, including special root–organism relationships, such as symbiotic mycorrhizal fungi and symbiotic nitrogen fixing organisms. The processes operating in the vicinity of the root are of particular importance in plant ecology and crop production, since good healthy plant roots generally result in good, healthy plants and crop yields.

Physical, chemical and biological processes combine to supply roots with adequate amounts of water and nutrients. Rhizotron studies have allowed monitoring of the distribution and rate of root growth. Detailed empirical laboratory studies have also measured ion diffusion rates in soils and soil materials, with and without the influence of active root uptake. There is still, however, some difficulty in answering a general question such as: 'do roots grow to exploit new sources of nutrients in the soil, or do nutrient solutes move towards roots down concentration gradients?' Part of the difficulty arises from uncertainty in separating the uptake of nutrients *per se* and uptake by associated micro-organisms, particularly mycorrhizae. Despite difficulties in developing successful techniques for studying delicate mycorrhizal hyphae under field conditions, Read and coworkers (see, for example, Read *et al.*, 1985) have presented some very exciting ideas concerning the interconnectivity of mycorrhizal systems and their importance in large scale nutrient transfers at the ecosystem scale.

CHAPTER 5

Solute Dynamics in the Rooting Zone

5.1 INTRODUCTION

Apart from carbon and oxygen, which are supplied to plants as CO_2 and O_2 in gaseous form and taken in through stomata in the leaf, all nutrients required by plants are taken up by roots as solutes from the soil solution. Some of the simpler forms of plant macronutrients (N, P, S, Ca, Mg, K, Cl) and micronutrients (Fe, Mn, Zn, Cu, Bo, Mo) which are found in the soil are given in Table 2.2 (p. 40). Some smaller amounts of carbon and oxygen may also be taken up by plant roots from the soil in more complex ionic or molecular form. Since plant nutrition relies on solute uptake, an understanding of the dynamics of soluble plant nutrients and other chemicals in the soil solution is imperative for ensuring plant health and encouraging maximum yields.

Within the zone of influence of plant roots, the concentration of soluble chemical nutrients required for uptake are controlled by production, transformation, storage and transport processes. The processes that produce simple plant nutrients from mineral and organic soil materials have been discussed in detail in Chapters 1 and 2, together with some of the more widely operative transformation processes such as oxidation and reduction (see Chapter 3). As well as conveying nutrients to plant roots, transport processes can cause nutrient loss or removal through drainage and leaching. Operating in opposition to these transport processes are storage mechanisms which retain plant nutrients against loss or movement out of the soil profile. These are the mechanisms of cation and anion adsorption and exchange. The distribution and equilibration of solute ions between the surfaces of soil solids and the aqueous solution depends on the amount and type of charge generated on the colloid surfaces and on the type and concentration of ions present in the soil solution. Equilibration virtually never exists in the ion exchange system due to continually changing colloid charges and the dynamics of soil solute concentrations. The soil thermal and moisture regime as well as root uptake control the concentration of ions in the soil solution. In the following sections, an introduction to ion exchange and solute transport processes will be given, together with an outline of the developments that have occurred in the modelling of solute transport at the field scale.

5.2   PROPERTIES OF COLLOID SURFACES AND EXCHANGE MECHANISMS

Soil colloidal particles, both mineral and organic, possess surface charges. In clay minerals this charge is due, first, to isomorphic substitution in individual layers of the crystal lattice, and second, to the presence of unsatisfied valencies at the broken edges of particles. In the case of organic colloids, dissociation of carboxylic and phenolic groups at particle surfaces cause surface charge to be generated. A comprehensive discussion of the sources and surface density of charge on soil inorganic and organic colloids is given in Talibudeen (1981). Surface charge due to isomorphic substitution in the aluminosilicate lattice of clay colloids produces negative charge which is *permanent* and independent of changes in solution pH. Surface charge due either to dissociation of chemical groups at colloid surfaces and broken particle edges, or to adsorption onto these groups, is *pH dependent* and is characteristic of most naturally occurring colloids such as the oxides of Fe, Al, Mn, Si and colloidal organic matter.

Dissociation of surface hydroxyl and organic groups varies with pH (Fig. 5.1). Only clay colloids can generate a positively charged surface when the soil solution is acidic, while both clay and organic colloids will have negatively charged surfaces when the soil solution is alkaline. In between these two extreme conditions lies the isoelectric point, which is the pH at which the surface charge on the colloid is zero. This value can be determined experimentally or calculated for pure mineral colloids such as naturally occurring aluminosilicates, oxides and hydroxides (Table 5.1). It is clear that the aluminium species and some iron species which have high isoelectric pHs, will predominantly exhibit positively charged surfaces with anion attraction and exchange under normal soil conditions where the pH < 7.0, while aluminosilicate clay minerals which have a much lower isoelectric pH will predominantly exhibit negatively charged surfaces with cation attraction and exchange. The magnitude of charge on soil colloid surfaces determines their ability to attract and exchange cations and anions from the soil solution. These are the cation and anion exchange capacities (usually abbreviated to CEC and AEC). When adsorption occurs at the colloid surface, as occurs with phosphate adsorption on Al and Fe hydrous oxides, the pH at which the surface charge is zero (the point of zero charge, or PZC) is different from the isolectric point. Hingston *et al.* (1968) suggested that sorption of anions on sesquioxide surfaces shifts the PZC to lower pH. Since more cations are required to balance the additional negative charge, the CEC is increased. This phenomenon has been widely reported in highly weathered, sesquioxide-rich tropical soils (see, for example, Mekaru and Uehara, 1972). The PZC for soil material generally can be considered as a simple measure of the relative magnitude of positive and negative charge on colloidal surfaces.

(a) *Mineral groups*

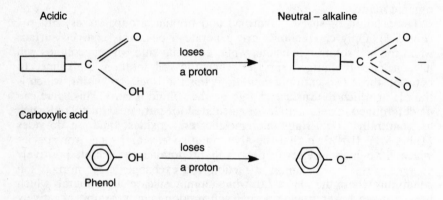

(b) *Organic groups*

Figure 5.1 Development of pH-dependent charge in **(a)** clay colloids; **(b)** organic colloids

5.2.1 Contributions of clay minerals and organic matter to soil CEC

Until recently, work on reactivity of solute ions and charged soil surfaces has mainly concentrated on temperate agricultural soils, where the bulk of soil colloids have permanent negative charge. Current and continuing interest in the fertility of weathered tropical soils and in organic soils has shifted attention to the reactivity of variable charge colloids. In temperate agricultural soils, many attempts have been made to assess the relative contributions made to the total soil cation exchange capacity (CEC) by clay minerals and organic colloids. Current research focuses more on the components of variable charge: (1) the hydrous oxides of Fe and Al,

Table 5.1 Experimentally determined isoelectric points (pH) for some naturally occurring minerals and organic compounds (from Drever, 1982* and Swift, 1980[†])

Mineral/compound	Isoelectric point (pH)
Quartz* (SiO_2)	2.0
Gibbsite* ($Al(OH)_3$)	~9.0
Corundum* (Al_2O_3)	9.1
Goethite* ($FeO(OH)$)	6–7
Magnetite* (Fe_3O_4)	6.5
Haematite* (Fe_2O_3)	6–7
Kaolinite*	~3.5
Montmorillonite*	<2.5
Histidine[†] (amino acid)	7.6
Lysozyme[†]	11.1
Haemoglobin[†] (proteins)	7.0
Gelatin[†]	9.0
Urease[†]	5.1
PVA[†] (polyvinyl alcohol)	uncharged

(2) different clay mineralogies, and (3) influence on CEC of organic matter adsorption on clay surfaces.

It has been known for a long time that the clay and organic matter content of temperate mineral soils control the CEC. Helling, Chesters and Corey (1964) found that in soils with uniform clay mineralogy, about 92 per cent of the variation in CEC could be attributed to their clay and organic matter contents. The results of several investigations have suggested that organic matter content mainly determines soil CEC in cultivated topsoils, with clay content mainly responsible for the CEC of subsoils (see, for example, Wright and Foss, 1972). Particle sizes within the clay faction appear to contribute differently, with the < 0.2 μm clay fraction being the main CEC contributor in B horizons (Wilding and Rutledge, 1966). Martel, De Kimpe and Laverdière (1978) have also shown that where the relative contribution of the clay fraction to soil CEC is 3.5 to five times greater than that of organic matter, nearly 50 per cent of the variation in CEC could be explained by clay mineralogical composition (per cent smectite + per cent vermiculite + per cent illite and chlorite).

The relative magnitude of permanent and pH-dependent charge in soil is also determined by type and amount of clay minerals and organic matter present. The contribution of clay mineral OH groups to pH-dependent charge in soil varies with the type of clay minerals present. Bohn, McNeal and O'Connor (1985) suggest a simple rule of thumb which states that about 5–10 per cent of the negative charge on 2 : 1 layer silicates such as smectites and vermiculites is pH-dependent while about 50 per cent of the charge on 1 : 1 clay minerals such as kaolinite can be pH-dependent. The main reason for this difference is due to the vertical stacking of kaolinite lattices which exposes a large vertical or edge surface area compared to the horizontal or planar surface area. Since smectites do not stack in this way, their horizontal or planar surface areas can be up to twenty times higher than their vertical or edge surface areas. The CEC of 1 : 1 type clay minerals is smaller and more dependent on soil pH than that of 2 : 1 type minerals (Table 5.2).

Table 5.2 Cation exchange capacities (CEC) for common soil minerals and organic matter

Soil material	CEC (in me 100 g^{-1})	Reference
Soil organic matter	150–300	White (1979)
Kaolinite	2–5	White (1979)
2:1 clays	40–150	White (1979)
Humus	200	Brady (1974)
Vermiculite	150	Brady (1974)
Montmorillonite	80–100	Brady (1974)
Hydrous mica	30	Brady (1974)
Chlorite		
Kaolinite	3–15	Brady (1974)
Hydrous oxides	4	
Illite	15–40	Brady (1974)
Kaolinite	2.3	Wiklander (1955)
Illite	16.2	Wiklander (1955)
Montmorillonite	81.0	Wiklander (1955)

Weight for weight, soil organic matter has a higher CEC at pH 7 than any other type of soil colloid (see Table 5.3). Not surprisingly, soils high in organic matter have a potentially large CEC which is pH-dependent. So one reason why many natural peats are infertile is that their low pH values of 3–4 result in low CEC. The pH-dependence of organic CEC is well illustrated by Helling, Chesters and Corey (1964) for a range of Wisconsin soils (Fig. 5.2). Tan and Dowling (1984) attempted to characterise the contribution made by soil organic matter to the permanent and pH-dependent CEC properties of several clay soils in the southern United States. They found that in montmorillonite-dominated soil, removal of organic matter caused an increase in CEC while in soils of mixed clay mineralogy, removal of organic matter resulted in a decrease in CEC. The lack of explanation for this finding may reflect gaps in our knowledge of (1) binding mechanisms between clays and different organic molecules in soil organic matter, and (2) interferences that binding may have with ion exchange on the clay and organic matter surfaces. The CEC of inorganic and organic soil components is not additive and it may be that adsorption of soil organic matter onto clay

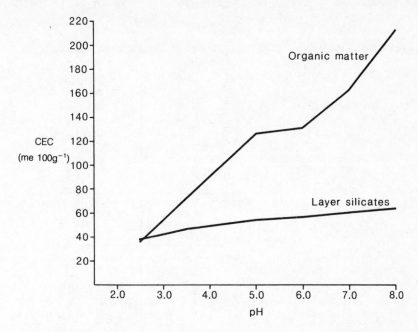

Figure 5.2 Effect of pH on CEC for 60 Wisconsin soils. (Source: drawn from the data of Helling *et al.* 1964)

surfaces causes blocking of ion exchange sites.

Many authors have simplistically considered the cation exchange properties of the so-called clay–humus complex in soil. That this is a gross oversimplification should be obvious from a knowledge of soil mineralogy and the heterogeneity of soil organic matter. In a comprehensive review, Mortland (1970) has described a wide range of possible mechanisms for the adsorption of different organic materials on clay surfaces. Few studies have, however, examined the effects of organic matter bonding on ion exchange capacity. Using synthetic organic materials, Swift (1980) found that the adsorption of an amino acid, histidine, and three proteins, lysozyme, haemoglobin and gelatin, caused a reduction in the CEC of montmorillonite, illite and kaolinite. Since the amino acid could be removed by leaching with a dilute salt solution, he reasoned that it was retained on clay surfaces by simple ion exchange. Proteins are held by a combination of ion exchange and physical adsorption. The use of ion exchange sites by these organic molecules reduced the overall surface charge. Swift compared these results from synthetic, commercially available organic compounds with a humic acid preparation, extracted from Haughley clay soil from Surrey, England. Not only did adsorption of the humic acid increase the CEC for all three clay minerals, but the increase was directly related to the amount of humic acid adsorbed and the measured CEC closely approximated the numerical sum of the mineral and organic components. These findings suggest either that the

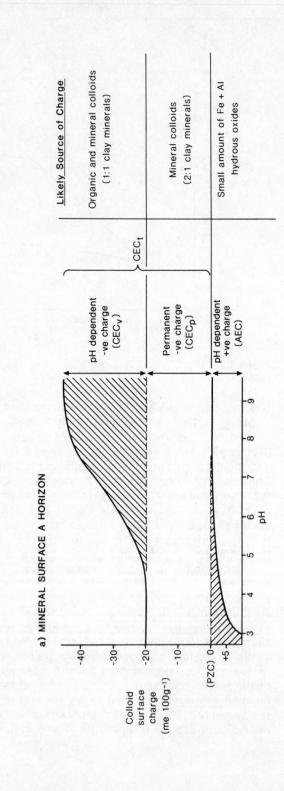

a) MINERAL SURFACE A HORIZON

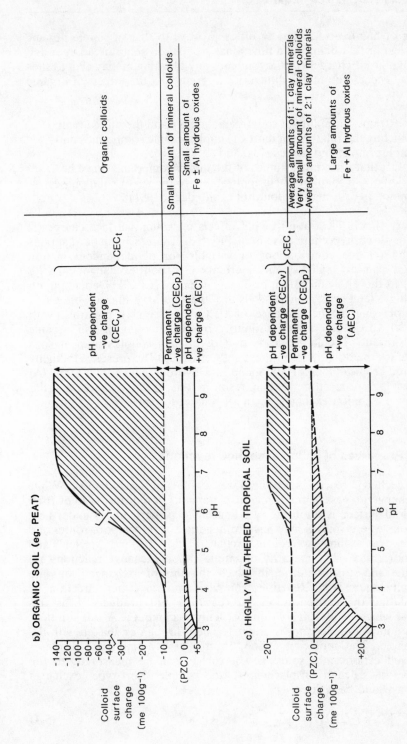

Figure 5.3 Schematic effect of pH on surface colloidal charge for temperate, mineral, organic and tropical soils

binding of humic acid occurs in different ways to that of simple organic molecules, or that humic acid functional groups can generate a larger and more significant charge which overcomes any blocking of clay charge sites that may occur. The major difficulties in predicting the effects of organic matter on soil CEC are

(1) the extreme variability of soil organic matter and the difficulty in predicting the relative contents of simple organic compounds and complex humic and fulvic acids; and
(2) the fact that since most studies of the humic compounds have been carried out in solution, their applicability to the gel or solid phases common in soils may be doubtful (Talibideen, 1981).

A very simple illustration of pH effects on colloid surface charges is given for three hypothetical soils in Fig. 5.3. These schematic diagrams show the relative contributions of variable, or pH-dependent, cation exchange capacity (CEC_v) and permanent cation exchange capacity (CEC_p) to the total soil cation exchange capacity (CEC_t). Depending on mineralogy, temperate mineral soils are likely to have measurable CEC_t at any pH due to the presence of CEC_p on 2:1 clay minerals, with additional CEC_v determined primarily by the amount of soil organic matter present. The large proportion of organic colloids in peat soils accounts for the dominance in these soils of CEC_v. The presence in highly weathered tropical soils of allophane and hydrous oxides of Fe and Al with very high isoelectric pHs may result in a net positive charge and the dominance of anion exchange in these soils at low pH.

5.2.2   Properties of cation exchange reactions

Cation exchange reactions are rapid, reversible and stoichiometric. While the exchange process is practically instantaneous, ion diffusion to or from the colloid surface is often the rate-limiting step, particularly under field conditions where solutes may have to negotiate tortuous pore routes and stagnant water films to reach exchange sites (Bohn, McNeal and O'Connor, 1985). The majority of simple cation exchange reactions are reversible and stoichiometric, in which amounts of exchanged ions are chemically equivalent. Reactions can be driven in either direction by manipulating the concentrations of reactants and products. It is this principal which is used in soil ion extraction techniques. A salt solution, such as ammonium acetate, is either leached through or shaken with the soil sample. The replacing cation, in this case ammonium, displaces exchanged cations into solution where they can be measured. In the example of calcium displacement, the simple Kerr-type exchange equation would be written as:

$$\boxed{}\text{--}Ca + 2NH_4^+ \rightleftharpoons \boxed{}\text{--}2NH_4 + Ca^{2+} \tag{5,1}$$

Since two ammonium ions are required to replace one calcium ion, the reaction coefficient for Ca exchange is:

$$K = \frac{[(NH_4)_2 -\boxed{}]^2 \, [Ca^{2+}]}{[Ca -\boxed{}] \, [NH_4^+]^2} \tag{5,2}$$

$$= \frac{\left[\begin{array}{c}\text{concentration of}\\\text{adsorbed replacement}\\\text{cation}\end{array}\right]^2 \left[\begin{array}{c}\text{concentration of}\\\text{Ca in solution}\end{array}\right]}{\left[\begin{array}{c}\text{concentration of}\\\text{adsorbed Ca}\end{array}\right]\left[\begin{array}{c}\text{concentration of}\\\text{replacement cation}\\\text{in solution}\end{array}\right]}$$

where $[\,] = $ concentration in moles per litre (mol l^{-1}).

The typical cation exchange equation is:

$$\frac{[(NH_4)_2 -\boxed{}]^2}{[Ca -\boxed{}]} = K \frac{[[NH_4^+]^2}{[Ca^{2+}]} \tag{5,3}$$

or: $\begin{array}{c}\text{ratio of } [NH_4]^2 \text{ to}\\ Ca^{2+} \text{ on the colloid}\end{array} = K \times \begin{array}{c}\text{ratio of square of } NH_4^+\\ \text{to } Ca^{2+} \text{ in soil solution}\end{array}$

When exchange occurs between ions of different valence, dilution of the equilibrating solution causes the more highly charged cations to be preferentially retained. The dependence of cation exchange reactions on cation valence is called the ratio, or valence dilution effect. This effect means that the ratio of monovalent replacement ion (NH_4^+ in this example) to exchanging ion (Ca^{2+}) will decrease during dilution of the bathing solution (Table 5.3).

The Kerr-type exchange equation assumes that ionic concentrations are directly proportional to their activities. This equation works quite well over small concentration ranges, but for wider applicability, the Gapon equation is used. For the reaction:

$$(Ca)_{\frac{1}{2}} -\boxed{} + NH_4^+ \rightleftharpoons NH_4 -\boxed{} + \tfrac{1}{2}(Ca^{2+}) \tag{5,4}$$

the Gapon equation has the form:

$$\frac{[NH_4 -\boxed{}]}{[(Ca)_{\frac{1}{2}} -\boxed{}]} = K_G \frac{[NH_4^+]}{[Ca^{2+}]^{\frac{1}{2}}} \tag{5,5}$$

in which exchangeable cation concentrations are in me g^{-1} and soluble cation concentrations are in mmol l^{-1}. Again, Bohn, McNeal and O'Connor (1985) indicate a limited applicability of the Gapon equation over wide concentration ranges, but suggest that the Gapon exchange

Table 5.3 Numerical examples of valence dilution effect during cation exchange

	Ratio of $[NH_4^+]$ to $[Ca^{2+}]$ on soil colloid	Ratio of $[NH_4^+]$ to $[Fe^{3+}]$ on soil colloid
Bathing solution concentration:	$[NH_4^+] = [Ca^{2+}] = 1$ mmol l^{-1} ratio $[NH_4^+]:[Ca^{2+}] = 1^2:1 = 1$ mmol l^{-1}	$[NH_4^+] = [Fe^{3+}] = 1$ mmol l^{-1} ratio $[NH_4^+]:[Fe^{3+}] = 1^3:1 = 1$ mmol l^{-1}
X10 dilution:	ratio $[NH_4^+]:[Ca^{2+}] = 0.1^2:0.1 = 0.1$ mmol l^{-1}	ratio $[NH_4^+]:[Fe^{3+}] = 0.1^3:0.1 = 0.01$ mmol l^{-1}

So, ratio of NH_4^+ to Ca or of NH_4^+ to Fe^{3+} on the soil colloid decreases during dilution

coefficient (K_G) for Ca^{2+}–Na^+ exchange, for example, is acceptably constant for the concentration ranges of interest in irrigated agricultural soils of the western United States.

5.2.3   Selectivity of soil colloid attraction for different cations

The relative attraction of cations to charged colloid surfaces depends partly on characteristics of the cation and partly on properties of the colloid surface. Cations with the same charge are not equally attracted colloidal surfaces. This is due to differences in (1) size and shape of cations of the same valency, and (2) the surface charge density and surface geometry of soil colloids. Marshall (1975), reporting the results of cation exchange work on montmorillonite clay by Schachtschabel (1940), gives the following order of cation exchange:

$$\text{Replaceability} \longrightarrow$$

monovalent cations: $Li^+ > Na^+ > K^+ > Rb^+ > Cs^+$
divalent cations: $Mg^{2+} > Ca^{2+} = Sr^{2+} > Ba^{2+}$

The order of cation replacability was not very different for kaolinite, muscovite, biotite and orthoclase feldspar systems. Bohn, McNeal and O'Connor (1985) suggest that in general the full sequence follows the order:

$$\text{Replaceability} \longrightarrow$$

$$Li^+ \approx Na^+ > K^+ \approx NH_4^+ > Rb^+ > Cs^+ \approx Mg^{2+} > Ca^{2+} > Sr^{2+}$$
$$\approx Ba^{2+} > La^{3+} \approx H^+(Al^{3+}) > Th^{4+}$$

This order of exchange is called the *lyotropic series* and expresses increasing cation replacability, or decreasing cation attraction, with increasing atomic radius, increasing atomic number and thus decreasing polarising power and ionic hydration. Generally, less hydrated ions with smaller hydrated radii, such as Na^+ and K^+, are more strongly attracted to charged mineral surfaces. Cation valency is a second major control on the order of exchange, with multivalent ions being more strongly retained than monovalent ions. Trivalent ions are thus more strongly attracted than divalent ions which are more strongly attracted than monovalent ions. For very simple ionic systems, it is possible to predict the composition of cations on the cation exchange complex, from a knowledge of their activities in the bathing solution. The *ratio law* states that in a simple two-ion system, in which one ion is monovalent and the other divalent, then the ratio of the amounts held by soil colloids is dependent on the ratio of the activity of the monovalent ion to the square root of the activity of the divalent ion in solution. Similarly, for a monovalent and a trivalent cation, the ratio held will depend on the ratio of the activity of the monovalent to the cube root of the activity of the trivalent ion in solution. These are simple examples of the Gapon equation:

for monovalent:divalent   for monovalent:trivalent

$$\frac{[K - \boxed{}]}{[Ca - \boxed{}]} = K_G \frac{[K^+]}{[Ca^{2+}]^{\frac{1}{2}}} \qquad \frac{[K - \boxed{}]}{[[Al - \boxed{}]} = K_G \frac{[K^+]}{[Al^{3+}]^{\frac{1}{3}}}$$

It is the extreme complexity and temporal variability of soil solution ionic compositions that makes predictions of the composition of cations on exchange sites virtually impossible.

An additional reason for cation selectivity is the exact matching of cation size to structural holes in the surfaces of soil minerals (Talibudeen, 1981). Three main clay colloid surface properties are considered by Talibudeen (1981) to be important factors in cation selectivity. He first identifies the generation of a greater density of surface negative charge, by a high degree of isomorphic substitution, as a possible reason for increased K-fixation. Second, the strong preference of micas for NH_4^+ over Li^+, Na^+, or K^+ may be due to the formation of $NH_4 - O$ links in the hexagonal holes of the $Si - O_2$ lattice layers. Talibudeen (1981) indicate, third, that the presence of cracks, cleavages or crevices in clay surfaces may be a reason for cation selectivity within mineral types. Due to their size and shape, some cations can be specifically held in structural holes whose shape and dimensions were created by particular isomorphic substitutions such as the internal pores in interlayer regions of chloritised micas, formed by islands of brucite $(Mg(OH)_2)$ and gibbsite $(Al_2O_3.3H_2O)$.

The cation selectivity of organic colloids depends on the configuration of carboxylic and phenolic groups on organic molecules because it is this which determines the surface negative charge density. Two or three adjacent end groups on aliphatic and aromatic compounds (see Fig. 4.4, p. 116) attract multivalent cations more selectively than when they are widely spaced (Talibudeen, 1981). According to Talibudeen, the influence of one carboxyl group on the dissociation of an adjacent carboxyl group is negligible when they are separated by more than four CH groups. He notes three general rules of organic colloid–cation selectivity: (1) multivalent cations are preferred to monovalent cations; (2) transition group metals (Fe, Mn, Cu, Zn) are preferred to strongly basic metals (K, Na, Ca, Mg), and (3) cation selectivity increases with CEC.

5.2.4 Modelling cation exchange

Most models of cation exchange at colloidal surfaces have adopted a diffuse double-layer (DLL) structure, representing the parallel alignment of (1) the negatively charged colloid surface, and (2) the positively charged layer or diffuse 'cloud' of counterions in solution immediately

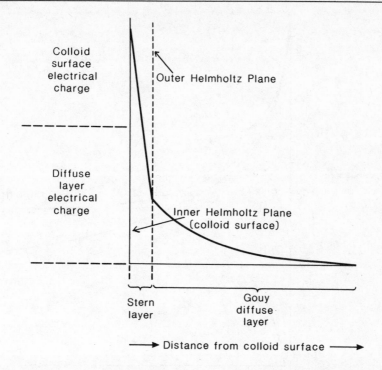

Figure 5.4 The Stern diffuse double layer model of cation exchange. (Source: van Olphen, 1977)

adjacent to the surface, with ionic concentration declining with distance from the colloid surface. A comprehensive introduction to the theory and assumptions of these models is given by Arnold (1978). The most widely accepted of these simple models is the Stern model (Fig. 5.4) which divides the soluble phase into two layers. Firstly, the Stern layer lies immediately adjacent to the colloidal surface. In this layer, electrical charge decays linearly with distance from the colloid surface. The Stern layer of cations is separated from the outer diffuse, or Gouy, layer by the outer Helmholtz plane. In the Gouy layer, electrical charge decays exponentially. The high negative charge generated at the colloid surface results in an excess of cations and a deficit of anions in the Stern layer, with concentrations approaching that of the bulk solution with distance from the colloid surface (Fig. 5.5). Neilsen *et al.* (1972) show in this figure that the main effect of a higher bulk solution ionic concentration is to decrease the 'thickness' of the DDL. When the soil dries out by drainage or evaporation, we can visualise a contraction of water films around soil particles, resulting in increased bulk solution ionic concentration and causing a contraction of the DDL at the surface of colloids. On further water removal, truncation of the DDL is envisaged by Bolt (1978), resulting in the tendency to absorb water by a process analogous to osmosis, until the full DDL is reinstated. As well as solution ionic

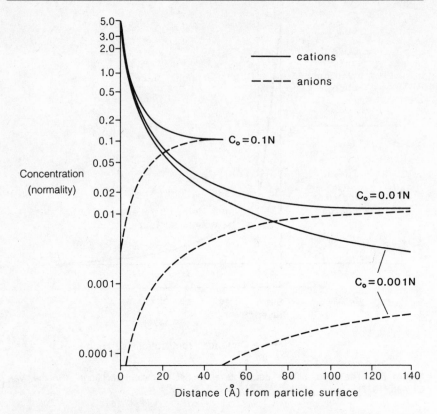

Figure 5.5 Distribution of monovalent cations and anions near the surface of a montmorillonite particle at three different concentrations in the bulk solution (Source: Nielson *et al.*, 1972)

concentration, a decrease in DDL thickness is also caused by increasing valency of exchanging cation. This results in a thicker DDL around clay colloids when they are bathed in soil solutions containing K^+ and Na^+ than when they are bathed in Ca^{2+} and Mg^{2+}. The calculation of DDL thickness was first carried out by Schofield (1947) by considering the negative adsorption of repelled anions. An increase in DDL thickness is caused by increasing temperature, increasing anion valency and increasing hydration of both cation and anion species. The theoretical treatment of these controls on DDL thickness is given in White (1979) and Arnold (1978).

5.2.5 Cation exchange reactions and equilibria

The fundamental reason for modelling cation exchange in soils is to aid the prediction of soil responses to management procedures such as fertilising or to external influences such as acid rain. These applications are particularly important when considering fertiliser efficiency in the soil–crop system. Although exchange reactions and equilibria were originally formulated for very simple exchange systems, it is now possible to develop thermodynamic equations applicable to the more complex, multication exchange systems found in soils.

Simplified equations and reaction constants for the equilibrium of cation exchange at the surface of a clay colloid have been given in Chapter 1 (Equations **1,17**, **1,18**, **1,19**, **1,20**). In an attempt to define equilibrium constants for cation exchange reactions, many authors have formulated empirical relationships similar to Equation (**1,18**). While a range of authors have shown that these empirical 'constants' can be derived from thermodynamic principles, none are constant over the whole range of exchange (Goulding, 1983a). They are thus usually defined as equilibrium 'coefficients'. A major confusion has existed in exchange equilibria work between thermodynamics and molecular theory (Sposito, 1981a). Although the earliest formulations did not consider a thermodynamic treatment of cation exchange equilibria, two thermodynamic formulations of cation exchange reactions have since been developed, their form depending on whether the exchangeable cations and anions on the exchange colloid (that is, the adsorbed ion activities) are expressed in molar or equivalent fractions. The main exponent of the mole fraction approach was Vanselow (1932). The second method, the Gaines and Thomas (1953) convention, represents an important development of the earlier Gapon (1933) formulation, expressing adsorbed ion activities in equivalent fractions.

In his 1932 example, Vanselow used an exchange reaction similar to Equation **1,17**, using the aqueous chloride salt as the exchanging solution phase:

$$\beta \boxed{}\!\!- A_\alpha + \alpha BCl_\beta \rightleftharpoons \alpha \boxed{}\!\!- B_\beta + \beta ACl_\alpha \qquad (5,6)$$

Simplifying this to use just the exchanging ion species, $A^{\alpha+}$ and $B^{\beta+}$, the thermodynamic equilibrium constant (K) is:

$$K = \frac{(\boxed{}\!\!- B_\beta)^\alpha \, (A^{\alpha+})^\beta}{(\boxed{}\!\!- A_\alpha)^\beta \, (B^{\beta+})^\alpha} \qquad (5,7)$$

where ( ) represents thermodynamic activities. In Vanselow convention, adsorbed ion activities are expressed in mole fractions (M) to give the Vanselow selectivity coefficient (K_v):

$$K_v = \frac{M_B^\alpha \, (A^{\alpha+})^\beta}{M_A^\beta \, (B^{\beta+})^\alpha} \qquad (5,8)$$

in which

$$M_A = \frac{\{\alpha \ \boxed{}\!\!- A\}}{\{\alpha \ \boxed{}\!\!- A\} + \{\beta \ \boxed{}\!\!- B\}} \quad \text{and}$$ (5,9)

$$M_B = \frac{\{\beta \ \boxed{}\!\!- B\}}{\{\beta \ \boxed{}\!\!- A\} + \{\beta \ \boxed{}\!\!- B\}}$$

are mole fractions of adsorbed $\alpha \ \boxed{}\!\!- A$ and $\beta \ \boxed{}\!\!- B$ respectively, and {} represent concentrations in moles per unit mass of exchanger (Sposito, 1981b). Only when the ion exchange reaction is 'ideal', as in the case of $Mg^{2+} - Ca^{2+}$ exchange for example, will K_v equal K. For all other exchange reactions, ion activities are related to mole fractions using activity coefficients (f), where:

$$f_A = \frac{(\alpha \ \boxed{}\!\!- A)}{M_A} \quad \text{and} \quad f_B = \frac{(\beta \ \boxed{}\!\!- B)}{M_B}$$ (5,10)

So K, expressed in Vanselow convention (K_{mol}), is given by:

$$K_{mol} = \frac{M_B^{\alpha} \ f_B^{\alpha} \ (A^{\alpha+})^{\beta}}{M_A^{\beta} \ f_A^{\beta} \ (B^{\beta+})^{\alpha}}$$ (5,11)

and

$$K_v = K \frac{f_A^{\beta}}{f_B^{\alpha}}$$ (5,12)

In an effort to express adsorbed ion activities in an experimentally accessible way, Gaines and Thomas (1953) defined the adsorbed ion activity coefficient (g) in terms of equivalent fractions (E) such that:

$$g_A = \frac{(\alpha \ \boxed{}\!\!- A)}{E_A} \quad \text{and} \quad g_B = \frac{(\beta \ \boxed{}\!\!- B)}{E_B}$$ (5,13)

The Gaines and Thomas selectivity coefficient (K_T) is given by:

$$K_T = \frac{E_B^{\alpha} \ (A^{\alpha+})^{\beta}}{E_A^{\beta} \ (B^{\beta+})^{\alpha}}$$ (5,14)

and the equilibrium constant (K_{eq}), expressed in Gaines and Thomas' convention, is given by:

$$K_{equiv} = \frac{E_B^{\alpha} \ g_B^{\alpha} \ (A^{\alpha+})^{\beta}}{E_A^{\beta} \ g_A^{\beta} \ (B^{\beta+})^{\alpha}}$$ (5,15)

This approach has subsequently been widely used in studies of ion exchange reactions in soils. For Equation (5,15), the Vanselow and Gaines and Thomas selectivity coefficients are related according to:

$$K_v = K_T \frac{\alpha^\alpha}{\beta^\beta} (\alpha E_B + \beta E_A)^{\alpha - \beta} \qquad (5,16)$$

(Goulding, 1983b).

Since these formulations, Sposito (1977) and Goulding (1983b) have drawn attention to problems associated with the expression of adsorbed ion activities based on equivalent fractions, as is the case in both Gapon and Gaines and Thomas conventions. Thermodynamic activity coefficients of ions in solution are a function of mole fractions and thus should be defined using Vaneslow convention. Sposito (1981b) suggests that the calculation of a selectivity coefficient based on equivalent fractions is only of use for homovalent ionic exchanges and is only a formal procedure for heterovalent ionic exchange since the ion activity coefficients, g_A and g_B have no strict thermodynamic meaning. Goulding (1983b) has examined this problem as it affects the interpretation of previous soil ion exchange work based on equivalent fractions. Most work on ion exchange in British soils is based on calculations of change in free energy, enthalpy and entropy. Goulding shows that the equivalent fraction definition results in alteration of ΔG and entropy of exchange by a constant factor which is independent of cations and exchanger. His major conclusions are (1) equivalent fractions correctly indicate the heterogeneity in the exchange process, and (2) comparisons of series of cations and exchanging soil colloids using these methods are not substantially different from those based on mole fractions.

5.2.6 Cation exchange and clay flocculation

As a result of the interactions of DDLs of adjacent clay micelles, soils exhibit flocculation or dispersion, depending on the relative strengths of attraction and repulsion. These effects have very important implications for permeability and rates of hydraulic conductivity in saline and irrigated soils, and hence influence solute transport routes and rates.

In the description of the DDL model given in Section 5.2.4, the inner clay lattice layer is envisaged as carrying only negative charge. This is an oversimplification since, in reality, clay surfaces with Al and Fe hydroxide coatings and the broken edges of clay micelles carry positive charges. At broken clay edges, this can result from the unsatisfied positive charges of Al^{3+} and Si^{4+} in octahedral and tetrahedral clay positions respectively. A clay can only form a stable deflocculated paste when the positive charges on broken platelet edges are suppressed. In field soils, there are two possible ways that this can happen. In the DDL, the lattice negative charge is neutralised by a layer of counterions immediately adjacent to the colloid surface. The cations in this layer are simultaneously attracted by electrostatic forces to the negatively charged clay lattice surface while also displaying a tendency to diffuse from the zone of high concentration at the lattice surface to zones of lower concentration in the bulk solution. An increase in concentration of ions in the bathing solution causes the thickness of this counterion layer to decrease. By increasing the concentration of solution Na^+ bathing an Na clay, by marine inundation for example, a condition is set up in which the diffuse layer of counter Na^+ ions at the lattice surface becomes very contracted, so that, according to Russell (1973), it no longer blankets positive charges at the platelet edges. When two platelets approach each other, a force of attraction is set up between the positive charges at the broken edges of one particle and the negative charge on the face of the other. The result is the very open structured edge-to-face packing that is characteristic of a thixotropic suspension. To deflocculate this suspension, the concentration of Na^+ in the bathing solution must be reduced. This results in thicker Na^+ ion diffuse layers on platelet surfaces. When two particles approach each other, the negatively charged diffuse Na layers repel each other.

In the field, the second cause of deflocculation is a rise in soil pH. Neutral to alkaline conditions cause charges at broken lattice edges to change from positive to negative, so that attractive forces between particles are reduced. In the laboratory, a polyphosphate such as Calgon (Na hexametaphosphate) causes flocculation when it is chemisorbed into broken clay edges where it reacts with exposed Al to produce a negative charge (van Olphen, 1977). Edge-to-face attraction cannot operate and higher electrolyte concentrations are required for flocculation.

Since divalent ions are attracted to lattice surfaces twice as strongly as monovalent ions, Ca-saturated clays exhibit a more compressed DDL then do Na-saturated clays. For this reason, the broken lattice edges of a Ca-clay retain their positive charges and can only be deflocculated at solution concentrations which are more than 100 times smaller than for sodium (Russell, 1973). Simply, the repulsive forces between adjacent particles decrease with increases in solution ionic concentration and valence of adsorbed ions. These tendencies are illustrated diagrammatically in Fig. 5.6. Summarising some of his early work, Jenny (1980) suggests that the flocculating efficiency of cations in solution follows the order:

$$Na^+ < Ca^{2+} < La^{2+}, Al^{3+}, Fe^{3+} < Th^{4+}$$

For the two most important flocculating cations in irrigated agricultural soils, Na^+ and Ca^{2+}, more recent studies have characterised their flocculating values for different clay types. The flocculating value, or the critical coagulation concentration (c.c.c.), is the minimum electrolyte concentration which causes flocculation. Van Olphen (1977) suggests that the flocculating values for monovalent ions lie between 25–150 mmol l^{-1}, for divalent ions, 0.5–0.2 mmol l^{-1} and for trivalent ions, 0.01–0.1 mmol l^{-1}.

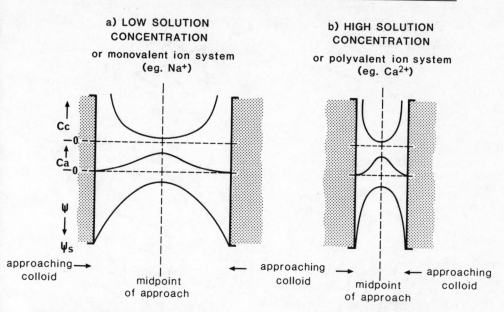

a) LOW SOLUTION
CONCENTRATION

or monovalent ion system
(eg. Na⁺)

b) HIGH SOLUTION
CONCENTRATION

or polyvalent ion system
(eg. Ca²⁺)

Figure 5.6 Diagrammatic comparison of electrical potential Ψ and ion concentration between two interacting negatively charged clay micelles at **(a)** low, and **(b)** high bathing solution concentration. Cc = concentration of cations; Ca = concentration of anions; Ψ_s = electrical potential at the colloid surface. (Source: modified from Bohn, McNeal and O'Connor, 1985)

As the bathing solution concentration is reduced by dilution, a dispersed clay suspension will become flocculated to produce a gel structure, characterised by edge-to-face attractive forces. With further dilution and flocculation, alignment of clay platelets occurs to form domains. When the bathing solution concentration is reduced further, water is taken up between clay plates and the domains swell (Fig. 5.7). This influence of electrolyte concentration on clay structure and subsequently on soil hydraulic properties has been widely studied because of its importance in saline and irrigated soils. In these soils it became apparent that the hydraulic conductivity of soil irrigated with water high in exchangeable sodium and high in soluble salts generally, was very much higher than when the same soil was irrigated with water of low salt concentration. Investigating the effect of electrolyte concentration on the hydraulic conductivity of soils saturated with different exchangeable cations, Quirk and Schofield (1955) defined the *threshold concentration* of the bathing solution which caused a 10–15 per cent decrease in soil permeability. they found that the threshold concentration was highest for Na⁺-saturated soils bathed in NaCl solutions of different concentrations and lowest for Ca²⁺-saturated soil, bathed in CaCl₂ solutions.

Irrigated soils containing > 15 per cent Na on the exchange complex are classified as alkali soils. This is termed the exchangeable sodium percentage (ESP). If these soils are irrigated with waters of low salt

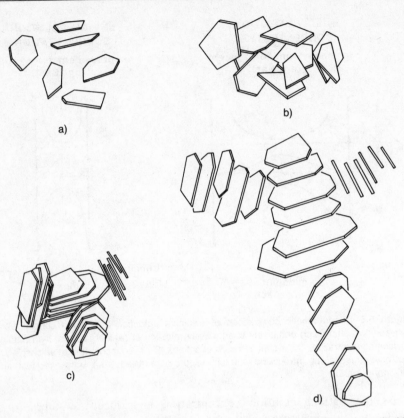

Figure 5.7 Characteristic clay structures produced during the dilution of the bathing solution concentration **(a)** dispersed; **(b)** 'gel' structure showing edge-to-face attractions; **(c)** well-aligned clay domains; **(d)** domain swelling as water is taken up between lattice layer (Reproduced with permission from Russell, 1973, *Soil Conditions and Plant Growth*, 10th edition, © Longman Group UK Ltd.)

concentration, the clays deflocculate. For a given soil, Quirk and Schofield (1955) showed that for solution concentrations less than the threshold concentration, hydraulic conductivity decreased with increasing ESP. A number of soil properties have subsequently been considered important in affecting the response of hydraulic conductivity to electrolyte concentrations. Of these, clay content and clay mineralogy seem to be the most important. McNeal *et al.* (1968) found that, in the presence of mixed salt solutions, the hydraulic conductivities of soils of uniform clay mineralogy, decreased markedly with increasing clay content. Frenkel, Goertzen and Rhoades (1978) further showed that montmorillonite and vermiculite clays dispersed more than kaolinitic clays when leached with solutions of low concentration.

Two main mechanisms have been suggested for the decrease in hydraulic conductivity observed with an increase in ESP and a decrease in solution concentration. Quirk and Schofield (1955) originally proposed that clay swelling partially blocked the conducting pores. Frenkel, Goertzen and Rhoades (1978) suggested that dispersed clay particles plugged soil pores. When these dispersed clay particles were washed out of the soil, Pupisky and Shainberg (1979) showed that the hydraulic conductivity was again increased.

Several models have been developed to predict clay swelling effects on hydraulic conductivity. Using the empirical relationship:

$$1 - k_s = \frac{fx^n}{(1 + fx^n)} \tag{5,17}$$

where: k_s = saturated hydraulic conductivity, relative to that of a divalent cation-saturated soil at high salt concentration

       f = empirical constant, characteristic of a given soil

       x = colloid swelling factor

       n = factor which varies with estimated soil ESP.
          (n = 1 for ESP < 25, 2 for ESP = 25–50, and 3 for ESP > 50)

McNeal and Coleman (1966) generated a series of curves (Fig. 5.8) to predict saturated hydraulic conductivity (k_s) changes with soil ESP and soil solution concentration.

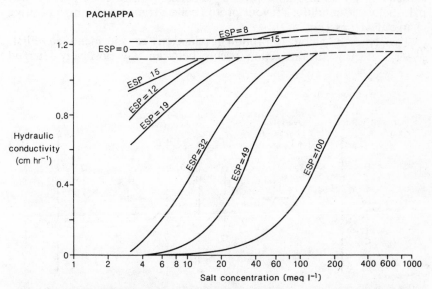

Figure 5.8 Hydraulic conductivity of Pachappa soil (western United States) in relation to salt concentration and exchangeable sodium percentage (ESP). (Source: from the data of McNeal and Coleman, 1966, redrawn by Bresler *et al.* (1982), *Water Resources Research*, 15, © American Geophysical Union)

5.3   ANION ADSORPTION

In soils dominated by organic and lattice-layer clay minerals, the high colloid surface negative charge means that cation attraction and exchange greatly exceeds anion attraction and exchange. Indeed, there is a greater tendency for anions to be repelled. Only in soils generating significant positive charge does anion adsorption assume dominant importance. Highly weathered tropical soils contain significant amounts of Fe and Al hydroxyoxides. Such soils are often termed 'variable charge' soils since their surface charge properties are strongly controlled by changes in soil pH. When acidic, they generate large amounts of positive charge and anion adsorption becomes important. In recent years, there have been two main reasons for increased interest in anion adsorption. First, clearance of tropical rainforest and the agricultural development of tropical soils has focused attention on the chemistry of variable charge soils. Second, more research has concentrated on the problems of soil contamination by heavy metals and other agrochemicals such as pesticides, some of which dissociate in the soil to form anionic species (see Section 9.4).

Anions are attracted to positively charged colloid surfaces either at broken clay lattice edges, where unsatisfied Al^{3+} groups are exposed, or on the surfaces of Fe, Mn and Al oxide and hydroxide films. In the aqueous soil solution, metal oxides bond with water molecules which subsequently lose or gain a proton, depending on solution pH. At low pH, the acidic conditions favour proton concentration and a high positive charge develops (Fig. 5.9).

Two types of anion adsorption mechanisms can be identified. First, general electrostatic attraction occurs between any positively charged

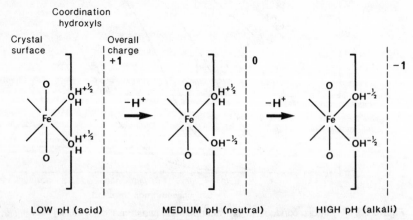

LOW pH (acid)   MEDIUM pH (neutral)   HIGH pH (alkali)

Figure 5.9   Effect of solution pH on the dissociation of unsatisfied hydroxyls at the surface of Fe or Al hydroxyoxides. (Reproduced with permission from Russell, 1973, *Soil Conditions and Plant Growth*, 10th edition, © Longman Group UK Ltd.)

surface and any negatively charged anion. The second type of adsorption is anion specific. The hydroxyoxides of Fe and Al have a particular affinity for phosphate, molybdate, silicate and arsenic in solution.

5.3.1 General electrostatic anion attraction and repulsion

Since the majority of organic and inorganic colloid surfaces exhibit negative charge, anion repulsion, or negative adsorption, is an important consideration, particularly in temperate soils. All factors controlling the configuration of the DDL (see Section 5.2.4) also control anion repulsion. Five main factors affect anion repulsion:

(1) it increases with increasing anion concentration;
(2) it increases with increasing anion valency – Mattson (1929) found the following order of increasing repulsion:

$$Cl^- = NO_3^- < SO_4^{2-} < Fe(CN)_6^{4-}$$

(3) it decreases with soil pH since this decreases the net negative charge on soil colloids;
(4) it decreases when the DDL is saturated with divalent or polyvalent cations – Mattson (1929) found that Cl^- repulsion decreased according to the sequence: $Na > K > Ca > Ba$ saturating the DDL of a montmorillonitic clay;
(5) it increases with increased density of negative charge on the colloid surface – 2:1 type clays such as montmorillonite thus repel anions more strongly than 1:1 type clays such as kaolinite.

$Cl^- NO_3^-$ and SO_4^{2-} are the main anions of importance in non-specific attraction. Mattson (1929), in Fig. 5.10, clearly showed how non-specifically adsorbed anions are repelled in montmorillonitic clay, even at low pH, since all negative charge is permanent. In kaolinitic clay with a significant amount of pH-dependent negative charge, Cl^- and SO_4^{2-} non-specific adsorption occur at pH < 7.

5.3.2 Specific anion adsorption

Fe and Al hydroxyoxides are complex ion compounds in which the central cation is closely surrounded by six oxygens or hydroxyls. These surrounding species are called coordinating groups or ligands and are shared by two adjacent cations in the crystal. The unsatisfied charges at crystal surfaces (see Fig. 5.9) occur because here ligands are coordinated with only one Al^{3+} or Fe^{3+} cation. During specific anion adsorption, the oxygen of the phosphate or molybdate anion enters into six-fold coordination with Fe^{3+} or Al^{3+}. For this reason, these specific anion adsorption reactions are often termed ligand exchange.

The chemistry of phosphate adsorption on iron oxides has been studied

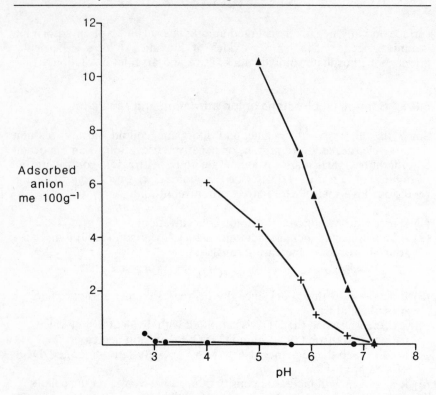

Figure 5.10 Electrostatic adsorption of Cl^- and SO_4^{2-} on kaolinitic and montmorillonitic clay soils x—x:Cl^- adsorption on kaolinitic clay soil; ▲—▲: SO_4^{2-} adsorption on kaolinitic clay soil; ●—●: Cl^- adsorption on montmorillonitic clay soil. (Source: Mattson, 1929)

extensively. The ligand exchange mechanism of phosphate on geothite has been elucidated by Parfitt and Atkinson (1976) using infrared spectroscopy. Their work has shown that a binuclear bridging complex is formed on the crystal surface (Fig. 5.11). This produces a particularly strong adsorption complex and explains why Hingston, Posner and Quirk (1974) failed to desorb phosphate from goethite by washing at low pH. Hingston, Posner and Quirk (1972) have also shown how borate, molybdate, and silicate are adsorbed on goethite by ligand exchange, as are humic and fulvic acids (Parfitt, Fraser and Farmer, 1977).

The mechanisms of specific anion sorption on aluminium hydroxides is suggested by Parfitt (1978) to be less well understood than on iron oxides because their surfaces are less well defined. Parfitt *et al.* (1977) found that edge $Al(OH)H_2O$ groups in the gibbsite crystal were responsible for phosphate adsorption. Their results also suggested that less crystalline forms of gibbsite adsorbed larger amounts of phosphate.

Parfitt (1978) cites a large number of studies whose results suggest that the adsorption of phosphate, borate, molybdate and arsenate on amorphous iron and aluminium hydroxides, forms of Fe and Al

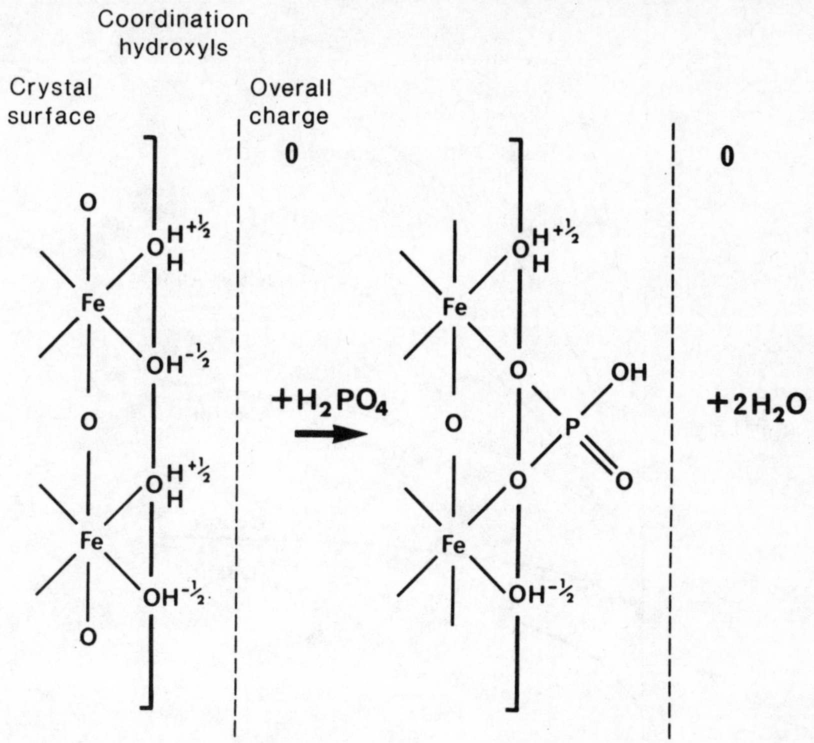

Coordination
hydroxyls

Crystal
surface

Overall
charge

Figure 5.11 Binuclear Fe.OP(O)$_2$O.Fe bridging complex, characteristic of phosphate absorption on goethite

particularly common in weathered soils, occurs by ligand exchange, probably by binuclear bridging mechanisms.

5.3.3 Adsorption isotherms

An ion adsorption isotherm is the temperature-dependent relationship between the amount of ion sorbed at colloid surfaces and the concentration of that ion at equilibrium in the bathing solution. An adsorption isotherm is constructed by shaking the soil sample with ionic solutions of different concentration and calculating the amount of phosphate ion sorbed in each case. Plots of this relationship are known as quantity/intensity (Q/I) plots, in which the quantity refers to the amount of ion sorbed per unit mass of soil and intensity refers to the equilibrium concentration in solution. Fig. 5.12 shows the isotherms of phosphate adsorption for range of soils, in the form of Q/I plots. In true isothermal relationships, the amount of ion sorbed depends only on solution concentration and temperature. By holding temperature constant, the

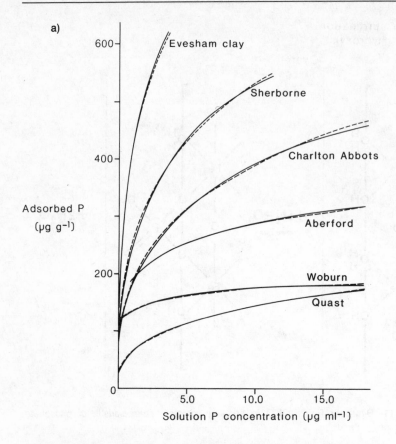

Figure 5.12(a) P absorption isotherms (Q/I) plots for four English soils, comparing observed results with predicted absorption using two forms of the Langmuir equation: a uniform surface model (---) and a two-surface model (——). (Source: Holford, Wedderburn and Mattingley, 1974). **(b)** P absorption isotherms for some English soils comparing the two-surface Langmuir model (——) with the Freundlich equation (---). (Source: calculated by Barrow from the data of Holford *et al*. (1975) and Holford and Mattingly (1975); from Barrow, 1978)

relationship between adsorption and solution concentration can be fully predicted.

By far the most widely studied adsorption process in soil is that of phosphate adsorption. Barrow and his co-workers have shown over a large number of experiments that three main factors affect the form of phosphate adsorption isotherms. For this reason, he suggests that the term 'isotherm' may be inappropriate for phosphate adsorption (Barrow, 1978). The influencing factors are

(1) a series of parameters relating to the shaking technique: the soil:solution ratio, the nature of the supporting solution electrolyte,

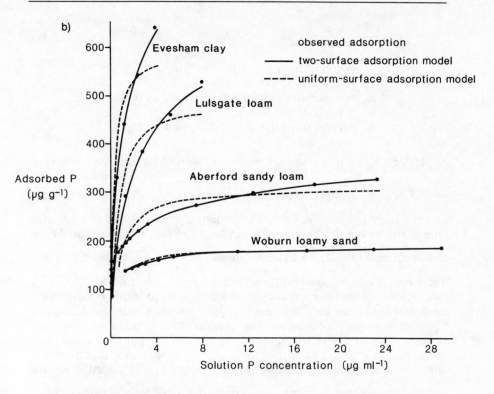

the length of shaking and the moisture content of the soil before shaking (Barrow and Shaw, 1979);

(2) any previous additions of phosphate or other specifically adsorbed anions (Barrow, 1974); and

(3) solution pH (Barrow, 1984).

Two types of equations, the Langmuir and the Freundlich, have been used to describe adsorption isotherms. In the original derivation of the Langmuir equation for the adsorption of gases on the surfaces of materials, Bohn, McNeal and O'Connor (1985) note that there were three main assumptions: (1) constant adsorption energy, irrespective of surface extent, (2) adsorption on specific sites, with no interaction between absorbate molecules, and (3) maximum adsorption was equal to a complete monomolecular layer on all reactive absorbant surfaces. In Q/I plots, assumption (3) is represented by the maximum adsorption plateau.

The original form of the Langmuir equation is:

$$\frac{x}{w} = \frac{Kcx_m}{1 + Kc} \qquad (5,18)$$

which is usually used in the linear form:

$$\frac{c}{x/w} = \frac{1}{Kx_m} + \frac{c}{x_m} \qquad (5,19)$$

where: c = equilibrium ion concentration
$\quad\quad x$ = amount of ion adsorbed
$\quad\quad x_m$ = maximum amount of ion that can be adsorbed
$\quad\quad w$ = mass of soil
$\quad\quad K$ = constant, related to adsorption energy

A plot of $\frac{c}{x/w}$ against c should be linear, with slope $1/x_m$ and intercept $\frac{1}{Kx_m}$. Since $\frac{c}{x/w}$ against c plots are frequently curvilinear over fairly large concentration ranges, the Langmuir equation is now considered to have limited use in the description of phosphate sorption relationships. Olsen and Watanabe (1957), for example, found that the $\frac{c}{x/w}$ versus c plot was linear only over the concentration range: 5×10^{-5} M to 5×10^{-4} M. A main reason for its failure relates to its inadequacy to 'explain' adsorption mechanisms. Olsen and Watanabe (1957) assumed that the Langmuir equation was only obeyed at low concentrations, other authors (for example, Ryden, McLaughlin and Syers, 1977) split the adsorption isotherm into separate linear portions, suggesting that each represented a different adsorption mechanism, while Gunary (1970) added a square root to the equation to correct the curved nature of his $\frac{c}{x/w}$ versus c plot.

Recently, the basic Langmuir and other adsorption equations have been modified in an attempt to account for (1) different adsorption sites on charged particle surfaces (see, for example, Sibbesen, 1981), and (2) the effect of desorbed ions in the equilibrating solution (Harter and Barek, 1977).

An early alternative to the Langmuir equation was the relationship proposed by Freundlich:

$$\frac{x}{m} = KC^{1/n} \qquad (5,20)$$

in which K and n are empirical constants.

The main implication of this relationship is that adsorption energy decreases exponentially as the extent of covered or reacted surface increases during adsorption (Bohn, McNeal and O'Connor, 1985). These authors also suggest that an advantage of the Freundlich equation is to liken the decrease in adsorption energy with increasing surface coverage to the effect of surface heterogeneity.

A suite of authors have subsequently examined the applicability of these two basic equations and of modified forms of these equations. A two-surface Langmuir model has been proposed by Holford (1982), while

Sibbesen (1981) has proposed extended versions of both the Langmuir and Freundlich equations; both authors seeking equations to realistically explain changing rates of adsorption with time. An additional problem in modelling adsorption is the effect of precipitation mechanisms at the adsorbing surface. Although Holford, Wedderburn and Mattingley (1974) examine the concurrent precipitation of calcium phosphate and P adsorption in soil, they do not specifically apply their two surface Langmuir model to this problem. It is clear that most isotherm data can be fitted mathematically to either two or more simple Langmuir equations (Posner and Bowden, 1980) or to a two-surface Langmuir model (Sposito, 1982). Fig. 5.11a illustrates this fitting procedure for a uniform surface and a two-surface Langmuir model (Holford, Wedderburn and Mattingley, 1974). Using the same data, Barrow (1978) found that the fit of the Freundlich equation gave virtually the same result as the two-surface Langmuir (Fig. 5.11b). He concluded that a multiple surface model for adsorption may be unnecessary. Ratkowsky (1986) statistically tested the fit of seven different adsorption equations to six previously published isotherm data sets and found that the basic Freundlich or the extended form proposed by Sibbesen (1981) gave the best fit and best described surface adsorption.

Barrow (1978) identified two main reasons for studying phosphate adsorption curves. First, they may aid in the identification of soil properties associated with adsorption and hence aid in the prediction of efficient fertilizer use. Second, they may be to help in understanding adsorption processes. Most authors currently advocate the cautionary use of single equations designed to fit empirical adsorption isotherms. Most suggest that they should be used simply to describe the effect of a particular treatment and in the comparison of treatment effects rather than to explain adsorption processes (see, for example, Goldberg and Sposito, 1984; Ratkowsky, 1986).

A more encouraging approach to understanding ion adsorption mechanisms has been the development of a much more complex mechanistic model to describe phosphate sorption and desorption (Barrow, 1983). Equations in the model account for characteristics of the charged surface, the rate of adsorption, diffusion within and from the adsorbed layer, feedback effects on the electrostatic potential generated at the charged surface and thermal effects. Barrow has subsequently shown the model to have much wider application by successfully testing it for the sorption of molybdate and fluoride (Barrow, 1986a) and also for the cationic sorption of zinc (Barrow, 1986b).

5.4 SOLUTE TRANSPORT PROCESSES

Two mechanisms for solute movement are used in describing and modelling solute transport through soil pores from one zone of soil to another: mass flow (or convective flow), and diffusion. The movement of water through the porespace during processes such as drainage and leaching carries dissolved solutes with it. This mechanism is called mass flow or convention. The dispersion or migration through soil water of a dissolved substance, such as common salt, from areas of high concentration to areas of low concentration is called diffusion. This process is analagous to the diffusion of a gas, for example CO_2, through the air of the soil porespace.

5.4.1 Solute movement by mass flow

Soil water movement occurs when a difference in water potential exists in different parts of the soil. The soil water potential (ϕ_t) consists of three major components: the gravitational potential (ϕ_g), the matric potential (ϕ_m) and the osmotic potential (ϕ_o), such that:

$$\phi_t = \phi_g + \phi_m + \phi_o$$

The rate of water movement under steady state conditions is given by Darcy's law:

$$q = \frac{-kd\ \phi_t}{dz} \qquad (5,21)$$

where: q = water flux (cm s^{-1}), defined as the quantity of water passing through a unit area in unit time

k = hydraulic conductivity (cm s^{-1})

ϕ_t = total soil water potential ($\phi_g + \phi_m + \phi_o$). Under saturated soil conditions, ϕ_t = the hydraulic head (H).

z = depth (cm)

Although steady flow conditions may be maintained artificially in laboratory experiments, water movement in field soils mainly occurs under unsaturated conditions. Transient flow can be expressed mathematically by combining the steady state equation of Darcy with the equation of continuity for water flow:

$$\frac{\partial\theta}{\partial t} = -\frac{\partial q}{\partial z} \qquad (5,22)$$

where: θ = volumetric water content (cm^3 cm^{-3})

t = time (s)

Combining Equations (5,21) and (5,22) yields the Richards equation, in which the full derivatives of Equation (5,21) become partial derivatives since water content is now a function of both depth and time:

$$\frac{\partial \theta}{\partial t} = \frac{\partial}{\partial z}\left[k\theta \ \frac{\partial \phi_t}{\partial z} \right] \tag{5,23)}$$

where: $k\theta$ = hydraulic conductivity dependent on volumetric
water content

Water and solutes move at different rates through soil. During their movement through soil, they may be subject to various reactions and interactions such as adsorption/exchange, chemical transformation, precipitation, uptake by plant roots, or changes of state such as volatilisation. When solutes are considered as 'ideal', where there are no gains or losses and no solute/surface or solute/solute interractions, the convective flow of solutes associated with water movement has been written by Raats (1984) as:

$$\frac{\partial \theta C}{\partial t} = -\frac{\partial F}{\partial z} \tag{5,24}$$

where: C = concentration of solute in solution (g cm^{-3})
  F = mass flux of the solute (g cm^{-3} s^{-1})
  θ = volumetric water content (cm^3 cm^{-3})
  z = depth (cm)

Since solute movement is also governed by the bulk solution concentration, Equation (5,24) alone is generally too simplistic to realistically describe even the simplest of ionic movements in the soil solution.

5.4.2 Solute movement by diffusion

All atoms, molecules and ions in the soil solution possess Brownian motion which is random in all directions. Any difference in concentration of any ion in the soil solution will quickly be lost as random motion results in more ions moving in the direction of low concentration than are moving out, until the ionic concentration of the solution is uniform. The speed of this equalisation process depends on the concentration difference between the two volumes of solution and is described for the simple one-dimensional case by Fick's first law in which the amount of ion crossing a unit sectional area in unit time is given by:

$$F = -D \ \frac{dC}{dx} \tag{5,25}$$

where: F = solut flux in g cm^{-2} s^{-1} across an area 2 cm^2
      perpendicular to the direction of flow (x) in a time of
      1 s
  C = solute concentration in g cm^{-3}

$\dfrac{dC}{dx}$ = concentration gradient, which is negative, indicating that diffusion is from areas of high concentration to areas of low concentration

D = diffusion coefficient $(cm^2 \ s^{-1})$

For unsaturated soil conditions this could be written as:

$$F = -D(\theta) \frac{\partial C}{\partial x} \tag{5,26}$$

where: θ = volumetric soil water content $(cm^3 \ cm^{-3})$

As written, Equation (5,26) indicates unidirectional diffusion in the x direction. Rewritten as:

$$F_x = -D \frac{\partial C_x}{\partial x} \tag{5,27}$$

implies that while diffusion is considered only in the x direction, diffusion may also be occurring in the y and z directions, giving:

$$F_y = -D \frac{\partial C_y}{\partial y} \ \text{and} \ F_z = -D \frac{\partial C_z}{\partial z} \tag{5,28}$$

The rate of diffusion down the concentration gradient between adjacent loci in the soil decreases through time as adjacent solute concentrations equilibriate. This implies that it is useful to consider rates of change of diffusion processes. In Fig. 5.13, the rate of solute flux across a unit section of area y by z is given by:

$$F-D(yz) \frac{\partial C}{\partial x} \tag{5,29}$$

Crank *et al.* (1981) then consider the solute flux across a second section, of area y'z', a distance of δx downstream and parallel to section yz. A difference in inflow of diffusing solute to the first section and outflow from the second will be manifest in a change in solute concentration in the volume between the two sections. Looking firstly at the change in rate of solute flux over distance δx, if Fx represents the flux through yz and Fx + δx represents the flux through y'z', then the mean change in rate of flux over δx is given by:

$$\Delta F = \frac{(Fx + \delta x) - Fx}{\delta x} \ \text{which} \ \approx \frac{\partial F}{\partial x} \ \text{when} \ \delta x \ \text{is very small} \tag{5,30}$$

The rate of solute increase within the volume is:

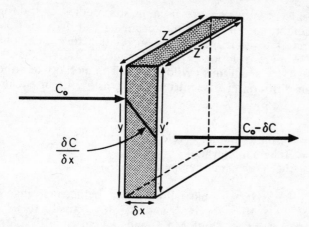

Figure 5.13 Diffusion across two parallel planes, yz and y'z' across distance δx from concentration C_O to concentration $C_O-\delta C$ (Reproduced with permission from Wild, 1981, In Greenland, D.J. and Hayes, M.H.B. (eds.), *The Chemistry of Soil Processes*, © John Wiley and Sons Ltd.)

$$Fx - (Fx + \delta x) \approx - \frac{\partial F}{\partial x} \delta x \qquad (5,31)$$

which = volume × rate of change of solute concentration (Crank *et al.*, 1981). So, second, looking at rate of change in concentration in the volume described by yzδx, Cδx represents the amount of solute in the volume and (∂C/∂tδx) represents the rate of change in concentration with time, so:

$$\frac{\partial C}{\partial t} \delta x = - \frac{\partial F}{\partial x} \delta x \text{ and for very small } \delta x, \ \frac{\partial C}{\partial t} = - \frac{\partial F}{\partial x} \quad (5,32)$$

Substituting for F using Fick's first law (Equation **5,25**) yields:

$$\frac{\partial C}{\partial t} = \frac{\partial}{\partial x}\left(D \frac{\partial C}{\partial x}\right) \qquad (5,33)$$

which, when expressed as:

$$\frac{\partial C}{\partial t} = D\left(\frac{\partial^2 C}{\partial x^2}\right) \qquad (5,34)$$

is called Fick's second law of diffusion. Here, the rate of change of concentration over time is described by ∂C/∂t and the rate of change of the concentration gradient with distance is given by $\partial^2 C/\partial x^2$ (Wild, 1981). Fick's second law also applies to heat conduction in soil, where C represents temperature and D represents thermal diffusivity.

Two coefficients of diffusion are used in soil studies. Self-diffusion refers to the equilibration of an ion interchanging with its own isotope. Bulk diffusion refers to ionic interchange in relation to a bulk solution concentration. In pure solutions, the difference between these two coefficients is small for small ions (Wild, 1981). The mobility of an ion is related to its self diffusion coefficient by the Nernst–Einstein equation:

$$D = ukT$$

where: u = absolute mobility (velocity in cm s^{-1})
      k = Boltzman constant
      T = thermodynamic temperature (°K)

In aqueous solution at 25°C, ionic self-diffusion coefficients range from 1.3 and 2.0 cm^2 s^{-1} $\times 10^{-5}$ for Na^+ and K^+ respectively to 0.7 and 0.8 cm^2 s^{-1} $\times 10^{-5}$ for Ca^{2+} and Mg^{2+} and 9.4 cm^2 s^{-1} 10^{-5} for H^+ (Robinson and Stokes, 1959).

As well as bulk solution concentration, diffusion coefficients and ion mobilities in soil are controlled by (1) the state of the substrate – solid, liquid or gas, (2) soil moisture content, and (3) tortuosity of the porespace. Crank et al. (1981) note typical diffusion coefficients ranging from 10^{-10} to 10^{-11} for solids to 10^{-5} to 10^{-8} for liquids and 10^{-2} to 10^{0} for gases. Data collated by Nye and Tinker (1977) indicate that the mobility of Na in montmorillonite clay (D = 4×10^{-6} cm^2 s^{-1}) is only one-third of its mobility in aqueous solution. The mobility of O_2 in the gaseous phase is about 10^4 times faster than in water. The retardation of ionic diffusion with reduced volume content is well illustrated by Rowell, Martin and Nye (1967) in Table 5.4. A two-fold to four-fold increase in diffusion of K^+, Ca^{2+} and Mg^{2+} was measured by Schaff and Skogley (1982) when soil moisture content was raised from 10 to 28 per cent. These results are probably due to increased tortuosity of the solute diffusion path as water films at particle surfaces become thinner during drying. Nye and Tinker (1977) introduce an impedance factor to modify the diffusion coefficient to take account of increased diffusion path due to increased tortuosity:

$$D_s = - D\theta f_1 \frac{\partial C}{\partial x} \qquad\qquad (5,36)$$

where: D_s = diffusion coefficient of solute through soil
      D = diffusion coefficient of solute in free solution
      θ   = fraction of the soil volume occupied by solution
           (cross-section of porespace available for diffusion)
      f_1 = impedance factor to account for tortuosity of diffusion
           pathway (increased length, decreased concentration
           gradient)

In Equation (5,36), no account is taken of ionic adsorption or exchange.

Table 5.4 Self-diffusion coefficients of Na^+, Cl^- and PO_4 in relation to soil moisture content in a sandy clay loam (from Rowell, Martin and Nye, 1967)

Ion	Moisture content (by volume) (%)	D_{soil} ($cm^2\ s^{-1}$)
Na^+	40	2.2×10^{-6}
Na^+	20	0.2×10^{-6}
Cl^-	40	9.0×10^{-6}
Cl^-	20	2.4×10^{-6}
PO_4	40	3.3×10^{-9}
PO_4	20	0.3×10^{-9}

This relationship was used by Hill (1984) to show that the diffusion of NO_3^-, Cl^- and SO_4^{2-} was related to the porosity of the chalk and to the extent of fissuring and cracking.

Nye (1979) nicely illustrates the significance of diffusion coefficient values for solute movement in the rhizosphere by comparing the distances moved by a single H^+ ion in water ($D \approx 10^{-6}\ cm^2\ s^{-1}$) and a single K^+ ion within an illitic clay ($D \approx 10^{-25}\ cm^2\ s^{-1}$). H^+ would move 1 mm in 50 s while K^+ would move only 1 nm in 16 years. These data illustrate that rates of ionic diffusion in soil are much slower than rates of convective or mass flow.

5.5 MODELLING SOLUTE TRANSPORT IN SOIL

Increased research activity on field scale solute transport studies and the development of solute transport models has been in response to the need for further information for two main thrusts of enquiry: the management of soil agrochemicals, particularly the leaching of surface applied fertilisers and pesticides, and to aid in the understanding of pollutant transport to and in groundwaters. It is apparent from the previous sections that the milieux of interacting solute processes in soil make for a particularly complex system which is virtually impossible to model realistically at the field scale. On the other hand, *in situ* field studies are expensive and results can be site specific with minimum general applicability. On the whole, the trend is currently towards model development with an as yet unsatisfied need for field verification studies.

The plethora of water/solute transport models which have been developed can be divided into *deterministic* models which reproduce the physical processes of mass flow and diffusion in response to a concentration gradient, and *stochastic* models which generate probabilities to cope with the heterogeneity and unknown variability of soil properties and processes (Addiscott and Wagenet, 1985).

Many modelling approaches use simple analogues such as chromatographic or electrical resistance systems. Earliest solute models, based on the principles of chromatographic separations in which solutes moved at different rates through an adsorbing medium, have now been developed into the coherent theory of miscible displacement. These deterministic approaches have also been termed convective/dispersive (CD) models since they are based on equations for convection (mass flow) and dispersion/diffusion. The study of solute elution, or breakthrough, curves (BTC) has facilitated the refinement of miscible displacement models.

CD models have achieved some success in predicting solute transport through soils with uniform transport properties, but suffer the same problems as water flow models in predicting flux where horizontal and vertical soil variability and preferential flow occur in field soils (Jury, 1982). The recognition that the incorporation of field scale soil variability into solute transport models was vital for more accurate predictions, has led to two developments: (1) the addition of scaling theory to deterministic mechanistic models, and (2) the introduction of stochastic models which treat the soil as a black box system whose internal mechanisms are unknown.

5.5.1 Convective-dispersive solute transport models

Miscible displacement models are based on solute input–output studies, usually in laboratory monoliths or columns. Solute ions added to the top of the column are flushed down the column and eluted from the base. A plot of elutant concentration (C_E) to the ratio of elutant concentration to input concentration (C_E/C_O) against the number of pore volumes passed through the column is called a breakthrough curve (BTC). Two general forms of BTC are (1) those caused by a pulse of solute (in Fig. 5.14a this is a pulse of NaCl), and (2) those caused by frontal displacement (in Fig. 5.14b this is the continuous addition of NaCl). In both BTC forms there are two components of convective flow: mean porewater velocity and local variations in velocity which cause solute spreading and mixing, often called hydrodynamic dispersion. It is from these two components that convective-dispersive transport models are named. Two other effects change the shape of the elution curve. These are anion exclusion and ionic adsorption.

Hydrodynamic dispersion

As a solution flows through a simple cylindrical pore, the frictional resistance of the pore walls cause slower flow than in the centre of the pore. Slow flow in very small soil pores and delayed flow along longer, more tortuous pathways add to solute dispersion (Fig. 5.15). The effect of dispersion on the BTC, to spread out the elution curve, is illustrated in Fig. 5.14. Three factors operate to cause mixing and a spreading out of

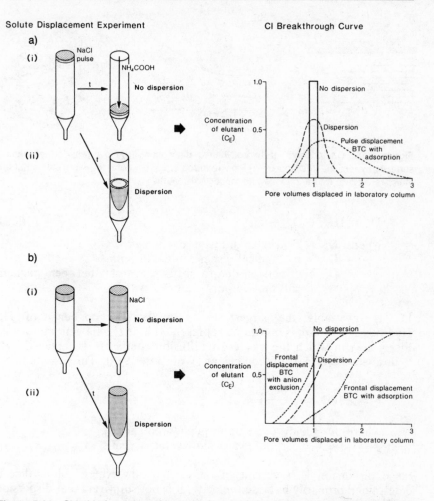

Figure 5.14 Chloride breakthrough curves characteristic of different solute displacement experiments. **(a)** pulse of NaCl with and without dispersion; **(b)** frontal displacement of NaCl with and without dispersion. Frontal displacement with anion exclusion and adsorption are also illustrated

the elution curve during hydrodynamic dispersion. An increase in the size of individual pores and an increase in porewater velocity cause increased dispersion (Fig. 5.16a). A similar BTC shape is caused by an increase in solute concentration in the displacing solution (Fig. 5.16b). Since solutes move from areas of high concentration to areas of low concentration during dispersion, the process is thus similar to diffusion in mode of operation. According to Rao *et al.* (1980), hydrodynamic dispersion (D_h) can be defined as:

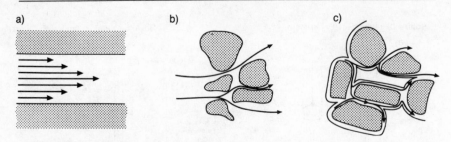

Figure 5.15 Components of hydrodynamic dispersion. **(a)** Flow retarded at particle surfaces by frictional resistance. **(b)** Flow retarded in narrow pore. **(c)** Flow retarded at low moisture content (more tortuous and longer flow path)

$$D_h = [D_e + D_m + D_s] \tag{5,37}$$

where: D_e = molecular diffusion coefficient
D_m = mechanical dispersion coefficient
D_s = dispersion due to diffusive transfer between stagnant and mobile pore water

D_m is invariably the largest of these three components of D_h. Hydrodynamic dispersion can be represented in the same general form as Fick's law, in which the molecular diffusion coefficient is replaced by a hydrodynamic dispersion coefficient (Wagenet, 1983). This results in:

$$F = - \theta D_h \, v \left(\frac{\partial C}{\partial x} \right) + v \, \theta C \tag{5,38}$$

where D = hydrodynamic dispersion coefficient
v = mean porewater flow velocity (cm s^{-1})

Since dispersion is a variant of convective flow where solute flux is determined primarily by water velocity, it is not surprising that dispersion coefficients are much higher than diffusion coefficients. In a series of chloride leaching experiments, Kirda, Nielsen and Biggar (1973) found that molecular diffusion was obscured by much larger dispersion effects at high water velocities in coarse textured soils. Their empirical studies showed that D_m was two orders of magnitude larger than D_e at porewater velocities > 0.002 cm s^{-1}. As porewater velocity decreases, the dispersion coefficient decreases (Rose, 1973) and molecular diffusion assumes increasing importance in solute mixing and dispersion effects (Kirda, Nielsen and Biggar, 1973).

Anion exclusion and ion adsorption

The negative charge of most particle surfaces causes anions in solution to be repelled and excluded from the volume of water immediately adjacent to the particle surface. Anions thus have a smaller effective water volume

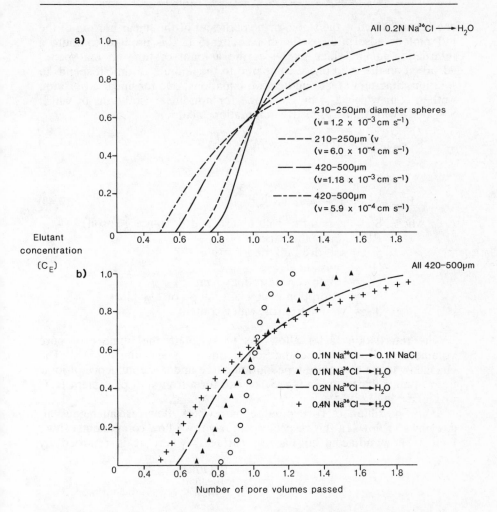

Figure 5.16 (a) Effect of two different sizes of particle diameters (and hence pore diameters) on the frontal displacement BTC for H_2O being displaced by ^{36}Na tagged NaCl. **(b)** Effect of three different solute concentrations on frontal displacement BTCs in 420–500 μm diameter particles at a pore water velocity of 1.18×10^{-3} cm s^{-1}. Treatment three (— —) 0.2N Na^{36} Cl → H_2O through 420–500 μm particles at a velocity of 1.18×10^{-3} cm sec^{-1}, is common to both graphs (a) and (b). (Source: Rose and Passioura, 1971)

available for transport than non-reactive ions (10–20 per cent less, according to Wild, 1981). However, since anion concentration is highest in the centre of pores where water is flowing fastest, net anion flux can thus exceed net water flux, resulting in forward displacement of the BTC in Fig. 5.14(b).

Ion adsorption at particle surfaces causes a retardation in solute flux as solute ions equilibrate between solution and solid phases. The adsorption coefficient, which is the slope, or b coefficient of the linear portion of the Q/I plot or adsorption isotherm (see Fig. 5.12), is used to compute a retardation factor which expresses the amount of time an ion spends adsorbed on the solid phase compared to the amount of time it spends in solution. Smettem (1986) quotes two equations, one for linear adsorption isotherms in which $S = bC$, and one for non-linear isotherms in which $S = bC^N$, with their respective retardation factors:

$$RF_{linear} = 1 + (\rho b/\theta) \qquad\qquad (5,39)$$

and

$$RF_{non-linear} = 1 + (\rho b/\theta)NC^{N-1} \qquad\qquad (5,40)$$

where: S = amount of solute adsorbed (g solute g^{-1} soil)
RF = retardation factor
ρ = soil dry bulk density (g cm^{-3})
N = constant
C = solute concentration (g cm^{-3})
b = adsorption coefficient (slope of Q/I plot)
θ = volumetric soil water content

The retardation factor allows us to calculate the number of pore volumes required to elute the solute out of a leaching column. The change in shape of a pulse-type elution curve under the influence of ionic adsorption is illustrated in Fig. 5.14a and for a frontal displacement BTC in Fig. 5.14b.

It is possible to conceptualise porewater flow as analogous to streamwater flow. In this respect it will have two flow components: slow, laminar flow adjacent to particle surfaces, where it is retarded by

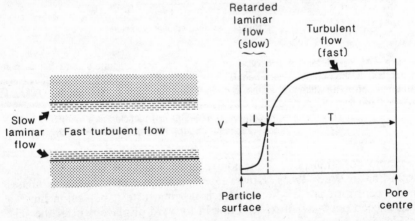

Figure 5.17 Diagrammatic representation of two-component porewater flow in a simple cylindrical pore

frictional resistance; and faster, turbulent flow in the centre of the pore (Fig. 5.17). In large pores, fast, turbulent flow dominates and the concentration of non-reactive solutes will be fairly uniform across the pore cross-section due to mixing. In very small pores, slower laminar flow adjacent to particle surfaces will assume a more important role. Solute transport will be made up of two convective or mass flow components: (1) slow convection adjacent to particle surfaces, with a concentration gradient set up in the zone of slow laminar flow, and (2) fast convection in the pore centre where solute mixing occurs due to turbulence (Fig. 5.18a). The concentration gradient so generated will set up diffusion towards the pore centre. In critically small pores where the diameter is less than twice the thickness of the zone of laminar flow, ionic diffusion, in relation to a concentration gradient, and mean porewater velocity will characterise transport for non-reactive solutes. Similar diagrams for anion exclusion and ion adsorption are given in Fig. 5.18b and c. The shape of the ion adsorption curve is caused by ion removal (adsorption) at the particle surface and by ion removal (transport) in the pore centre by fast porewater flow. The same concepts apply to preferential water flow and solute transport in soil macropores.

The simple one-dimension convective-dispersive solute transport equation for steady flow through a uniform soil is given by:

$$\frac{\partial C}{\partial t} = D_h\left(\frac{\partial^2 C}{\partial x^2}\right) - v\left(\frac{\partial C}{\partial x}\right) \qquad (5,41)$$

Ion adsorption during solute transport is commonly handled by incorporating an adsorption factor to account for solute partitioning between solution and adsorbed phases:

$$T = \theta C + \rho S \qquad (5,42)$$

where: T = total mass of solute ($T = \theta C$ for non-reactive solutes)

Substituting into Equation (5,41) gives:

$$\frac{\partial C}{\partial t} = D_h\left(\frac{\partial^2 C}{\partial x^2}\right) - v\left(\frac{\partial C}{\partial x}\right) - \frac{\rho}{\theta}\left(\frac{\partial S}{\partial t}\right) \qquad (5,43)$$

and by substituting a retardation factor (either Equation (5.39) or (5,40) into Equation (5,43) above; Smettem, 1986) to give:

$$\frac{\partial C}{\partial t} = \frac{1}{RF}\left[D_h\left(\frac{\partial^2 C}{\partial x^2}\right) - v\left(\frac{\partial C}{\partial x}\right)\right] \qquad (5,44)$$

A large number of models have been suggested to describe cation adsorption/desorption during solute transport. Wagenet (1983) classifies these models into three types:

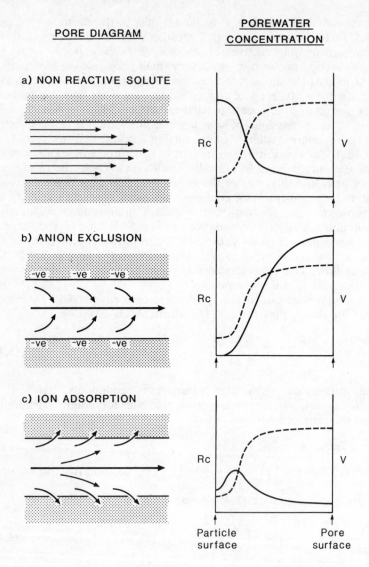

Figure 5.18 Diagrammatic representation of porewater ionic concentration Rc (—) in relation to porewater velocity V (---) in a simple cylindrical pore for **(a)** non-reactive ions; **(b)** anion exclusion; and **(c)** ion absorption

(1) Linear equilibrium models (such as Lapidus and Amundson, 1952).
(2) Non-linear and hysteretic adsorption/desorption isotherms:

    (a) at low velocities: kinetic, non-equilibrium models have been shown to improve adsorption/desorption predictions (see, for example, Lindstrom and Boersma, 1970).

(b) At higher velocities, predictions are improved by combining mobile/immobile water theory with diffusive transfer between the two liquid phases (see, for example, van Genuchten and Wierenga, 1976).

(3) Time-dependent differential adsorption models; the use of a two-rate adsorption process: an instantaneous phase, with the balance of adsorption on remaining sites being time-dependent (see, for example, Cameron and Klute, 1977).

The inevitable problem encountered in incorporating cation exchange into models of transport for reactive solutes relates to the complexity of a true soil system and the numerous interactions of competing ions. Most approaches are simple, based on only two to four cations and often omitting anion exchange. Many use cation exchange selectivity coefficients (such as those described in Section 5.2.5) to describe the replacement of one cation by another in a simple one-dimensional transport system. Ion exchange can be incorporated into transport models in several ways. Firstly, the $\partial S/\partial t$ term in Equation (**5,43**) can be used directly for ion adsorption, or it can be replaced by a cation selectivity coefficient. Secondly, a cation exchange subroutine can be used to modify solution concentrations according to calculated proportions of cations in solution and adsorbed on the solid phase.

Robbins, Jurinak and Wagenet (1980) use a series of simple two-ion exchange equations similar to that in Equation (**1.17**) to formulate cation exchange coefficients according to the Gapon equation (see Section 5.2.2). For Ca^{2+}, Mg^{2+}, Na^+ and K^+, they develop six selectivity coefficients, one to represent each pair of ionic exchanges:

For the $Ca^{2+} - Mg^{2+}$ exchange couple: (**5,45**)

$$K_{Ca/Mg} = \frac{(Ca)^{\frac{1}{2}}\ \boxed{}\!\!-\ \tfrac{1}{2}(Mg)}{(Mg)^{\frac{1}{2}}\ \boxed{}\!\!-\ \tfrac{1}{2}(Ca)}$$

For the $Ca^{2+} - Na^+$ exchange couple:

$$K_{Ca/Na} = \frac{(Na)\ \boxed{}\!\!-\ \tfrac{1}{2}(Ca)}{(Ca)^{\frac{1}{2}}\ \boxed{}\!\!-\ Na}$$

If total CEC is assumed to be $= \Sigma\ \boxed{}\!\!-\ \tfrac{1}{2}(Ca)\ +\ \boxed{}\!\!-\ \tfrac{1}{2}(Mg)\ +\ \boxed{}\!\!-\ Na\ +\ \boxed{}\!\!-\ K$ then exchangeable Ca is calculated from:

$$\boxed{}\!\!-\ \tfrac{1}{2}(Ca) = \frac{\text{total CEC}}{\left[\dfrac{(Mg)^{\frac{1}{2}}\ K_{Ca/Mg}}{(Ca)^{\frac{1}{2}}} + \dfrac{(Na)}{(Ca)^{\frac{1}{2}}\ K_{Ca/Na}} + \dfrac{(K)}{(Ca)^{\frac{1}{2}}\ K_{Ca/K}} + 1\right]}$$

(**5,46**)

with similar calculations for exchangeable Mg^{2+}, Na^+ and K^+. As well as successfully predicting Ca^{2+}, Mg^{2+}, Na^+ and K^+ migration, this approach was also successful in predicting solution electrical conductivity and sodium adsorption ratio (Robbins, Jurinak and Wagenet, 1980). Cho (1985) used a generalised selectivity coefficient (K), based on two replacing ions with modifiable valencies, to examine cation adsorption effects on the shape of the displacing front. He found that as K increased, the front of the displacing ion became sharper and the maximum concentration of the displacement ion in the effluent increased.

5.5.2 Laboratory and field testing of solute leaching models

Simple one-dimensional solute transport models have been widely developed for solute leaching studies, particularly of non-reactive ions, such as Cl^- and NO_3^- in irrigated soils. The aim in most studies is to predict solute penetration depth, or solute peak, for ions whose transport is assumed to depend only on convection, dispersion and diffusion. Convective-dispersive (CD) models have been adapted in various ways to account for more realistic field soil conditions. In particular, attempts have been made to model solute transport (1) in layered soils, and (2) under the influence of macropore flow. Selim (1978) uses a CD model with adsorption to plot depth – concentration and depth – adsorption for solute transport through soils with layers of different textures. A first order kinetic adsorption routine was used to generate the profile illustrated in Fig. 5.19. It is clear from these simulations that the order of textural layering in a soil profile plays a major role in determining the distribution of solute retention and the rate of solute loss by drainage. Although this model has not yet been field tested, the layered approach appears to offer a simple way of modelling depth variability due to changes in pore geometry.

The influence of macropores on both hydraulic conductivity and solute leaching has received much attention recently (see, for example, Bevan and Germann, 1982; White, 1985). Differential rates of soil matrix flow has been modelled by subdividing soil water into 'mobile', 'immobile' and 'stagnant' phases (see, for example, van Genuchten and Wierenga, 1976). If we consider the concepts illustrated in Figs 5.17 and 5.18, it is doubtful whether the simple CD transport equation is sufficient to model solute transport in soils containing a wide range of pore diameters and porewater velocities. White, Thomas and Smith (1984) used laboratory cores to show that antecedent soil moisture conditions have an important influence on preferential routing of solutes. They found that the drier the soil initially, the more marked was the bypassing flow, with high C_E/C_O ratios in the early part of breakthrough curves. The influence of pore geometry on rates of solute transport was examined by Scotter (1978) who compared breakthrough curves simulated for idealised soils containing only cylindrical channels, and for idealised soils containing only planar voids. He found that the minimim channel diameter for

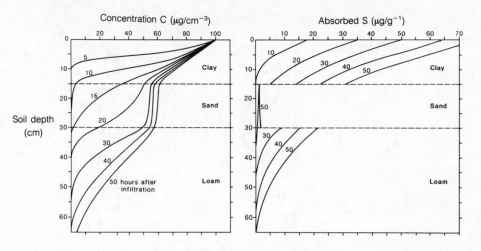

Figure 5.19 Solution and absorbed concentration distribution in three-layered soils 5–50 h after infiltration; a first-order kinetic model is used. (Source: Selim, 1978)

preferential flow of both reactive (PO_4^{3-}) and non-reactive (Cl^-) ions was 0.2 mm. The minimum planar crack width for preferential flow was 0.1 mm. Although Scotter (1978) warns about the problems of applying simulated results from idealised soil models to field soils, his results indicate that if soils containing cylindrical channels > 0.2 mm and planar cracks > 0.1 mm are wetter than field capacity, both reactive and non-reactive solutes will exhibit 'bypassing' flow.

Although laboratory validations of leaching and solute breakthrough simulations have generally proved quite successful (see, for example, van Genuchten and Cleary, 1979), less success has been achieved with field testing.

Cameron and Wild (1982), comparing the ability of three models to predict chloride leaching in a chalk soil in southern England, found that most accurate results were obtained by the CD method of Rose *et al.* (1982) (Fig. 5.20). Leaching predictions for actual rainfall proved to be less accurate than those for regular irrigation applications. Cameron and Wild (1982) report that accurate leaching predictions using the Rose method could only be made once estimates of field diffusivity were available. This result highlights the major problem in applying simple transport models to heterogeneous field conditions. Field measurements have shown that there is considerable spatial and temporal variability in soil water parameters and solute transport processes (Nielsen, Biggar and Erh, 1973; Biggar and Nielsen, 1976). Many attempts to predict field leaching of solutes have concluded that the lack of success was due to failure to account for the often unknown local field variability in soil hydrological parameters such as infiltration (Jury *et al.*, 1976) and porewater velocities (Van der Pol, Wierenga and Nielsen, 1977).

Disappointing results from deterministic leaching models have resulted

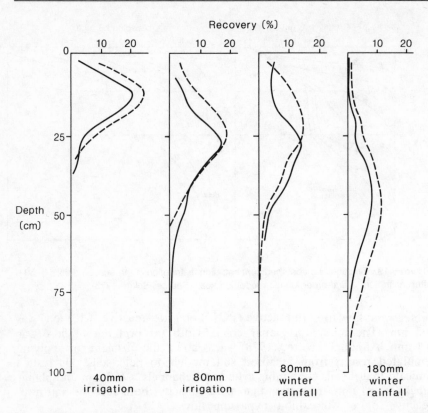

Figure 5.20 Chloride leaching in a chalk soil, as measured (—) and predicted (---) using the convective-dispersive model of Rose *et al.* (1982). (Source: Cameron and Wild, 1982)

in two main developments. First, several mechanistic models have been developed to include a stochastic routine for handling spatial variability in soil hydraulic parameters. The most popular approach has been to include a scaling factor to account for the field variability of soil moisture retention ($h\theta$) and unsaturated hydraulic conductivity ($k\theta$). Peck, Luxmoore and Stolzy (1977) and Warrick, Mullen and Nielsen (1977) used scaling theory to suggest that different zones of a heterogeneous porous medium were scale magnifications of a simple, uniform reference porous medium. In a Cl^- leaching model for irrigated soil, Bresler, Bielorai and Laufer (1979) used a single, dimensionless scaling factor to estimate the range of field values of $h\theta$ and $k\theta$. The results for Cl leaching using three values of the scaling factor (Fig. 5.21) agree quite well with the 95 per cent confidence limits of observed values.

Jury (1982) abandoned a deterministic approach to modelling solute transport and developed a quite different simulation procedure. He used a lognormal transfer function to estimate the solute travel times from the soil surface to a reference depth. The probability of a surface applied

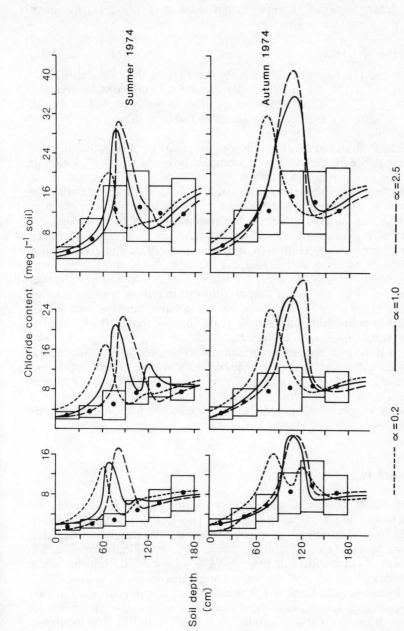

Figure 5.21 Chloride leaching in irrigated soil. Observed data (●) with 95% confidence limits (rectangles) compared to computed results using a convective-dispersive model with three different values for the scaling factor (α). (Source: Bresler, Bielorai and Laufer, 1979)

solute reaching a depth (L) after a net amount of water (I) has been applied at the surface is given by:

$$P_L(I) = \int_0^I f_L(I)dI \qquad (5.47)$$

where: $f_L(I)$ = probability density function for the probability (P_L) that a solute added to the soil surface will arrive at depth (L) as the quantity of water applied at the soil surface increases from I to (I + dI).

Field testing this model, Jury, Stolzy and Shouse (1982) found good agreement between measured and predicted bromide (Br^-) migration at four depths in a loamy sand.

Almost without exception, both laboratory and field validation of solute transport models have concentrated on non-reactive species such as Cl^-. While considerable attention has also been paid to modelling the transport of non-decaying molecular solutes such as organic pesticides (see, for example, van Genuchten and Wierenga, 1974) and groundwater pollutants such as heavy metals (see, for example, Anderson, 1979), there have yet been virtually no field validation studies. Attempts to model NO_3 transport with the added complication of microbial transformations during leaching, have so far been less successful, perhaps because of difficulties in simulating microscopic scale soil O_2 status which controls microbial nitrification and denitrification. Considering the range of mixed inorganic compounds applied to agricultural soils as fertilisers, the absence of field-verified solute transport capable of handling cation and anion interactions is, perhaps, surprising. There is thus a clear need for further direct field validation of solute transport models for a wider range of ionic species with greater attention paid to *in situ* field soil hydraulic properties and pore geometries.

5.6 SUMMARY

The dynamics of solute ions within the soil rooting zone are controlled by both exchange and transport processes. The cation exchange capacity (CEC) of soil is a combination of permanent and pH-dependent negative charge on the surfaces of soil clay and organic colloids. Weight for weight, soil organic matter at pH 7 has the largest CEC, smectite clays have the next highest, and non-expanding clays have the lowest CEC. Cation exchange reactions are reversible, stoichiometric and rapid. Cation exchange reaction coefficients are used to describe the exchange of cations from the bathing solution onto soil colloids. The lyotropic series indicates the order of cation replacability on cation exchange sites. Generally, multivalent cations are more strongly retained on exchange sites than are monovalent cations. In very simple chemical systems, it is possible to predict the composition of cations on exchange complexes if their valencies, ionic radii and activities in the bathing solution are known. The extreme complexity and temporal variability of soil solution ionic compositions in field soils makes the prediction of the composition of cations on exchange sites virtually impossible. Simple models of cation exchange use a diffuse double layer (DDL) structure, representing the parallel alignment of the negatively charged colloid surface and the positively charged 'layer' of exchanged and exchanging cations, with cation concentration declining with distance from the colloid surface.

Variable charge soils, such as highly weathered tropical soils containing significant amounts of Fe and Al hydroxyoxides, have particularly high anion adsorptive capacity when they are acidic. This property has important implications for the efficient use of fertilizer phosphorus. Phosphate is specifically adsorbed onto the surfaces of Fe and Al hydroxyoxides by ligand exchange. Quantity/intensity (Q/I) plots of anion adsorption are used to study rates of adsorption and desorption under the influence of changing environmental factors such as temperature and concentration of anion in the soil solution.

Transport of solute ions through soil occurs by two mechanisms: mass flow (convection) and diffusion. These processes are most commonly simulated using the convective-dispersive (CD) model which is based on equations for convection and dispersion/diffusion. The study of solute breakthrough curves (BTCs) in laboratory leaching columns has greatly aided the refinement of these miscible displacement models. While the miscible displacement approach and CD models have been successful in predicting solute transport through uniform media, they suffer the same problems as water flow models in predicting flux where horizontal and vertical soil variability and bypassing flow occur in field soils.

CHAPTER 6

Biological Processes in the Rooting Zone

6.1 INTRODUCTION

While the physicochemical controls on soil solute dynamics outlined in Chapter 5 may seem complex enough, under field soil conditions a whole series of biological interactions between soil and root, root and micro-organisms and micro-organisms and soil, can be superimposed to complicate this dynamic picture of the rooting zone even further. The aim in this chapter is to examine the ways in which plant roots and associated micro-organisms alter the rooting zone to make its fabric and processes different from the rest of the soil.

Within the rooting zone of plants we can identify several scales at which different chemical and biological processes operate. The term 'rhizosphere' has been used by soil ecologists to mean the 1–2 mm zone of soil surrounding the surfaces of plant roots, or rhizoplane. It is the zone altered by root exudation and the development of a microbial population associated with, and feeding on, this highly organic substrate. During nutrient uptake, a zone of nutrient depletion develops in the soil around individual roots. The largest depletion zones occur for mobile ions such as NO_3^- and Cl^-, the next largest for exchangeable ions such as K^+, and the smallest zones of depletion are for immobile ions such as phosphate (Fig. 6.1). Three main factors control the size and shape of the root depletion zone:

(1) the diffusion rate of the ion in question and the controlling influences of soil moisture content and the tortuosity of the porespace (see Section 5.4.2);
(2) mass movement of water carrying nutrient solutes to the root; and
(3) the root absorbing power, which depends on the concentration of nutrient ion in the soil porespace and at the root surface.

Bhat and Nye (1973) suggest that when the zone of depletion around the root is very small, as in the case of phosphate, root hairs begin to play an important role in nutrient depletion (Fig. 6.2). If roots absorb solute ions faster than water, as in the case of phosphate and potassium (Nye and Tinker, 1977), the concentration of these ions in the soil solution at the root surface must fall. If roots absorb water faster than solute ions, as illustrated by Nye and Tinker (1977) for SO_4^{2-} ions, then the solute ions must accumulate at the root surface. Ion depletion in the vicinity of the root may promote ion desorption from adjacent soil solids, while ion

Distance from the centre of the root (mm)

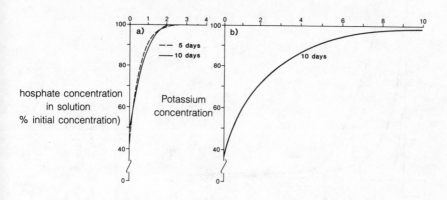

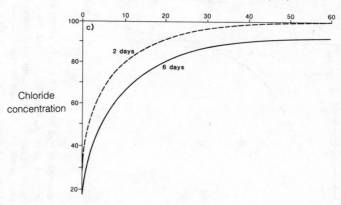

Distance from the centre of the root (mm)

Figure 6.1 Observed solute ion concentration profiles in soil around roots. **(a)** Phosphate concentrations around a 5-day-old (---) and 10-day-old (—) onion root. (Source: modified from Bhat and Nye, 1974 Copyright 1974 by Martinus Nijhoff Publishers. Reprinted by permission of Kluwer Academic Publishers.). **(b)** Potassium concentration around a root after 10 days. (Source: Nye, 1968). **(c)** Chloride concentration around an onion root in moist soil after 2 days (---) and 6 days (—) (Source: modified from Dunham and Nye, 1974)

accumulation may promote diffusion away from the root vicinity. The competition between adjacent roots for available soil nutrients creates overlaps in their nutrient depletion zones (Fig. 6.3), which results in potentially very heterogeneous nutrient concentrations and concentration gradients in the volume of soil exploited by roots. In this section, the total volume of soil exploited by the roots, or the rooting volume of plants, will be considered.

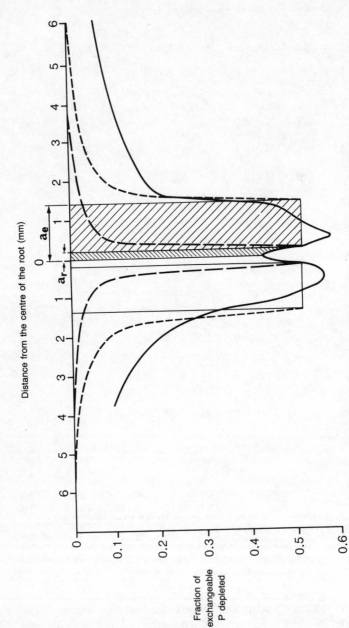

Figure 6.2 The influence of root hairs on the zone of phosphate depletion around a rape root; a_r = radius of root axis; a_e = radius of root hair cylinder; —— calculated, assuming root hairs as inactive; --- calculated, assuming intense root hair activity and uniform depletion from within the root hair cylinder; —— measured in the experiment. (Source: Bhat and Nye, 1973. Copyright 1973 by Martinus Nijhoff Publishers. Reprinted by permission of Kluwer Academic Publishers.)

Distance from the centre of the root (mm)

Fraction of exchangeable P depleted

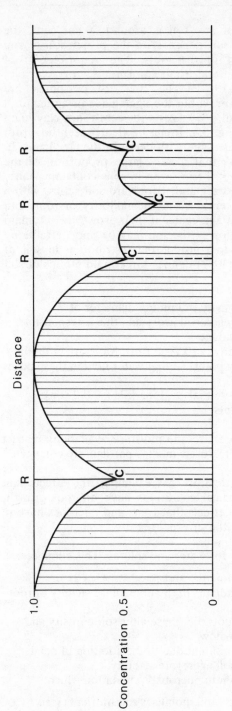

Figure 6.3 Concentration-distance profile to show effect of overlap of diffusion zones on concentration at the root surface; R denotes a root position, C = concentration in solution at the root surface. (Source: Nye and Tinker, 1977)

Although the total porosity of any soil is around 50 per cent, the ability of roots to penetrate the soil depends on the proportion of soil pores smaller than the diameter of the roots. The majority of root diameters are > 60 μm, with lateral roots in cereals, for example, around 200–400 μm (Russell, 1977). The majority of pores are < 0.2 μm in clay soils and < 1 μm within the aggregates of agricultural soils generally. The distribution of pore sizes with depth in a clay loam and in a sandy loam is given in Fig. 6.4. It can be clearly seen how root penetration becomes physically restricted with depth in the clay loam. In Fig. 6.5, Greenland (1979) shows that a very large proportion of the pores in 5 cm soil aggregates are < 1 μm, even in soils containing fairly low clay contents (for example, Rosemaud, Bromyard soil series, with a < 2 μm clay content of 24 per cent). Since the majority of roots can penetrate only the cracks between aggregates if their growth is to remain unrestricted, a significant proportion of the soil may be impenetrable by plant roots. It is also pertinent to consider the distribution in soil of different function pores. Russell (1977) defines three main groups of soil pores:

(1) Pores draining freely under gravity (minimum diameter of pore = 30–60 μm); air filled when soil is dry and this is responsible for soil aeration as well as drainage.
(2) Pores holding water against gravity (0.2–60 μm). Moisture potential > 1585 kPa (that is, $>$ wilting point) so that water held in these pores is unavailable for root uptake.
(3) Fine pores holding water at potentials > 1585 kPa (< 0.2 μm); moisture is inaccessible to plants.

These data indicate that both pore size and moisture availability restrict root growth in dense and finely textured media, particularly compacted clays of subsoil horizons.

While a large literature in agriculture has examined the effects that particular soil physical and chemical conditions have on root growth, attention here will be paid to the effects that roots and microbes have in altering soil conditions. Some of the effects are:

(1) Roots act as absorbing organs for water and solutes. This enhances:

    (a) soil moisture potential gradients, and
    (b) nutrient concentration gradients in the immediate vicinity of root surfaces.
(2) Root channels remain when roots die; these alter soil porosity and possibly encourage macropore flow.
(3) Enhanced soil organic matter content due to the addition of dead roots and root exudates aids soil aggregate stability.
(5) The development of a rhizosphere microbial population which:

    (a) decomposes organic matter and mobilises soil nutrients (see Chapter 2).

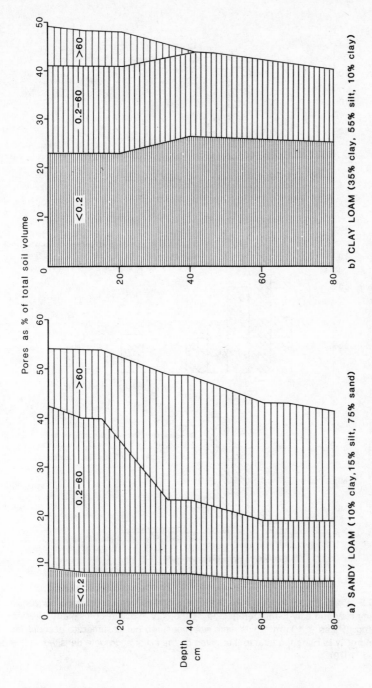

Pores as % of total soil volume

a) SANDY LOAM (10% clay, 15% silt, 75% sand)

b) CLAY LOAM (35% clay, 55% silt, 10% clay)

Figure 6.4 Percentage of total soil volume occupied by pores < 0.2 μm, 0.2–60 μm and greater than 60 μm in diameter in (a) sandy loam, and (b) clay loam soils; particle size distributions averaged for an 80 cm depth. (Source: Russell, 1977)

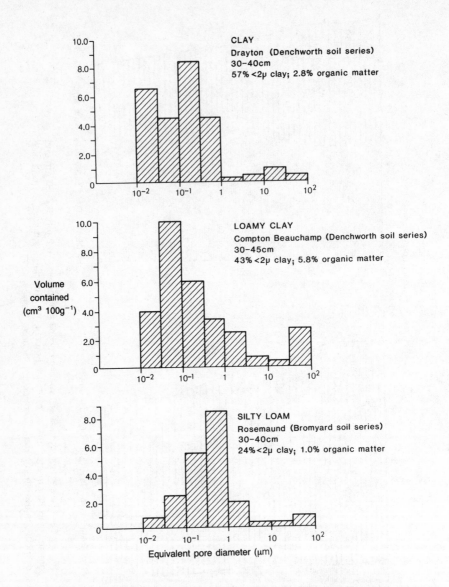

Figure 6.5 Pore size distributions of 5 cm³ aggregates from subsurface horizons of a clay, loamy clay and silty loam, determined by mercury injection porosimetry after critical point drying. (Pores < 10 μm radius were not measured but assumed to account for the residual porosity to bring the total to that given by the initial aggregate densities). (Source: Greenland, 1979)

(b) enhances aggregate stability through the production of gels and mucilages.

(c) forms root/microbial symbiotic relationships which influence plant nutrition.

Some authors have formulated water and solute transport models for soils with actively growing roots, incorporating the effects of enhanced moisture potential gradients and solute concentration gradients at the root/soil interface. Models of this type aim to predict water uptake (see, for example, Neuman, Feddes and Bresler, 1975) and nutrient uptake (see, for example, Cushman, 1982; 1984) by plant roots, but not the effect of uptake on other rooting zone conditions and processes, such as cation and anion exchange, pH, moisture content and aeration. Another important aspect of the rooting zone which until recently has received little attention, is the influence of roots and earthworms on the provision of macropores for bypassing flow. These routes are now thought to be vital in the rapid transmission of soil water and solutes through the rooting zone of surface soils.

Three main aspects of biological activity in the rooting zone will be examined here:

(1) the organic matter input to soil through underground parts of plants;
(2) the importance of roots, root exudates and soil organisms on organic matter dynamics, aggregate stability and macropore formation;
(3) relationships between plant roots and rhizosphere organisms such as mycorrhizas and symbiotic N-fixing bacteria.

6.2 ROOT ORGANIC MATTER TURNOVER

In any discussion of rhizosphere processes, it is useful to start by illustrating the magnitude of the belowground root biomass. A rather useful comparison to make is between the biomass of the above-ground and below-ground parts of plants, called the root/shoot ratio. This is not strictly a root mass to shoot mass ratio since litter, non-shoot and non-root materials (for example, rhizomes) are included. In the above-ground and below-ground biomass budget produced by Duvigneaud and Denaeyer-de Smet (1970) for a mixed oak forest in Belgium, it can be seen that the total below-ground biomass is very much smaller than the above-ground biomass (Table 6.1). Whittaker and Marks (1975) list root/shoot ratios ranging from 0.05 to 6.23 for a range of plant and ecosystem types, including herbaceous dicots such as bedstraw (*Galium aparine*) and ragweed (*Ambrosia artemisifolia*) at 0.09, shrubs such as blueberry (*Vaccinium vacillans*) and *Rhododendron maximum* at 2.52 and 1.62 respectively and forest trees such as white oak (*Quercus alba*) and pitch pine (*Pinus rigida*) at 0.91 and 0.28. Their review of root/shoot data reveal that shrubs, such as shrubby oak, and prairie/steppe grasses, have the highest ratios, reaching 6.23 and 3.4 respectively. These figures

Table 6.1   Above and below-ground annual primary productivity (biomass) in a mixed-oak forest community (from Duvigneaud and Denaeyer-de Smet, 1970)

Forest component	Biomass (kg ha^{-1})	Per cent contribution to above-ground or below-ground biomass
Above-ground		
Trees	6120	50.00
Shrubs	216	1.76
Ground flora	658	5.38
Leaf litter	3165	25.86
Non-leaf litter	2082	17.01
Total above ground biomass	12,241	
Below-ground		
Tree and shrub roots	2000	93.72
Ground flora (underground parts)	134	6.28
Total below ground biomass	2,134	

indicate that a substantial volume of soil can be occupied and modified by plant roots.

A large amount of organic matter is also added to soils every year through root turnover. A range of data, accumulated by Fogel (1983a), on fine root annual dieback in forest trees, indicates annual losses (per cent total weight) of fine root biomass ranging from 40 to 92 per cent. The actual weight of annual fine root and mycorrhizal loss in a Douglas fir ecosystem ranged from 14.6 to 18.8×10 kg ha^{-1} (Fogel and Hunt, 1983). Only recently has enough quantitative data been available to show, particularly in boreal forests, that the below-ground fine root and mycorrhizal turnover is around two to five times more important in returning organic matter to the soil than is above-ground litterfall, including branches (Fogel, 1983b; Vogt *et al.*, 1983). As with litterfall inputs of organic matter to soil, larger amounts of fine root mass and fine root turnover have been recorded in tropical forests compared to cold temperate and boreal forests (Vogt, Grier and Vogt, 1986). The relationships between organic matter input and latitude are illustrated in Figs 6.6a and b. Although lowest total fine root masses are measured in cold temperate forests, tropical rainforests with particularly fast organic matter decomposition and nutrient cycling also have low fine root masses.

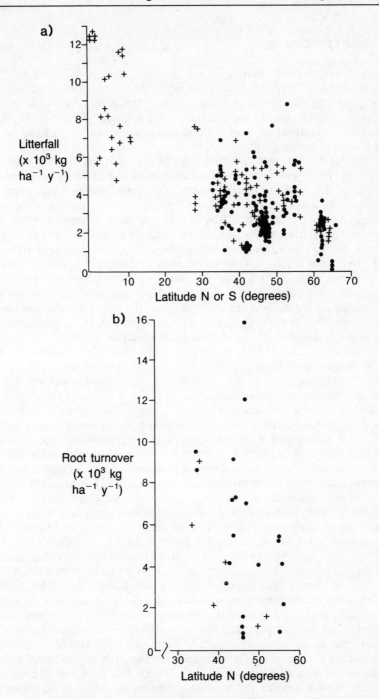

Figure 6.6 Relationship between **(a)** litterfall mass and **(b)** root turnover and latitude for broadleaved (+) and needleleaved (•) forests. (Source: Vogt, Grier and Vogt, 1986)

6.3 ROOT EXUDATES

Apart from organic matter input due to fine root dieback, roots contribute to soil organic matter through various types of exudates. The magnitude and distribution of these carbon inputs has been examined by applying $^{14}CO_2$ to plant tops and measuring organic ^{14}C and $^{14}CO_2$ in the rhizosphere adjacent to roots. Root exudates generally include soluble and insoluble mucilage produced by the root cap, sloughed fragments of root tip and root cell material, moribund root hairs and root hair secretions. Root 'exudate' is sometimes considered to be an inappropriate term since much of the organic matter lost from roots is tissue material, sloughed off from the root cap or root cortical cells during growth and passage through the soil. Rovira, Foster and Martin (1979) have suggested that four terms should be used to describe the exudate gel component: *exudate* refers to low molecular weight compounds leaked from live root cells; *secretions* are metabolically mediated release; *mucigels* refer generally to the gels covering root surfaces in non-sterile soils; and *lysates* are released from within live root cells when they rupture. In addition, Tinker (1980) suggests the general term *root carbon release* for the total ^{14}C losses measured in $^{14}CO_2$ experiments. A significant proportion of the carbon synthesised by plants is lost through root exudation and as CO_2 from root respiration. In a study of wheat roots, Bowen and Rovira (1973) found that 0.8–1.6 per cent of root carbon and 80 per cent of total carbon released into the soil was derived from insoluble mucigels and sloughed cells from the root cap.

Root exudates are composed of a range of organic compounds, including polysaccharides and other carbohydrates, proteins, amino acids, lipids and sometimes even enzymes and vitamins (see, for example, Oades, 1978). It is interesting to note that different species, different ages of the same species and annual versus perennial plants appear to produce different amounts and compositions of exudates. W. H. Smith (1977) shows that tree species, such as *Pinus radiata*, release greater amounts of root exudates overall, with higher quantities of amino acids and other organic acids than do herbaceous crops such as wheat, sorghum and tomato. The presence of root exudates in the zone immediately adjacent to root surfaces stimulates the development of a rhizosphere microbial population whose biomass and diversity depend on exudate composition. The presence of rhizosphere micro-organisms also has an important positive feedback on exudation. Vancura *et al.* (1977) examined root exudation from maize roots and from roots inoculated with *Pseudomonas putida*. They showed that the presence of micro-organisms increased the quantity of exudates by a factor of 2 to 2.6 compared with sterile roots. They suggest that the uptake of compounds by the growing bacterial culture increases the concentration gradient between the root and the nutrient solution, which results in increased exudation. In root/*Pseudomonas* cultures in which exudates were the sole source of both carbon and energy, the bacteria metabolised 73–76 per cent of maize exudates and 80–91 per cent of wheat exudates.

6.3.1   Effects on soil of root exudates

Apart from being a significant source of soil carbon as indicated above, there are several other important effects of root exudates on soil physical and chemical properties: (1) they facilitate root–soil contact, particularly for nutrient solute uptake from soil or losses from the root; (2) enhance soil CEC; (3) stabilise soil aggregates; and (4) influence the development of symbiotic root microflora. The influences of root exudates on aggregate stability and microbial symbiosis are discussed in Section 6.4.1 (pp. 193–197).

6.3.2   Soil–root contact

Several authors have suggested that during rapid transpiration, as the soil dries out, the shrinkage of roots can cause the development of a gap between the root surface and the soil, thus reducing root–soil contact. Although Tinker (1980) confirms that partial contact at low water inflow rates may be nearly as efficient as total contact, the presence of mucilage around root surfaces allows thorough contact with the irregular surfaces of soil particles which aids water and ion exchange between the gel and clay particles (Oades, 1978). The role of exudate gels as links in transferring nutrients between soil and root and between adjacent roots has been implied. Woods and Brock (1964), for example, applied ^{32}P and ^{45}Ca to stumps of *Acer rubrum* (red maple) and subsequently detected these radioisotopes in the leaves of nineteen other species at distances of up to 8 m from donor trees. Root exudation and/or mutually shared mycorrhizae (see Section 6.4.1) were thought to be responsible for these ionic transfers.

The concentration of major cations and anions in root exudates from mature trees in the Hubbard Brook Experimental Forest in New Hampshire were measured by Smith (1976). In Fig. 6.7, the net movement of these ions (balance between root uptake from solution and loss through exudation) indicates highest cationic movement of Na^+ and highest anionic movement of SO_4^{2-}. Root exudates from *Betula alleghaniensis* consistently show ionic compositions about three times higher than *Acer* and *Fagus*. Although the inorganic component appears to be the most important fraction in these root exudation results (Table 6.2), Smith (1976) suggests that the disproportionately high Na concentrations may indicate experimental contamination. Despite this, Smith's results indicate that root exudates in forest ecosystems may play an important role in nutrient cycling; the release of inorganic ions by intact roots constituting an internal cycle pathway between roots and the

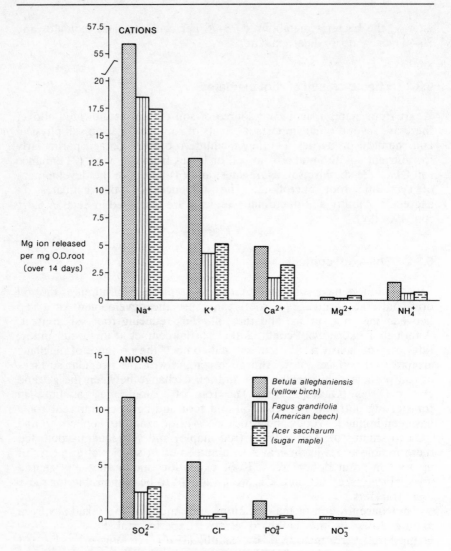

Figure 6.7 Cationic and anionic content of root exudates from three dominant tree species in the Hubbard Brook Experimental Forest, New Hampshire (mg ion released during 14 days/mg oven dry root). (Source: drawn from the data of Smith, 1976)

available nutrient pool in the soil. W. H. Smith (1977) also suggests that the organic fraction of root exudates may enhance the solubility of plant nutrients and hence aid availability. In particular, this mechanism may be important in the complexing and uptake of Ca, Al, Fe and Mn (Tinker, 1980). Bhat, Nye and Baldwin (1976) found that the phosphate depletion zone around a rape root was larger than predicted, even accounting for root hair effects. They suggest that root exudates may in some way alter

Table 6.2   Composition of root exudates from three dominant tree species in the Hubbard Brook Experimental Forest, New Hampshire (from Smith, 1976)

	Betula alleghaniensis	*Fagus grandifolia*	*Acer saccharum*
Per cent composition of exudate			
Cations	70	70	76
Anions	17	8	10
Organic acids	7	16	10
Carbohydrates	5	5	4
Amino acids	1	1	<1

the phosphate desorption isotherm in the vicinity of the root. It is clear that root exudates can hold an important reserve of plant nutrients but much further information is required on the magnitude, composition and seasonality of root exudation, as well as better quantification of the fine root distribution in different ecosystems, before we can properly assess the importance of these organic materials in nutrient budgeting and cycling.

6.3.3   Enhancement of Soil CEC

Since organic acids containing carboxylic groups are a major component of root exudates, they act as cation exchangers at the root–soil interface. Oades (1978) gives several examples of low pH measurements (2.5–3.0) in root exudates, indicating the influence of the organic acid component on the chemical characteristics of the rhizosphere. He also points out that since the root tip has the highest CEC of any region of the root, its movement through soil and distribution of mucigel over particle surfaces probably results in active ion exchange between root, carboxyl groups in the exudate and soil minerals.

6.3.4   Environmental controls on root exudation

A large number of environmental parameters are thought to influence root exudation. Hale and Moore (1979) group these effects under three main headings. First, environmental influences on plant foliage, including temperature and light, not only alter photosynthesis, but also the rate of translocation of photosynthates to the roots, the rate of enzyme degradation of photosynthates and the permeability of cell membranes. Smith (1971) found that defoliated sugar maple trees exuded higher quantities of fructose, glutamine and lysine and smaller quantities of sucrose, glycine and acetic acid than did trees with intact foliage. More recent work has indicated that the application of chemicals, particularly pesticides, to plant leaves affects the amount and composition of root exudates. The effect of pesticides on soil biological processes is discussed in Chapter 9. Herbicide effects on root exudation appear to be rather inconsistent. 2,4-D reduced the rhizosphere microbial population by suppressing the root exudation of ribose, maltose and raffinose (Jalali, 1976), while chloramben enhanced root exudation overall and increased amino acid concentration in root exudate from soybeans by a factor of 5.4 (Lee and Lockwood, 1977). Perhaps of more concern was the discovery that foliarly applied pesticides could be exuded from roots. Leaf-injected glyphosate has been detected in root exudates of *Agropyron repens* (Coupland and Caseley, 1979). Reid and Hurtt (1970) also demonstrated that the herbicides dicamba, picloram and 2,4,5-T were exuded from roots in quantities high enough to affect neighbouring plants.

Soil conditions exert a second environmental control on root exudation. Soil moisture, anaerobic conditions and texture effects have been recorded. Several authors have shown that temporary reversible wilting of plants causes enhanced root exudation. Higher amounts of amino acids (Katznelson, Rouatt and Payne, 1955), amino nitrogen and reducing sugars (Vancura and Garcia, 1969) have been found in root exudates collected from reversibly wilted plants. Since roots under field soil conditions are likely to experience a continually changing range of moisture stresses and surfeits, their exudation rate and composition may be continually changing. The main influence of anaerobic conditions, due to soil waterlogging, is also to alter the composition of exudates. Young, Newhook and Allen (1977), for example, found that ethanol was produced in the rhizosphere of waterlogged lupin seedlings. Different amounts of root exudates are produced when plants are cultured in soil compared to solution. More abrasive rooting media appear to cause greater exudation (Kepert, Robson and Posner, 1979), as testified by Boulter, Jeremy and Wilding (1966) who found that the exudation of certain amino acids could be increased by a factor of seven when roots were grown in quartz sand. It has also been shown that exudate composition is affected by mechanical impedance. Barber and Gunn (1974) used ballotini culture media to examine the effect of increased pressure on exudation by barley and maize roots. With increasing soil pressure, they found that roots exuded more amino acids and carbo-hydrates.

As indicated earlier in the work of Vancura *et al.* (1977), the third important influence on root exudation is the presence of rhizosphere micro-organisms that stimulate the production of mucilage. They may do this by:

(1) altering the permeability of root cells through damage, altering root metabolism, preferentially using particular exuded compounds or excreting toxins;
(2) altering nutrient availability to the root;
(3) affecting root metabolism; or

(4) physically blocking root surfaces (Rovira and Davey, 1974; Bowen and Rovira, 1976).

Many studies have shown that substantially greater amounts of exudates are released under non-sterile compared to sterile conditions. Bacterial and fungal colonisation of mucigel on root surfaces has been illustrated by Rovira and Campbell (1974), who also show that the surface gels contain materials of both root and microbial origin. Hale and Moore (1979) review a wide range of root colonising organisms, including saprophytes, pathogens and symbionts, whose activities are stimulated by the presence of root exudates. Many of these were shown to increase $^{14}CO_2$ release in the rhizosphere when experimental plants were supplied with labelled $^{14}CO_2$.

6.4   STABILISATION OF SOIL AGGREGATES BY ROOTS AND MICRO-ORGANISMS

6.4.1   Aggregate stabilisation

In an early paper, Emerson (1959) proposed a model of a soil crumb in which organic polymer compounds linked clay domains to the surfaces of quartz particles. Since then, electron microscopy has aided the study of the microbiology of soil structure and aggregation. In the electron micrographs of Foster *et al.* (1983), for example, many soil aggregates can be seen to contain amorphous organic matter which binds mineral particles together. Campbell and Rovira (1973) used scanning electron microscopy to show the intimate relations between soil particle surfaces, root mucigels and soil micro-organisms such as actinomycetes. Many authors have also studied quantitative links between soil organic matter content and aggregate stability. In Fig. 6.8, for example, Chaney and Swift (1984) illustrate the regression of aggregate stability (measured as the mean weight diameter of wet sieved aggregates) against soil organic matter content for two Scottish soils whose characteristics are given in Table 6.3a. For a wide range of British soils, they found highly significant ($P > 0.001$ and $P > 0.01$) correlations between aggregate stability and (1) soil organic matter, (2) soil carbohydrate, and (3) soil humic materials. The results for the Stirling and Humbie soils used in Fig. 6.8 are given in Table 6.3b. The addition of organic matter alone is not enough to create stable pores and aggregates. For good aggregation it is mainly the byproducts of microbial decomposition, such as gums, waxes, resins and gels, which stick mineral soil particles together. Root exudates and soil microbial or earthworm secretions operate in the same way to stabilise pore channel walls and surrounding soil aggregates.

The difference in aggregate stability and soil organic matter between permanent pasture and arable soils can be seen in Table 6.3a. Low (1972), in Fig. 7.1 (p. 234), also illustrated the drastic decline in water stability of soil aggregates when permanent pasture was ploughed for

Table 6.3(a) Aggregate stability, organic matter and texture of two Scottish soil series. **(b)** Correlation coefficients between mean weight diameter and various organic matter constituents for two Scottish soil series (from Chaney and Swift, 1984)

(a)

Soil series		Aggregate stability	Per cent organic	Per cent particle size distribution			Textural category
				sand	silt	clay	
Stirling (n = 27)	Arable	118	3.6	10	57	33	Silty clay loam
	Permanent pasture	225	9.5	9	59	32	Silty clay loam
Humbie (n = 14)	Arable	132	3.7	51	25	22	Sandy clay loam
	Permanent pasture	225	5.6	54	25	23	Sandy clay loam

(b)

Soil series	Organic matter	Carbohydrate	Pyrophosphate extract	Humic materials	
				sodium hydroxide[a] extract	Sodium hydroxide[b] extract
Stirling (n = 27)	0.8674***	0.7916***	0.6068***	0.7642***	0.7018***
Humbie (n = 14)	0.7743***	0.7106**	0.7034**	0.7463**	0.7668***

*** significant at 0.01 per cent level; ** significant at 1 per cent level
[a] extracted sequentially from the same soil sample
[b] extracted from a separate soil sample

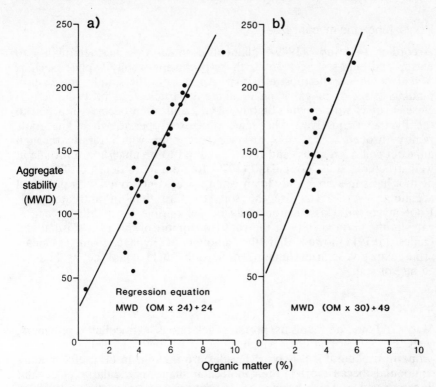

Figure 6.8 Relationship between aggregate stability mean weight diameter (MWD) after wet sieving and organic matter (OM) content for **(a)** Stirling series soils, and **(b)** Humbie series soils. (Source: Chaney and Swift, 1984)

arable crops. The main reasons for the difference is the proliferation of a dense, adventitious root system by grasses that not only contributes a large amount of organic matter to the soil, but also acts to enmesh soil particles, forms stable root channels and enhances the soil organism population. When pasture is ploughed, improved aeration and temperature changes speed up the rate of organic matter decomposition. Soil organic matter content declines and structural stability is lost.

Apart from the growth of a densely branching root system of very fine roots and root hairs, the most effective biological methods of binding soil particles are the production of filamentous and woolly hyphae by fast-growing soil fungi, the production of gelatinous algal sheaths composed of organic polymers and the ingestion and subsequent egestion of soil by earthworms in the form of casts. Tisdall and Oades (1982) suggest that aggregation caused by enmeshment by roots and fungal hyphae can be considered as a *temporary stability*, while aggregation due to the effects of soil polymers such as polysaccharides is only a *transient stability*, since these compounds are easily biodegraded. The physical binding of soil particles and the glueing action of soil polymers will be considered in turn.

Roots, fungi and mycorrhizas

According to Swaby (1949), filamentous fungi vary in their ability to entrap and bind soil aggregates; those producing woolly hyphae, such as *Mucor sp.*, being the most effective. Apart from the simple, mechanical binding effect of fungal hyphae, mineral particles can be brought into closer contact with organic binding agents and often become attached to the hyphae themselves. The same processes operate when fine roots ramify through soil. Roots produce root hairs which ramify through adjacent soil aggregates, and root hairs tend to be much more prolific in well structured soils. Greenland (1979) found a more dense proliferation of root hairs in a fine sandy loam when it was conditioned with polyvinyl acetate (PVA) to stabilise the soil structure than in untreated soil. Root–mycorrhizal associations may be responsible for stabilising aggregates in the close vicinity of the root. In support of this idea, Tisdall and Oades (1979) showed that the amount of hyphae from vesicular-arbuscular mycorrhizal fungi (see Section 7.3.1) could be as high as 55 m/g of soil.

Soil polymers

Many soil micro-organisms produce gelatinous extracellular polymers, such as polysaccharides, which appear to have particle cementing properties. Natural soil polysaccharides are thought to be useful binding compounds because of (1) their molecular shape, particularly length and linear structure, which bridge spaces between mineral particles; (2) their flexibility, allowing many points of contact – van der Waals forces can be more effective; (3) a large number of hydroxyl groups allowing hydrogen bonding; and (4) acid groups allowing ionic bonding (Martin, 1971). Actual bonding mechanisms between polysaccharides and soil particles are reviewed by Lynch and Bragg (1985). Research attention on microbially mediated aggregate stability has focused primarily on the role of polysaccharides. If polysaccharides are responsible for stabilising aggregates, then sodium periodate treatment, which oxidises and cleaves component sugars, should result in disaggregation. Several authors have reported that aggregate stability is little affected by sodium periodate treatment, unless it is continued over a period of several days. This suggests that while some forms of polysaccharide are easily destroyed, others are much more resistant (Cheshire, Sparling and Mundie, 1983). Olness and Clapp (1975) confirmed that polysaccharides attached to expanding clays, such as montmorillonite, are resistant to periodate treatment. Another explanation for their resistance to degradation could be that they are located on internal surfaces of microaggregates and are hence unavailable for microbial decomposition. Adu and Oades (1978) found that > 90 per cent of the surfaces in soil are protected from microbial attack inside soil aggregates.

It has been suggested that the manipulation of algal populations in agricultural soils may be a method of improving soil structure. Barclay and Lewin (1985), however, found that algal polysaccharides were restricted to the top 2 mm of soil, and that rotavation would be required for their incorporation. The use of artificial soil conditioners, such as synthetic polyvinyl alcohols (PVA), polyvinyl acetates (PVAc) and polyacrylonitriles, aim to imitate the cementing action of natural soil polymers.

Lynch and Bragg (1985) envisage an hierarchical sequence in aggregate stabilisation, in which aggregates of < 0.2 μm are built up sequentially into structures > 2000 μm. They suggest that three mechanisms of aggregation may be responsible.

(1) Polymers produced by bacteria may become adsorbed to soil particles.
(2) Micro-organisms may bind soil particles by becoming adsorbed to mineral surfaces.
(3) Groups of micro-organisms may interact with each other or with roots to stabilise aggregates.

These processes operate at two levels: (1) and (2) produce micro-aggregates, while (3) leads to a higher level of organisation. All the evidence points to improved aggregate stability initiating around roots. The first step in the process appears to be the adsorption of fine clay particles to mucigel at the root surface (Turchenek and Oades, 1978). Fungi, including mycorrhizas, can enmesh particles around the root and these can be further cemented by polymers either produced directly by the fungus or by bacteria associated with the hyphae (Lynch and Bragg, 1985). Once a root 'channel' or zone of aggregate stability has been established, soil drying out and root shrinkage may form a gap between the root surface and adjacent soil particles. Tinker (1976) has suggested that the development of a root–soil gap may be an important factor in limiting mass flow and diffusion of water and solutes to the root, but that mucigel bridging the gap may act to conduct ions between the soil matrix and the root surface.

Earthworms

Many authors have reported higher proportions of water stable aggregates in earthworm casts compared to surrounding soil (reviewed by Edwards and Lofty, 1972). The larger species which produce casts within the soil, such as *Allolobophora longa* and *Lumbricus terrestris*, produce the largest aggregates while surface casting species, such as *L. rubellus*, produce small aggregates. It seems unlikely that aggregate stability is due to cementing by internal mucilaginous secretions alone, since worm casts in forest and grassland soils have a larger proportion of water stable aggregates than those in arable soils (Edwards and Lofty, 1972). These authors suggest that stability may be due to the presence of calcium humate, with calcium secreted from the calciferous glands, or that bacteria or fungi produce stabilising compounds within the casts.

6.4.2   The production of soil macropores

The formation of soil macropores, or 'biopores', by plant roots and the larger soil fauna, such as earthworms, is now attracting the attention of hydrologists and solute chemists who are interested in characterising bypassing flow in soils. The concentration of water and solute flow through continuous macropores in unsaturated soils has important implications for the rapid transmission of solutes and pollutants through soils. Macropores can be considered to be continuous pores which allow non-equilibrium channeling flow (Bevan and Germann, 1982). This occurs when structural pores are large in relation to those in surrounding soil, so that the movement of water through the macropores can be much faster than the equilibration of moisture potentials in the soil matrix. It is now widely recognised that dead root channels and earthworm or soil fauna burrows can provide such routes in surface soils.

Perhaps the most obvious source of continuous macropores are the root channels cemented by root exudates and mucigels, particularly the channels of deep rooting species such as *Taraxicum officinale* (dandelion) and *Cirsium arvense* (field thistle), which can root to depths of 2–3 m. Root channels of annual species may be rather transient in soil, but roots of trees and shrubs are protected by bark which takes longer to decay than internal root tissues. 'Hosepipe' macropores are formed which contain loosely packed organic matter, derived from decaying root tissues. According to Aubertin (1971), such macropores may make up 35 per cent of the volume of a forest soil, decreasing with depth.

Earthworm channels, particularly in uncultivated and direct drilled soils have, for some time, been implicated in macropore flow (see, for example, Ehlers, 1975). It would be wrong to assume that all earthworms behave similarly and that all earthworm channels are similar. The shape, depth, density and longevity of burrows depends on earthwork species. Surface dwelling species, such as *A. caliginosa*, *L. rubellus* and *Octolasium cyaneum*, make temporary burrows, while larger species, such as *L. terrestris* and *A. longa*, have fairly permanent burrows, extending to 1–2 m and 45 cm respectively (Edwards and Lofty, 1972). Springett (1983) compared the relative abilities of five common earthworm species to produce and maintain burrows opened to the soil surface. All species, apart from *O. cyaneum*, produced 'open' burrows within 4–5 days (Fig. 6.9) and 70–100 per cent of these burrows were maintained for longer than 12 days. Springett (1983) concluded that *L. terrestris* burrows were generally less stable than the burrows of *Allolobophora sp.* Earthworms of the genus *Lumbricus* do not burrow extensively. Using a laboratory system of soil sandwiched between two sheets of glass, Evans (1947) found that *A. caliginosa* formed an extensive burrow network in the surface 20 cm in just 2 days, while *L. terrestris* took 4–6 weeks.

Depth of macropore penetration, their interconnectivity and the extent to which they are open to the soil surface, will determine their effectiveness as routes for rapid bypassing flow. Using fluorescent dyes to

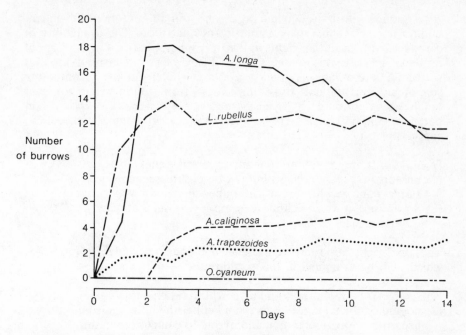

Figure 6.9 Mean cumulative total of burrows opening to the soil surface over a 12-day period for five common earthworm species. (Source: Springett, 1983)

trace the pattern of macropores in the surface layers of a sandy loam, Omoti and Wild (1979) noted about a hundred earthworm channels per square metre, with a range in diameter of 2–10 mm. Nearly all channels were continuous to 15 cm and about 10 per cent were continuous to 70 cm depth. In a study of the effect of earthworm macropores on water infiltration, Ehlers (1975) stressed the importance of macropore continuity throughout the soil profile. He noted that while there were a large number of earthworm channels in the subsoil of both tilled and untilled loess soil, they were not effective in water transmission, nor did they influence surface infiltration, because they were not connected with the soil surface. It is clear that the enhanced earthworm numbers and activity in reduced tillage and directly drilled soils (see, for example, Edwards and Lofty, 1980) may have an important influence, not only on macropore formation and soil aeration, but also on infiltration and drainage. Edwards and Lofty (1980) also found that the roots of directly drilled winter wheat preferentially exploited existing earthworm channels.

6.5 BIOLOGICAL ASSOCIATIONS IN THE ROOTING ZONE

Roots growing in soil are always intimately associated with micro-organisms and it is at the scale of a few cubic millimetres around the root–soil interface that most biological action in the soil solution and pore space occurs. Both free-living organisms and those totally dependent upon a host plant may play a role in root nutrition. The availability of carbon substrates and nutrients in the rhizosphere, perhaps in the form of exudates, increases the numbers of free-living micro-organisms at the root surface. These organisms operate as normal heterotrophs, decomposing soil organic matter and providing simple forms of nutrients for root uptake. Symbiotic micro-organisms obtain their carbon in whole or in part from the host plant. Three broad grades of root–microbial relation can exist:

(1) *symbiosis*, in which both root and organism benefit;
(2) *parasitism*, or *pathogenesis* in the case of diseases, in which the plant suffers damage while the organism benefits; and
(3) *neutralism*, in which neither plant nor organism is affected (Dommergues, 1978).

The two most important and beneficial plant–microbial relationships which shall be examined in this section are

(1) symbiosis between roots and mycorrhizal fungi; and
(2) nodule symbiosis between roots and either the nitrogen fixing bacterium, *Rhizobium* or actinomycete-like micro-organisms.

As shall be seen, in these relationships, both plant and micro-organism (symbiont) are thought to benefit. The intention in this section is not to concentrate on the physiology and biochemistry of these relationships, but to focus attention on soil and environmental factors that control microbial efficiency in aiding the provision of nutrients for plant uptake.

6.5.1 Mycorrhizal associations

Mycorrhizas are specialised symbiotic fungi that grow in intimate contact with the roots of most healthy plants. A wide range of mycorrhiza–root relationships are found in nature, ranging from mutualistic symbiosis, in which the fungus and host mutually support one another, through to those fungi that are parasitic and pathogenic. Some mycorrhizal fungi, particularly vesicular–arbuscular types, are facultative parasites and are able to switch from one nutritional mode to another, depending on soil environmental conditions. Two main groups of mycorrhizal fungi occur. They have traditionally been classified into ectomycorrhizas, or sheathing mycorrhizas, and endomycorrhizas, including vesicular–arbuscular (VA) and ericaceous types. VA mycorrhizas (VAM) are the most ubiquitous and economically important group, since they infect the majority of grass and herbaceous species of temperate and semiarid grassland ecosystems, as well as some tropical tree species. Ectomycorrhizas infect the trees of boreal and temperate forests. VAM and ericaceous mycorrhizas ramify through the root cortical cells, producing a much branched system of arbuscles, with storage organs called vesicles. There is also an extensive network of single hyphae in the external soil. Ectomycorrhizas produce a mass of mycelium which forms a sheath around lateral roots. While hyphae do not penetrate cortical cells, they do ramify between cell walls to produce an internal hyphal system called the Hartig net. A very extensive network of hyphae is produced outside the root, ramifying widely through the soil.

Root and soil conditions which affect mycorrhizal infection

High carbohydrate concentration in the root, as a result of high photosynthetic activity, is one of the main factors enhancing the host's susceptibility to infection. Mycorrhizal infection also depends on soil fertility and nutrient content. Marx, Hatch and Mendicino (1977) found that high N and P concentration in the soil caused enhanced root growth and protein synthesis, causing a decrease in carbohydrates in feeder roots of *Pinus taeda*. This caused a dramatic reduction in their susceptibility to ectomycorrhizal infection by *Pisolithus tinctorius* (Fig. 6.10). Since

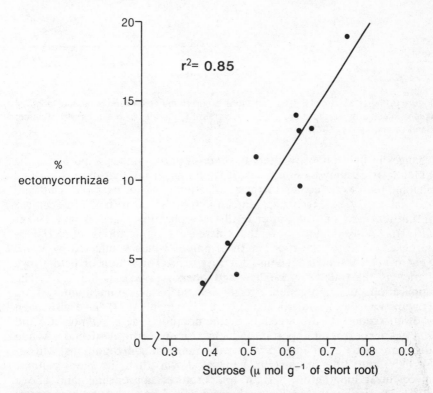

Figure 6.10 Correlation between sucrose content of short roots of 10–week old loblolly pine seedlings and percentage ectomycorrhizas formed after innoculation with *Pisolithus tinctorius*. (Source: Marx, Hatch and Mendicino, 1977)

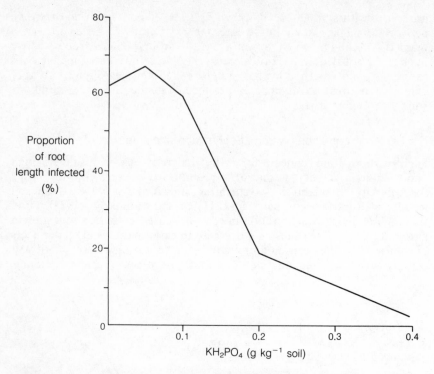

Figure 6.11 The effect of KH_2PO_4 additions to soil on the percentage of the root length of onions infected by vesicular-arbuscular mycorrhizal fungi after 8 weeks growth. (Source: Sanders and Tinker, 1973)

changes in light intensity alters the rate of photosynthesis and hence the amount of carbohydrate in roots, it is not surprising that defoliation and shading tend to reduce mycorrhizal infection (see, for example, Daft and El-Giahmi, 1978). Mycorrhizal infection has also been shown to decrease in nutrient rich soils and after fertiliser application. Sanders and Tinker (1973), for example, show that increasing applications of KH_2PO_4 fertiliser decreases the root length of onions which is infected by VAM (Fig. 6.11). A similar response for mycorrhizal infection of field-grown barley is illustrated by Sanders and Sheikh (1983) in Fig. 6.12. The application of nitrogenous fertilisers such as ammonium nitrate (Lanowska, 1966) and calcium nitrate (Hayman, 1970) have also been shown to reduce VAM infection. Soil temperature, aeration/moisture and soil pH also exert an important influence on mycorrhizal infection. While there are fungi adapted to a very wide range of soil conditions, Mosse, Stribley and Le Tacon (1981) have shown that, in pure culture, mycorrhizal fungi grow best at temperatures approaching and above 30°C. A similar picture, of enhanced initial infection at higher soil temperatures, was observed by Smith and Bowen (1979), who studied infection of *Trifolium subterraneum* and *Medicago truncatula* whose roots

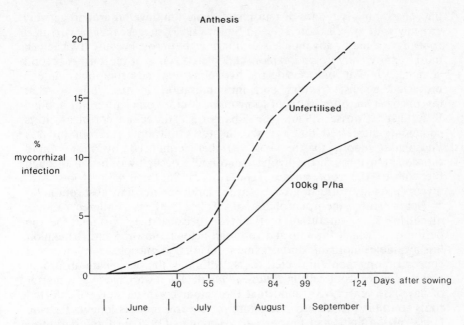

Figure 6.12 The progress of mycorrhizal infection in field grown barley (anthesis = flowering and setting of seed). (Source: Sanders and Sheikh, 1983. Copyright 1983 by Martinus Nijhoff Publishers. Reprinted by permission of Kluwer Academic Publishers.)

were kept in soils at 12°C, 16°C, 20°C and 25°C. Mycorrhizal activity may also be enhanced at higher temperatures since root exudation is also greater under these conditions. Both deficits and excesses of soil water can severely decrease mycorrhizal infection. The effect of waterlogging is to restrict oxygen diffusion to the fungus. Jackson and Mason (1984) suggest that the ability of *Pinus contorta* to survive partially anaerobic, waterlogged conditions by transporting oxygen from shoots to roots via air spaces in the stele, may be the reason for more successful ectomycorrhizal colonisation in wet forest soils. Soil pH also influences mycorrhizal activity. Ectomycorrhizas and ericaceous mycorrhizas need acid conditions for good growth. Reduced mycorrhizal infection in alkaline soils has been ascribed to an increased nitrogen availability due to enhanced nitrification, rather than the high pH itself (Jackson and Mason, 1984).

The effect of mycorrhizas on phosphate uptake

It is now well established that, compared to uninfected roots, those infected with mycorrhizas can take up greater amounts of phosphate and possibly other nutrients such as trace metals. In an early paper by Harley and McCready (1950), ectomycorrhizal roots of beech were shown to absorb as much as five times more phosphate than non-infected roots. Since phosphate ions diffuse very slowly in soil while plant roots adsorb phosphate rapidly, zones of phosphate depletion develop around actively growing root systems. Harley and Smith (1983) suggest that mycorrhizal roots have a major advantage over non-infected roots because hyphae can more rapidly colonise and exploit undepleted zones of the soil. A second advantage is that, once inside the hyphal system, adsorbed phosphate is protected against fixation and immobilisation in the soil, so that mycorrhizas are often thought to promote a more efficient use of applied P fertilizer (Mosse, 1986) (see Chapter 8). In ectomycorrhizas, it is commonly suggested that the sheath stores nutrients for root uptake. Phosphate supply to the host is then enhanced by more rapid translocation through the hyphae than could occur by diffusion through the soil to the root surface (Harley and Smith, 1983). Roots with mycorrhizas also offer a greater surface area for nutrient absorption.

Apart from the phosphate absorption effects outlined above, mycorrhizas are important in P uptake because they: (a) increase the period over which the infected root can actively absorb; and (b) exploit less available, inorganic and organic forms of soil phosphate at lower soil concentrations than uninfected roots. Although the phosphate absorbing power of apical root cells in *Pinus radiata* declined with age over a matter of days, Bowen (1973) found that the ectomycorrhizal fungi of infected roots continued to absorb phosphate for several months longer. Persson (1982) in the field, and Ferrier and Alexander (1985) in the laboratory, have even found that mycorrhizal activity persists for as long as 4–9 months after roots die or are excised. Clearly this temporal extension in absorption activity could be as advantageous for plant nutrition as the spatial extension of ramifying hyphae.

During ion uptake by roots, the relationship between external concentration and flux across the root membrane is characteristic of Michaelis–Menten type kinetics, which are used to describe and model rates of enzyme-catalysed reactions (Fig. 6.13). The rate of flux or uptake is usually directly proportional to external concentration at lower concentrations, levelling off at higher concentrations. This is given by:

$$f = \frac{C \, F_{max}}{K_m + C} \qquad \qquad (6,1)$$

where: f = rate of flux across the root membrane
$\quad \quad \; F_{max}$ = maximum rate of flux
$\quad \quad \; C$ = external solution concentration
$\quad \quad \; K_m$ = Michaelis constant

Since the flux of ions into the root is determined by both plant and ion type, it is standard to characterise the ion uptake capacity of roots using the Michaelis constant (K_m) for different types of ions. Values of K are given in units of concentration (M or mM). Nye and Tinker (1977) list K_m values for a variety of whole plants and different ions, ranging from 2 for K^+ uptake by common cocksfoot grass (*Dactylis glomerata*), to 10 for H_2PO_4 uptake by onions, 53 for NO_3^- uptake by onions and 170 for NH_4^+

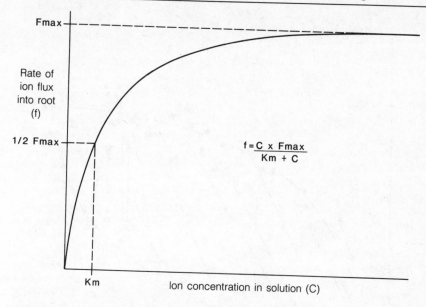

$$f = \frac{C \times F_{max}}{K_m + C}$$

Figure 6.13 Relationship between solution concentration and ion flux into the root (f) given by Michaelis-Menten kinetics in Equation **(6,1)**

uptake by maize. Soil microbiologists now think that the low K_m value for the uptake of phosphate by mycorrhizal hyphae may allow them to compete with other soil organisms in nutrient deficient soils for the low supplies of soil phosphate available (see, for example, Cress, Thronberry and Lindsey, 1979).

It has been suggested that mycorrhizas can exploit insoluble or sparingly soluble P sources which are not available to roots. Tinker (1975), for example, suggested that mycorrhizal roots exude more chelating acids, which may enhance P uptake. Barrow, Malajczuk and Shaw (1977), however, were unsuccessful in showing that mycorrhizal roots could use forms of 'fixed', or immobilised, soil phosphorus. Experimental evidence has shown that mycorrhizal plants grow better and take up more P than non-mycorrhizal plants when supplied with inorganic sources such as rock phosphate (Hayman and Mosse, 1972; Murdoch, Jakobs and Gerdemann, 1967), or bonemeal (Daft and Nicholson, 1966). Pairunan, Robson and Abbott (1980) have more recently suggested that this effect may not be seen over a wide range of rock phosphate application rates, but that mycorrhizal plants do appear to have a clear P-extracting advantage at realistic rock phosphate application rates of around 0–0.8 kg P/kg soil.

The suggestion that mycorrhizas can utilise organic forms of soil phosphorus has also received some attention. Both roots and mycorrhizal fungi produce phosphatases that can hydrolyse organic forms of phosphorus. Unpublished studies by Paterson and Bowen (1968), in Bowen (1973), showed that ectomycorrhizas in cultures could use sugar

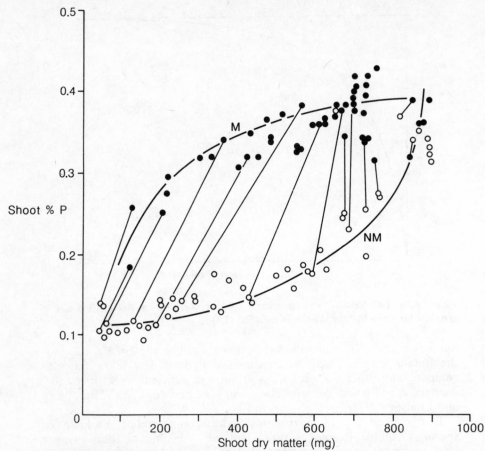

Figure 6.14 Relationship between per cent P in shoots and shoot yield per pot for plants infected (●) and uninfected (○) with *Glomus mosseae*, grown in soils of different P content, receiving five levels of added P: thin lines join data for mycorrhizal and non-mycorrhizal plants grown on the same soil at the same P level; line NM is typical for non-mycorrhizal plants generally. (Source: Stribley, Tinker and Rayner, 1980, Relations of internal phosphorus concentration and plant weight in plants infected by vesicular arbuscular mycorrhizas. *New Phytologist*, 86, © Cambridge University Press)

phosphates and nucleotides as sources of energy and phosphates. In subsequent experiments to examine the P nutrition of birch and pine, no correlation was detected between phosphatase production and PO_4-P concentration in pine, while mycorrhizal infection appeared to suppress phosphatase production by birch roots (Dighton, 1983). Since both roots and fungi produce phosphatases, it is difficult to quantify their relative contributions to plant P nutrition, but clearly any suppressive effects that mycorrhizas may have on organic P hydrolysis require further study.

That mycorrhizal plants take up more soil phosphorus than non-mycorrhizal plants has been effectively demonstrated by Stribley, Tinker and Rayner (1980). In Fig. 6.14 it can be clearly seen that mycorrhizal

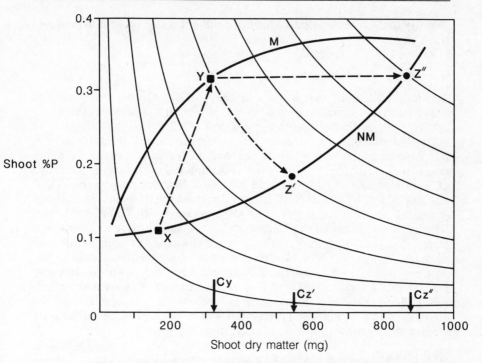

Figure 6.15 Model to explain higher per cent P in mycorrhizal plants (M) compared to uninfected plants (NM) in terms of carbon loss. Thin lines are isopleths of P concentration. The mycorrhizal plant at Y contains the same P content as the non-mycorrhizal plant at Z'. If an NM plant at X had become infected earlier, then it would occupy position Y, with a net increase in growth. The infected plant at Y contains as much P as the non-infected plant at Z'. Hypothetical dry weight 'loss' is $C_{z'}$-C_y/$C_{z'}$ or about 40%. If it is assumed that the infected plant at Y would have maintained the same per cent P content if it had remained non-infected, then the weight would have been the same as at Z", and the dry weight 'loss' would be $C_{z''}$-C_y/$C_{z''}$; or about 60% (Source: Stribley, Tinker and Rayner, 1980, Relations of internal phosphorus concentration and plant weight in plants infected by vesicular-arbuscular mycorrhizas, *New Phytologist*, 86, © Cambridge University Press)

plants, grown at the same P levels as non-mycorrhizal plants, show both increased shoot P content and increased shoot biomass. Stribley, Tinker and Rayner (1980) used these results to develop a model of P nutrition in mycorrhizal plants (Fig. 6.15). The losses in crop biomass predicted by this model are admitted by the authors to be crude, since:

(1) if an uninfected plant grew fast enough to reach point Z'', it could not have maintained the same shoot P concentration as at point Y in Fig. 6.15, hence the weight loss is an overestimation; while
(2) the larger root surface of a non-mycorrhizal plant becoming infected (position X to Y in Fig. 6.15) should be able to absorb more P than predicted, so the weight loss is an underestimation.

Despite these problems, the model shows much potential for examining not only mycorrhizal effects on plant growth and P nutrition, but perhaps also the study of mycorrhizal effects on the uptake of other nutrients and trace elements.

Mycorrhizas and nutrient cycling

The involvement of mycorrhizas in the uptake of other nutrients and water has been discussed by a range of authors. Although the faster uptake rates of nitrogen, both as NO_3^- and NH_4^+, imply the formation of nutrient depletion zones around active roots in exactly the same way as for phosphates, there is virtually no good evidence to support the hypothesis that mycorrhizal fungi aid root uptake of nitrogen. Only an indication of an effect is given by Haines and Best (1976) who found that leaching of soil NO_3^- and NH_4^+ from soil around the roots of *Liquidambar styriciflua* was reduced if the roots were infected with *Glomus mosseae*. Rhodes and Gerdemann (1978), studying the sulphur nutrition of onion plants, found a five-fold to ten-fold increase in sulphur concentration in roots and shoots of mycorrhizal compared to non-mycorrhizal plants. The positive influence of mycorrhizal fungi on trace metal uptake has also been shown for zinc (Tinker, 1978) and for copper (Gildon and Tinker, 1983). There have now been many indications that mycorrhizas aid the ability of plants to withstand drought. Two main effects of mycorrhizas on plant–water relations have been suggested:

(1) a larger absorbing root surface area, combined with reduced resistance to water uptake and flow in infected roots (Hardie and Layton (1981); and
(2) reduced soil potential at which stomata close, decreased stomatal resistance to water and CO_2 flux and hence increased rates of transpiration and photosynthesis (Allen *et al.*, 1981).

Enhanced photosynthetic capacity may allow increased competitive ability under drought conditions. Although Ellis, Larsen and Boosalis (1985) were able to show that VAM infection of wheat could produce a long-term drought resistance effect, they were unable to identify the mechanism.

In any symbiotic relationships, such as that between a mycorrhizal fungus and its host plant, it is important to evaluate any enhanced nutrient uptake against the carbon cost to the host. There have been reports of reduced plant growth as a result of VAM infection. Stribley, Tinker and Rayner (1980) explain the higher phosphate concentrations in VAM plants (Fig. 6.14) by suggesting that the fungus causes a carbohydrate drain on the host photosynthate. Studying the fate of ^{14}C in $^{14}CO_2$ fumigated faba beans (*Vicia faba*), Pang and Paul (1980) found that mycorrhizal plants transferred 47 per cent of the fixed carbon to the soil while uninfected plants transferred 37 per cent. The difference was attributed to enhanced rhizosphere respiration in the mycorrhizal root system. Whole plant dry weights were similar for both mycorrhizal and non-mycorrhizal plants, indicating that their increased ability to fix CO_2

and increased photosynthesis compensated for any symbiont carbon drain. Snellgrove *et al.* (1982) calculate the carbon cost to mycorrhizal leek plants to be of the order of 5–10 per cent of the total photosynthate, but also suggest that the infected plants appeared to be able to compensate for the drain by increasing leaf size.

Some very interesting ideas concerning the interconnectivity of mycorrhizal systems and their importance in large-scale nutrient transfers at the ecosystem scale have been presented by D.J. Read and coworkers (see Brownlee *et al.*, 1983; Francis and Read, 1984; Read, Francis and Finlay, 1985). Their ideas are substantiated by laboratory observations of interplant connections by both ectomycorrhizal roots (pine 'donor' to pine, spruce and birch 'receivers') and VAM roots (*Plantago lanceolata* 'donor' to *Festuca ovina* 'receiver') (Read, Francis and Finlay, 1985). Autoradiographs of root systems of 'donor' plants fed with $^{14}CO_2$ show that hyphal strands act to transfer ^{14}C throughout the fungal mycelium of the host and into the mycorrhiza of adjacent infected plants within 24 h of fumigation (Brownlee *et al.*, 1983). Duddridge, Malibari and Read (1980) have shown that water moves through interconnected hyphae in the opposite direction. Chiariello, Hickman and Mooney (1982) have also successfully demonstrated the transfer of P between living mycorrhizal plants in the field. Read and coworkers envisage a dynamic mycorrhizal system in which three main nutrient cycling functions operate:

(1) external hyphae which ramify through soil act to capture nutrients, particularly phosphates, which are in short supply;
(2) hyphal strands act as 'pipelines' to transport assimilates, water and nutrients between mycorrhizally interconnected plants; and
(3) the sheath and hyphae act as readily available stores of nutrients, particularly P, for plant uptake in poor soils.

Pot experiments showing that the transfer of materials, from nutrient rich mycorrhizal plants to starved, receiver seedling, could induce significant growth responses in the receiver plants (Whittingham and Read, 1982), appears to substantiate some of these ideas. Clearly, the importance of interplant connections in redistributing and conserving nutrient resources, particularly in nutrient deficient and economically important agricultural ecosystems, requires more study.

6.5.2 Associations with nitrogen fixing organisms

The primary source of soil nitrogen is the atmosphere. It is estimated that annually, some 12.2×10 tonnes of combined nitrogen are made available to the biosphere through processes of biological nitrogen fixation (Burris, 1980). The fixation of atmospheric nitrogen (N_2, or dinitrogen) is expensive in terms of energy consumption and dinitrogen fixing organisms obtain their energy either from solar radiation and photosynthesis, or from the oxidation of carbon compounds. In Chapter 2 these groups of organisms were termed phototrophs and chemotrophs respectively. The organisms responsible for dinitrogen fixation can be grouped into one of three possible life modes:

(1) freeliving, in soil or water
(2) freeliving, in close association with the rhizosphere of plant roots, or
(3) symbiotic, with other plants.

There are two types of freeliving dinitrogen fixers. First, photosynthetic bacteria and the cyanobacteria (blue-green algae) and photoautotrophs whose N_2-fixing activity is confined to the soil surface which is capable of light penetration, usually under warm and wet conditions, such as in rice paddy fields. The second group of freeliving N_2-fixers in the soil are chemoheterotrophic bacteria, primarily of the genera *Clostridium* (anaerobic conditions), *Azotobacter* (aerobic conditions) and *Bacillus* (facultative anerobe, fixing nitrogen only under anaerobic conditions). A more recent discovery was that some N_2-fixing chemoheterotrophic bacteria preferentially exploit the nutrient and organically rich zone of the rhizosphere, at the root surface of higher plants such as grasses and cereals. These are the organisms belonging to group (2) above.

The final group – symbiotic dinitrogen fixers – includes some simple associations, such as that between the cyanobacterium *Anabaena* and the water fern *Azolla*, in which symbiosis is restricted to extracellular contact. More important commercially are the more elaborate intracellular symbiotic relationships between the bacterium *Rhizobium* and the roots of legumes and between the actinomycete *Frankia* and the roots of alder (*Alnus*). The prime difference between N_2-fixers that are freeliving and those which are symbiotic chemoheterotrophs is the source of organic compounds for oxidation. Freeliving organisms oxidise soil organic compounds, rhizosphere freeliving organisms oxidise root exudates and other organic root and microbial debris, while symbiotic N_2-fixers oxidise organic compounds obtained from the host plant in the form of assimilates such as carbohydrates. An outline of these groups of dinitrogen fixing organisms is given in Table 6.4.

The relative dinitrogen fixing importance of the groups outlined in Table 6.4 relates to: (1) the location and economic importance of the habitats in which they operate; and (2) the rates at which they are able to convert N_2 into combined forms of N, primarily NH_3. N_2-fixing organisms are capable of producing the enzyme nitrogenase, which cleaves the triple bond in the dinitrogen molecule ($N \equiv N$) to produce two molecules of ammonia ($2NH_3$). Nitrogenase also cleaves the triple bond in the acetylene molecule ($HC \equiv CH$), converting it into ethylene ($H_2C = CH_2$). Acetylene reduction can thus be used to estimate the nitrogenase activity and hence the N_2-fixing potential of organisms and whole microbial populations. Since the reduction of N_2 to $2NH_3$ requires six electrons while the reduction of C_2H_2 to C_2H_4 requires two electrons, a theoretical conversion factor of 3 C_2H_2 reduced per N_2 fixed has usually been used in estimating dinitrogen fixation from acetylene reduction

Table 6.4 Groups of organisms and biological associations capable of nitrogen fixation in soils (modified from Smith and Rice, 1986; and Havelka et al., 1982)

Group	Mode of nutrition	N-fixing species (prokaryote)	Host (eukaryote)	Source of energy
(1) Freeliving	Photoautotrophic	Photosynthetic bacteria Cyanobacteria (blue-green algae)	– –	Solar radiation and photosynthesis
Freeliving	Chemoheterotrophic	Clostridium Azotobacter Bacillus + others	– – –	Oxidation of organic carbon compounds (from soil organic matter)
(2) Freeliving (in the rhizosphere)	Chemoheterotrophic	Azotobacter Azospirillium + others	Root surface of tropical and temperate grasses and cereals	Oxidation of organic carbon compounds (from root exudates in the rhizosphere)
(3) Symbiotic (prokaryote extracellular to eukaryote)	Chemoautotrophic	Cyanobacteria (blue-green algae)	Fungi Liverworts Sphagnum Azolla	Oxidation of organic carbon assimilates from host plant
Symbiotic (prokaryote within cells of eukaryote)	Chemoheterotrophic	Rhizobium Frankia (actinomycete)	Legumes Alnus Myrica	Oxidation of organic carbon assimilates from host plant
	Chemoautotrophic	Nostoc (Cyanobacteria)	Gunnera	

assays. Some variation in this ratio for different plant and environmental conditions are discussed by Hardy *et al.* (1977).

Nitrogen fixation by freeliving organisms

The factors influencing microbial activity and soil organic matter decomposition, which were outlined in Chapter 2, including resource amount and quality, temperature, moisture and pH, are all important factors influencing freeliving N_2-fixation. The availability of easily oxidisable organic compounds and suitable environmental conditions for their degradation, are important controls on the rate of N_2-fixation by heterotrophic bacteria. Alexander (1977) suggests that the addition of simple sugars, cellulose, straw or plant residues with wide C:N ratios, can often increase N_2-fixation (see Chapter 2 for a general discussion of C:N ratios). Some organisms, such as *Clostridium*, fix dinitrogen under anaerobic conditions, gaining energy for the process from the fermentation of soil organic matter. Soil moisture content determines the partial pressure of O_2 and hence the relative activities of facultative anaerobes, such as *Bacillus* and *Klebsiella*, as well as aerobic bacteria such as *Azotobacter*. *Azotobacter* is particularly pH-sensitive and is rarely found in soils whose pH lies below 6.0. The N_2-fixing bacterium *Biejerinckia*, on the other hand, grows well in aerated tropical soils under acid conditions, often as low as pH 3.0 (Alexander, 1977). As with all biological processes, N_2-fixation by freeliving bacteria is temperature-sensitive, with optimal conditions around 35°C (Brouzes and Knowles, 1973) and rates falling rapidly above this temperature. A substantial amount of N_2-fixation occurs in the tundra at temperatures around 0°C, but under these conditions, it is the cyanobacteria and lichens that are primarily responsible for N_2-fixation. Alexander, Billington and Schell (1978) found that the cyanobacterium *Nostoc* and the lichen *Peltigera* showed maximum acetylene reduction at 17.5°C–20°C, but that both were as sensitive to low light levels as to temperature.

Freeliving autotrophs such as the cyanobacteria may be only seasonally important in some habitats, due to drying out and desiccation of the soil surface during summer months. Since the cyanobacteria operate optimally in warm and wet conditions, they are agriculturally important in paddy rice culture where these conditions can be maintained. Under such conditions, N_2-fixing capacities of around 10 kg N ha^{-1} y^{-1} are common, although Balandreau *et al.* (1976) obtained a total N_2-fixing capacity of 70 kg N ha^{-1} y^{-1} in rice paddy culture for the combined effects of rhizosphere heterotrophs as well as freeliving and symbiotic cyanobacteria. N_2-fixation values of between 0.35 and 12.5 kg N ha^{-1} y^{-1} have been attributed to the activities of cyanobacteria on the forest floor of coniferous ecosystems (Granhall and Lindberg, 1978; Silvester and Bennett, 1973). Their contribution to the N budget of deciduous woodland is of a similar magnitude, measured by Todd *et al.* (1978) at around 8.53 kg N ha^{-1} y^{-1} and varying seasonally between 0.85 and 3.67 kg N ha^{-1} y^{-1}, depending on light and temperature.

Nitrogen fixation in association with rhizosphere organisms

Associative N_2-fixation is now known to occur in a wide range of grasses and cereals, including rice, maize, millet, sorghum, wheat, sugar cane and in a range of common higher plants, including species of the genera *Anthriscus*, *Convolvulus*, *Cyperus*, *Heracleum*, *Rumex*, *Stachys* and *Viola* (Alexander, 1977; Boddey and Dobereiner, 1984). Rhizosphere associations were originally detected in tropical grasses and quantified only once the acetylene reduction technique had been introduced. Dobereiner, Day and Dart (1972) noticed that, apart from rice, most associations were confined to plants possessing the C4-dicarboxylic acid photosynthetic pathway, and suggested that the more efficient solar energy conversion of these plants was connected with N_2-fixation on their roots. Dobereiner and Day (1975) found that several important cultivated forage grasses in the tropics could obtain most of their N requirements through associative dinitrogen fixation, estimating N_2-fixation rates of around 55 kg N ha^{-1} y^{-1}. These grass species also showed a marked seasonality in their ability to fix nitrogen, with fixation rates falling at, or just after, flowering. Dobereiner and Day (1975) have also found that sorghum has a diurnal cycle of nitrogenase activity, with peak N_2-fixation during peak photosynthetic activity. They suggest that the link between plant photosynthesis and bacterial N_2-fixation, together with the finding that root washing fails to remove nitrogenase activity, indicates that the N_2-fixers are located on the root, possibly in the mucigel layer.

The fixation of dinitrogen in wetland rice culture is due to both freeliving and rhizosphere organisms. The cyanobacteria are active in surface aerated soil, the facultative anaerobic bacteria, such as *Clostridium*, are active in the oxygen deficient layer below, while aerobic N_2-fixing bacteria are associated with the aerated rhizosphere zone at the surface of the rice root (Buresh, Casselman and Patrick, 1980). These authors quote N_2-fixing rates of 0.2–72 kg N ha^{-1} y^{-1} for paddy rice culture.

Symbiotic nitrogen fixation

The most widespread and widely studied symbiotic N_2-fixing relationship is that between legumes and the bacterium *Rhizobium*. It is estimated that this relationship accounts for between a third to a half of biological nitrogen fixation worldwide (Smith and Rice, 1986). Before considering legume symbiosis in more detail, two other, non-leguminous symbiotic N_2-fixing relationships are worth some discussion:

(i) the simple relationship between the cyanobacterium *Anabaena* and lichens, liverworts, *Sphagnum* mosses and the water fern *Azolla*; and

(ii) the more complex relationship between the actinomycete *Frankia* and plants such as *Alnus*, *Myrica* (bog murtle), *Hippophae* (sea buckthorn), *Ceanothus* (snowbrush) and *Dryas* (mountain avens).

Simple symbiotic nitrogen fixation

It was noted earlier that relationships between both lichens and *Sphagnum* and cyanobacteria were responsible for significant amounts of N_2-fixation in the tundra and in boreal and temperate coniferous and deciduous forests. In tropical watercourses and rice paddy fields, the *Anabaena–Azolla* symbiosis plays a particularly important role, fixing as much as 100 kg N ha^{-1} y^{-1} (Talley, Lim and Rains, 1977). Watanabe and Roger (1984) list five main factors which control rates of N_2-fixation by the *Azolla* symbiosis: temperature, light, water and wind, pH and nutrient supply. Species of *Azolla* are differentially temperature sensitive, with the cold-tolerant *A. rubra* operative between −5 and 38°C (optimum 20°C) and the heat-tolerant *A. microphylla* operative at 5–45°C (optimum 25–30°C). To maintain normal N_2-fixing rates in densely growing *Azolla* whose leaves overlap, higher light intensities are required. Talley and Rains (1980) also found that lower temperatures resulted in lower light intensity requirement for growth and nitrogenase activity. Apart from drought which desiccates the plant, *Azolla* growth is disrupted by wind which causes water turbulence and fragments the leaves. It is not found in fastly flowing water or large bodies of water. In terms of nutrition, *Azolla* is intolerant of acid water and soil conditions (< pH 3.5) and benefits greatly from phosphate additions (Watanabe and Roger, 1984). *Azolla* is now used as a valuable N-supplying green manure in rice culture (Talley and Rains, 1980).

Actinorhizal nitrogen fixation

Actinomycete infection of the roots of alder by the genus *Frankia* usually produce root nodules of varying shape and size which become the sites of active N_2-fixation. Nodulation in alder and other *Frankia* infected temperate hosts, such as *Myrica* and *Hippophae*, appears to vary with locality and plant age. The dry weight of the nodules of naturally growing *Alnus incana* and *A. glutinosa* varies from 1 to 7 per cent of the total dry weight of the trees, with highest percentages in young trees up to five years old, thereafter declining with increasing age (Akkermans and Houwers, 1979). Although the percentage dry weight decreases with age, these authors showed that under optimal growing conditions the number of nodules can reach several thousand per plant at only 6 years of age. Nodules can grow up to 8 cm in diameter. In 7 and 30-year-old stands of *Alnus rubra* in Oregon, Zavitkovski and Newton (1968) estimated nodule dry weights to be 117 and 244 kg ha^{-1} respectively. It is pertinent to note that the turnover rate of root nodules in *Alnus* can be high, amounting to as much as 35 per cent of the actual nodule biomass at any time, or about 10.6 g m^{-2} (van Dijk, 1979). This not only represents a substantial C and N input to the soil, but also indicates the scale at which the *Frankia* endophyte is 'recycled' back to the soil for the infection of new hosts.

The main environmental controls on actinorhizal nodulation are soil moisture content, pH and nutrient supply, while the factors controlling nitrogenase activity are the partial pressure of oxygen, together with seasonal factors that control plant photosynthetic activity, primarily photoperiod and temperature. Examination of the distribution of nodules on the roots of alder growing in waterlogged soil, wet peat and on river banks, indicates that nodules can only grow in surface soils above waterlogged zones, with adequate aeration. Deficits of soil water also have an important control on nodulation. Wheeler and McLaughlin (1979) suggest that moisture stress may be an important factor in controlling the shedding and formation of new nodules. Soil moisture also plays an important role in mediating the supply of oxygen to the roots. This is important for nitrogenase activity rather than nodulation.

Soil nutrition, particularly adequate supplies of phosphorus (Benecke, 1970), calcium (Youngberg and Wollum, 1976) and molybdenum (White, 1967) appear to be necessary for good nodulation. Since *Alnus* and *Myrica* grow on peats and gleys, it is not surprising to find that nodulation does not seem to be sensitive to high NH_4-N concentration, since ammonium does tend to accumulate under waterlogged conditions. Benecke (1970), however, found that nodulation of *Alnus viridis* was inhibited even at low levels of NO_3-N. Soil pH does not seem to be particularly critical for nodulation. Wheeler and McLaughlin (1979) noted that although nodulation in *Alnus* and *Myrica* could continue at pH as low as 3.3, good nodulation occurred between pH 5.5–7. The nitrogen fixation process is actually likely to cause acidification; Franklin *et al.* (1968) reporting soil pH values under 30–40-year-old *Alnus rubra* to be a whole pH unit lower than soil under adjacent conifers. A short discussion of acidification effects is given later.

As with rhizosphere nitrogenase activity, N_2-fixation in *Alnus* is related to, but probably not regulated by, photosynthetic activity. In evergreen species, such as *Ceanothus*, nitrogenase activity can continue through the winter, unless inhibited by low temperatures. In deciduous species on the other hand, nitrogenase activity ceases with leaf fall and reappears around bud burst. Wheeler and McLaughlin (1979) found that the onset of nitrogenase activity in the spring was independent from the supply of new photosynthates, but was strongly related to photoperiod and temperature changes. In plantation forestry it is important to understand the relationship between leaf senescence and dormancy and nitrogen fixation since this could indicate a mechanism for prolonging the seasonal accretion of fixed nitrogen (Wheeler and McLaughlin, 1979).

Nitrogen fixation in legumes

Numerically and economically, the legumes are the most important group of plants to host symbiotic N_2-fixation. Bacteria of the genus *Rhizobium* infect the roots of legumes to produce nodules. *Rhizobia* exist in soil as freeliving organisms, are attracted to the root by exudates at the root surface, and invade the root via infection of the root hairs. Different species of *Rhizobia* are known to be host specific. *R. japonicum* infects the roots of *Glycine max* (soya bean), *R. trifolii* infects the roots of

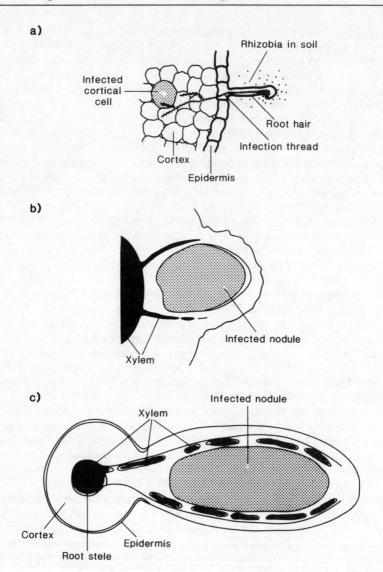

Figure 6.16 Development of root nodules in a legume by *Rhizobium* infection. **(a)** Root hair infection; **(b)** early nodule development; **(c)** mature nodule. (Source: Stewart, 1966, by permission of the Athlone Press)

Trifolium spp (clovers), *R. leguminosarum* infects the roots of *Pisum* spp (garden pea), *R. meliloti* infects the roots of *Medicago* spp (lucerne) and *R. phaseoli* infects the roots of *Phaseolus* spp. (field bean). The source of this specificity is now thought to be the presence of a protein lectin on the surface of the host root which binds with carbohydrates in the cell walls of the bacteria (Bohlool and Schmidt, 1974). Once the bacteria penetrate the root hair, they negotiate the root tissues as a fine infection thread, which is a thin strand of end-to-end bacteria. They finally penetrate an inner cortical cell where they are liberated and change into bacteroid form, stimulating cortical cell division and the formation of a cell mass, with further differentiation into a young nodule (Fig. 6.16). The different forms of *Rhizobium* have quite clearly distinct functions. The bacterial form reproduces rapidly and is responsible for colonisation and infection of the host. The bacteroid stage cannot multiply, but does possess nitrogenase and hence is the N_2-fixing stage of the organism. When nodules are shed, bacteria are rereleased into the soil for further multiplication and reinvasion of new host plants. The central tissue cells of a mature soya bean nodule can contain up to 500,000 bacteroids, with anything from 10,000 to 40,000 such cells in each nodule (Richards, 1974). In some studies, root nodule mass alone has been used as a useful estimation of nitrogenase activity. Wadisirisuk and Weaver (1985), for example, found much better correlations between root nodule mass in cowpeas (*Vigna unguiculata*) and the quantity of dinitrogen fixed than between the total number of bacteroids and N_2-fixation.

Environmental factors influencing N_2-fixation in legumes

Seasonal and diurnal cycles in N_2-fixation by soybeans have been observed by Sloger *et al.* (1975) who showed that the main period of N_2-fixing activity (as measured by acetylene reduction) occurs between flowering and seed development (Fig. 6.17). In the early stages of growth, tap root nodulation exceeds lateral root nodulation and this is reflected in slightly higher N_2 fixation by tap roots before flowering. After flowering, the proportion of N_2-fixed by lateral roots increases from a preflowering value of < 20 per cent to about 60 per cent. This change is directly related to an increase in fresh weight of lateral root nodules compared to tap root nodules. Sloger *et al.* (1975) also found strong relationships between diurnal fluctuations in both N_2-fixing activity and root respiration in soybeans, and solar radiation (Fig. 6.18). Cloudy periods caused a decrease in nitrogenous activity. Other evidence to suggest that shading can cause reduced nitrogen fixation is provided by Lawn and Brun (quoted by Sprent, 1976) who found that the self-shading which occurs when soybeans grow in size, depresses nitrogen fixation in older plants. Gibson (1976) found that *T. subterraneum* plants transferred to low light intensity (8600 lux) showed 40 per cent reduced nitrogenase activity compared to controls, while plants transferred to high light intensity (32,000 lux) showed a 50 per cent increase compared to controls. Defoliation has also been shown to detrimentally affect nitrogen fixation. Moustafa, Bell and Field (1969) found that nitrogenase activity in white clover (*T. repens*) took seven to ten days to recover after defoliation. Although this is a short-term effect, it may have important implications for nitrogen accretion rates in periodically mown silage and hay swards.

Several authors have demonstrated that increased N_2-fixation activity

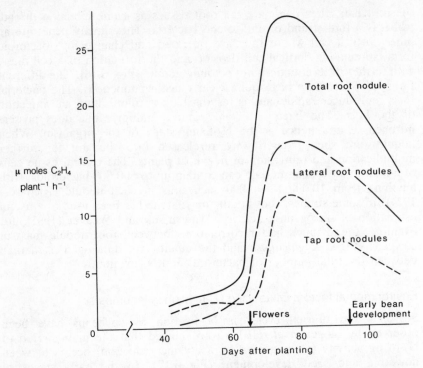

Figure 6.17 Seasonal profile of $N_2(C_2H_2)$-fixing activity for nodulated York soybeans. (Reproduced with permission from Sloger *et al.*, 1975, Seasonal and diurnal variations in N_2 (C_2H_2)-fixing activity in field soybeans. In Stewart, W.D.P. (ed.) *Nitrogen Fixation by Freeliving Microorganisms.* © Cambridge University Press.)

in legumes is directly related to photosynthetic activity. Only four hours after fumigating pea and clover plants with $^{14}CO_2$, Small and Leonard (1969) found that 5 per cent of the ^{14}C had been translocated to the root nodules of pea (*Pisum sativum*) and 9 per cent had reached the nodules in clover (*Trifolium subterraneum*). Using pea and lupin, Minchin and Pate (1973) further demonstrated that 74 and 71 per cent respectively of the fixed $^{14}CO_2$ was translocated to the roots; 16 per cent and 9 per cent being recycled within the plant, with 47 per cent and 40 per cent being respired to the soil. Hardy and Havelka (1976) report large increases in soybean N_2-fixing capacity when they were supplied with a CO_2-enriched atmosphere. The increase was due both to increased nodulation and to increased nitrogenase activity within the nodules. For peanuts grown under CO_2 enrichment, N_2-fixation increased by 59 per cent from 157 to 248 mg N plant^{-1} (Hardy, Criswell and Havelka, 1977). Havelka, Boyle and Hardy (1982) suggest that these results indicate the potential of genetic or chemical manipulation to alter rates of photosynthesis and hence N_2-fixation.

Apart from photosynthesis, soil temperature, moisture, pH and nutrient supply affect *Rhizobia* and N_2-fixation. Different species and strains of *Rhizobium* are adapted to live optimally in arctic, temperate

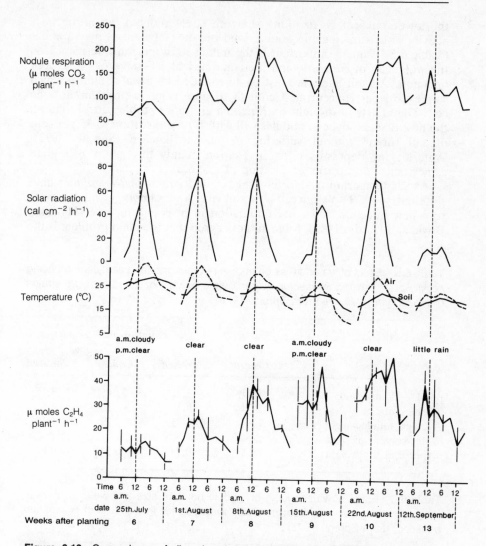

Figure 6.18 Comparisons of diurnal variation in $N_2(C_2H_2)$-fixing activity, soil and air temperatures, solar radiation, and nodule respiration in soybeans in Maryland, USA. Solid and dashed lines show the standard deviations and midday respectively. Each point is the average of 4 replicates (Reproduced with permission from Sloger *et al.*, 1975 Seasonal and diurnal variations in N_2 (C_2H_2)-fixing activity in field soybeans. In Stewart, W.D.P. (ed.) *Nitrogen Fixation by Freeliving Microorganisms.* © Cambridge University Press.)

and tropical conditions, but for any climatically adapted strain, heating and cooling can reduce nitrogenase activity. Laboratory grown (24°C) soybeans showed a rapid decline and recovery in N_2-fixation after cooling and reheating (24→14→24°C) and a slow decline and very slow recovery after heating and recooling (24→34→24°C) (Hardy, Criswell and Havelka, 1977). At low temperatures, Gibson (1976) notes that plants respond by producing a greater mass of nodule tissue to compensate for the lower rates of N_2-fixation. Moisture stress reduces the nitrogenase activity of existing legume nodules and the development of new nodules (Table 6.5). Equally important is the influence of moisture deficiency on the wellbeing of *Rhizobia* and their mobility at the stage of initial root infection. All soil bacteria must be surrounded by a water film in which solutes are not concentrated enough to pose osmotic problems for the cell. The effects of drought on *Rhizobia* are more problematical than on the host plants. Soybean nodules will withstand desiccation to 85 per cent of full turgor without suffering permanent damage (Sprent, 1972). Waterlogging depresses nitrogen fixation mainly through oxygen deficiency and reduced root respiration (Table 6.6).

As for temperature, different species and strains of *Rhozobium* have different soil pH tolerances. Low soil pH, for example, is inhibitory to root hair infection and early nodulation in peas and lucerne (Havelka, Boyle and Hardy, 1982). Most authors agree that the main problem is the

Table 6.5 Effect of water stress on nodule number, size and acetylene reducing activity of 44-day-old *Phaseolus vulgaris* (field bean) inoculated with two strains of *Rhizobium phaseoli*. (from Sprent, 1976)

| | *Rhizobium strain* | | | |
| | *3601* | | *3605* | |
	Control	*Stressed*	*Control*	*Stressed*
pmol C_2H_2 mg^{-1}min^{-1}	16.45	1.75	37.45	3.15
Nodule number	28.4	8.3	18.50	4.5
Average nodule weight (mg)	1.44	0.95	1.72	1.16
Water content of sand (per cent dry weight)	6.54	0.71	6.34	0.81

Table 6.6 Effect of waterlogging on nodule number, size and water content of *Phaseolus vulgaris* (field bean); values averaged for five strains of *Rhizobium phaseoli*. (from Sprent, 1976)

Age	*Treatment*	*Nodule number per plant*	*Average nodule size (mg)*	*Water content (fresh weight: dry weight)*	*pmol $C_2H_2(mg^{-1}$ fresh weight $min^{-1})$*
29 days	Waterlogged	15.1	3.63	2.72	4.10
	control	53.5	3.76	2.54	11.73
37 days	Waterlogged	44.2	3.65	7.46	6.09
	control	58.4	5.36	5.42	15.47
43 days	Waterlogged	94.5	4.49	9.67	6.30
	control	77.5	6.38	6.74	15.44

adverse effect of acidity on the survival of *Rhizobium* rather than on the symbiosis itself. Schmidt (1978) notes, for example, that *R. meliloti* is acid-sensitive, *R. trifolii* is less so, while *R. japonicum* is acid-tolerant Wheelan and Alexander (1986) studied the effects of low pH and high levels of Al, Fe and Mn on the survival of *Rhozibium trifolii* and on nodulation in *Trifolium subterraneum*. They found that nodulation could continue at a slow rate at pH values as low as 4.8, but that low pH in combination with low concentrations of Al (50 μM) or high concentrations of Fe (> 200 μM) inhibited *Rhizobium* growth and root nodulation.

Much attention has been paid to the effect of soil nutrient supply on N_2-fixation in legumes. Of particular interest is the supply of nitrogen, especially in the form of NO_3^-–N. A depressing effect of nitrate fertilizing on nitrogen fixation has been reported by several authors. Small and Leonard (1969) demonstrated that the amount of ^{14}C translocated to the nodules of subterranean clover and field bean plants supplied wth $NaNO_3$ for five days, was reduced by 60–75 per cent compared to NO_3^--free controls. Gibson (1976) found that nitrogenase activity in soybeans grown in NO_3^--free media, declined by 60 per cent within 48 h of receiving KNO_3. After elution of the NO_3^-, nitrogenase activity and nodulation resumed. Gibson (1976) suggests that soil NO_3^- affects N_2-fixation in two ways. First, NO_3^--N appears to have an adverse effect on root hair infection and early nodulation. Second, the additional NO_3^--N supply may alleviate N-deficiency in the plant at times of peak demand, particularly in young seedlings, when shoots and root (including nodules) are competing for the available N. Additional NO_3^--N supplied at this time should increase photosynthesis rates and the supply of soluble sugar photosynthates needed for nodule development (Gibson, 1976). The decline in N_2-fixation associated with NO_3^- additions is suggested by Oghoghorie and Pate (1971) to be due to photosynthate deprivation, resulting from competition for photosynthate from sites of active nitrate assimilation in both shoots and roots. From an extensive series of NO_3^- addition experiments, Oghoghorie and Pate (1971) concluded that the circulatory system of the plant (field pea) failed to completely mix the NO_3 it was receiving from direct supply and from N_2-fixation. This meant that any large-scale reduction or increase in either source could upset the nutritional balance of the plant. Thus, whether soil NO_3^- or any NO_3^- fertiliser application has a negative or positive effect on N_2-fixation depends on soil environmental factors, such as pH and supply of other nutrients, as well as the stage of development of both the host plant and the symbiosis. The complexity of these interrelationships led Gibson (1976) to suggest that until they were better understood, N-fertilising of legumes will 'remain a hit or miss affair'.

Lynd, Hanlon and Odell (1984) found that both P and K fertilising, both singly (KCl or $CaH_2 (PO_2)_4$) and in combination, increased mass of nodules, yield of shoots and nitrogenase activity in arrowleaf clover (*Trifolium vesiculosum* Sav.) (Fig. 6.19). As for plant growth generally, phosphorus additions appear to be most beneficial to nitrogenase activity

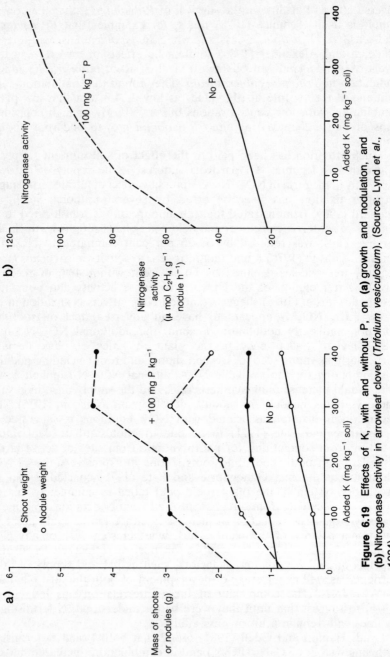

Figure 6.19 Effects of K, with and without P, on (a) growth and nodulation, and (b) Nitrogenase activity in arrowleaf clover (*Trifolium vesiculosum*). (Source: Lynd et al., 1984)

in particularly P deficient soils. Sharpe, Boswell and Hargrove (1986) showed that phosphorus fertilising caused a greater increase in rate of N_2-fixation, as measured by acetylene reduction in soybean, in soil containing 4–17 mg P kg^{-1}, with a rather smaller increase in nitrogenase activity in soils containing 17–39 mg P kg^{-1}. The effects of P and K are to stimulate growth of the host. Both molybdenum and cobalt are now known to be important for the symbiotic relationship and the efficiency of N_2-fixation by *Rizobium* (Alexander, 1977). Liming is frequently reported to have positive effects on dinitrogen fixation, perhaps by raising soil pH to improve molybdenum availability.

Soil acidification effects of N_2-fixation

Nitrogen assimilation by plants often involves changes in pH since, according to Raven and Smith (1976):

(1) NH_4^+ assimilation produces at least 1 H^+ per NH_4^+ assimilated
(2) N_2-fixation generates 0.1–0.2 H^+ per N assimilated
(3) NO_3^- assimilation produces almost 1 H^+ per NO_3^- assimilated.

During plant assimilation of NH_4^+ and NO_3^-, evolved H^+ and OH^- are liberated into the soil solution and hence do not disrupt pH balance within the plant. During N_2-fixation, it is now known that most organisms also produce an 'uptake' hydrogenase enzyme which allows the oxidation of evolved H^+. Hydrogenase activity not only allows the recycling of hydrogen, but also aids the synthesis of adenosine triphosphate (ATP) which provides the energy for further N_2-fixation. Most legume and non-leguminous symbionts as well as non-symbiotic dinitrogen fixers possess uptake hydrogenase. Examples of these are given by Schubert and Evans (1977) and Evans *et al.* (1979). The relative rates of nitrogenase and hydrogenase activity determine whether any excess H_2 will be lost to the soil solution, while the soil buffering capacity determines any subsequent soil pH change. More detailed field studies of these processes are required to allow prediction of conditions under which acidification can occur and to examine the complex effects of liming on dinitrogen fixation and crop yields.

Carbon cost of N_2-fixation

We have already seen that N_2-fixation is closely linked to the photosynthetic capacity of the host. Since rates of N_2-fixation can be reduced by limited energy supply (photosynthates) from the host, much attention has recently been directed towards assessing and improving the efficiency of energy usage during N_2-fixation. As for mycorrhizal symbiosis, the most useful measure of system efficiency is the relative carbon cost to the host plant compared to, in this case, nitrogen provision. This is measured ultimately as the cost in terms of plant yield. Does a N_2-fixing legume crop yield less than a non-fixing crop which obtains all of its N from the soil? Conflicting opinions on the efficiency of dinitrogen fixation has been due, in part, to the generation of a fairly wide range of carbon consumption values (0.3–20. mg C/mg N) through the use of a range of techniques on different plant species, with sometimes intact, though frequently excised, roots or nodules (Phillips, 1980). Recent data on intact roots indicates values of between 6.3 and 6.8 mg C per mg N fixed in a variety of legumes (see, for example, Ryle, Powell and Gordon, 1979a). Minchin and Pate (1973) suggested in an early paper that these values are no different from the energy requirement for nitrate assimilation (6.2 mg C/mg NO_3^--N reduced). Ryle, Powell and Gordon (1979b), however, have subsequently found that the respiration rate of nodulated roots fixing their own N was sometimes as much as twice that of plants lacking nodules and using supplied nitrate. They calculated that plants fixing their own nitrogen respire 11–13 per cent more of their fixed carbon each day than equivalent plants lacking nodules and using NO_3^--N. Using an average carbon consumption value of 6.5 mg C per mg N, Tinker (1984) estimated that for legume plants containing 1–2 per cent N, some 13–26 per cent of the plant dry weight may be needed for N_2-fixation. Despite this, there is little field evidence to suggest that crop yields may be reduced by this much. Since legume crops are often more demanding of phosphorus fertiliser and water than cereals, and the commercial inoculation of legume seed or soil with genetically improved *Rhizobium* are extra costs to the farmer, the true economics of symbiotic N_2-fixation are yet to be properly assessed.

N_2-fixation in ecosystem nutrient cycling

Despite the large literature on measured biological N_2-fixation, LaRue and Patterson (1981) point out the paucity of good data for N_2-fixation under field conditions. Table 6.7 lists some estimated values of N_2-fixation in different ecosystems by different organisms, both freeliving and symbiotic. All values are expressed in kg N ha^{-1} y^{-1} to allow crude comparisons to be made between different systems. These data, however, frequently summarise the results of short-term experiments, usually carried out under optimal, controlled conditions (lysimeters, glasshouses, growth chambers), and ignoring diurnal and seasonal variations in fixation rates. Even tentative comparisons indicate that many ecosystems have significant nitrogen accretion rates, decreasing under arctic and increasing under tropical conditions. Increased temperature, fertilisation and inoculation appear to be the main reasons for enhanced N_2-fixation, as testified by Agboola and Fayemi, 1972 for three different Nigerian pulses (see Table 6.7). Although legume seed inoculation is recommended for forage crops in the United Kingdom, Frame and Newbould (1986) report that inoculation of white clover seed effectively improved yields only on deep peat and wet peaty podzols.

Assessments of per cent plant N requirement supplied by symbiotic N_2-fixation indicate highest values (80 per cent) on nutrient deficient sites

(LaRue and Patterson, 1981), declining on more fertile sites. The value of using legumes such as tree lupin (*Lupinus arboreus*) and white clover (*Trifolium repens*) for vegetation establishment on nitrogen-deficient derelict sites such as china clay spoil or mine spoil from open cast coal extraction has been illustrated by Skeffington and Bradshaw (1980) and Jeffries, Bradshaw and Putwain (1981). They have shown that as much as 295 kg N ha^{-1} y^{-1} could be accumulated by a range of leguminous species on these sites. Significant transfers of nitrogen from legumes to companion grasses (up to 76 kg N ha^{-1} y^{-1}) also occurred within 2 years of sowing. In white clover–grassland forage crops, the predominant route for N transfer between legume and grasses is via grazing and recycling of urine and faeces, together with decomposition of leaf and root/nodule litter.

Two further areas where nitrogen transfers from N_2-fixing host to other vegetation is of prime importance are in intercropping systems and in silviculture. Intercropping, using both woody and herbaceous species, is particularly important on nutrient-poor tropical soils that are low in organic matter and hence low in nitrogen. Most useful systems include the use of a woody species such as the actrinorhizal *Casuarina* and the legume tree *Leucaena* which provide a vital source of firewood over 5–10 year rotations, usually intercropped with pulses or cereals. In temperate silviculture, two main approaches have been used to use biologically fixed nitrogen to supplement conifer N requirements. First, leguminous plants, particularly lupins, have been planted as an understory in nutrient-deficient conifer forests. Sprent and Silvester (1973), working on N_2-fixation by tree lupin (*Lupinus arboreus*) in thinned *Pinus radiata* in New Zealand, found that canopy shading reduced N_2-fixation to about 40 kg N ha^{-1} y^{-1}, or about 25 per cent of that in the open. Rehfuess (1979) found that *L. polyphyllus* gave more efficient N_2-fixation under dense Scots pine canopies and also altered the mor humus into a more mull-like surface organic layer which could be more readily decomposed. A second silvicultural management technique is to plant alder with conifers in mixtures, or to plant alder under thinned conifers. Miller and Murray (1979) suggest that such mixtures could provide anything from 14–300 kg N ha^{-1} y^{-1}, depending on the density of alder planting. Two main problems are associated with the management of alder mixtures:

(1) alder, planted with conifers such as sitka spruce (*Picea sitchensis* Bong. (Carr.)) and Douglas fir (*Pseudotsuga menziesii*), outgrows the conifer in the early years, so it must be pruned to prevent dominance, growth suppression and sometimes canopy damage of the conifer; and

(2) sufficient alder must be included in the mixture to eliminate N as a growth limiting factor, but since each alder replaces a conifer which would probably provide a greater economic return, alder–conifer mixtures tend to be economically viable only on upland sites which are wet and nutrient-poor.

Table 6.7 Rates of dinitrogen fixation in non-symbiotic and symbiotic systems (for comparative purposes, all results reported in mg N ha^{-1} y^{-1}); these frequently relate to growing season data, or peak activity periods only and take no account of diurnal or seasonal differences in N$_2$-fixation

N$_2$-fixing species	Environment	Annual N$_2$-fixation kg N ha^{-1} y^{-1}	Reference
(1) FREELIVING			
Nostoc (cyanobacterium)	Alaskan tundra	2.60–7.24	Alexander, Billington and Schell (1978)
Nostoc (cyanobacterium)	Swedish field soil	15–51	Henriksson (1971)
Four different cyanobacteria	Swedish waterlogged lakeside	4–44	Henriksson (1971)
Aerobic bacteria	Norway tundra meadow	6–6.4	Granhall and Lid-Torsvik (1974)
	Norway tundra heath	3	Granhall and Lid-Torsvik (1974)
Anaerobic bacteria	Norway tundra meadow	3.4–4.1	Granhall and Lid-Torsvik (1974)
	Norway tundra heath	2.6	Granhall and Lid-Torsvik (1974)
Soil bacteria	Canadian grassland	2	Vlassak, Paul and Harris (1973)
Cyanobacteria	Ivory Coast, paddy rice soils	180–240	Rinaudo, Balandreau and Dommergues (1971)
Bacteria – e.g. Azotobacter and Clostridium	Ivory Coast, paddy rice soils	26–108	Rinaudo, Balandreau and Dommergues
Freeliving soil bacteria and cyanobacteria	Coniferous forest floors (Wyoming and Montana)	0.35–1.9	Jurgensen et al. (1979)

(2) *RHIZOSPHERE ASSOCIATIVE*

Plant	Microorganism	Location/conditions	Rate	Reference
Paspalum notatum	*Azotobacter paspali*	Tropical savanna	91.25	Döbereiner and Day (1975)
Sugarcane (*Saccharum officinarum* L.)	*Azotobacter* and *Clostridium*	Controlled greenhouse conditions	0.16–8.69	Ruschel et al. (1978)
Wheat	NA	Broadbalk, Rothamsted, UK	115	Jenkinson (1977)
Mixed herbaceous vegetation	NA	Broadbalk, Rothamsted, UK	49	Jenkinson (1977)

(3) *SYMBIOTIC*
(a) *Non-legume*

Plant	Microorganism	Location/conditions	Rate	Reference
Peltigera (lichen)	Cyanobacteria	Alaskan tundra	7.65–9.55	Alexander, Billington and Schell (1978)
Azolla	*Anabaena* (cyanobacteria)	Philippines – rice culture	103–162	Becking (1979)
Azolla	*Anabaena*	New Zealand lake	164	Kellar (1979)
Gunnera dentata	*Nostoc* (cyanobacteria)	New Zealand	72	Silvester and Smith (1969)
Gorse (*Ulex europaeus*)	NA	United Kingdom reclaimed china clay waste	26	Skeffington and Bradshaw (1980)
Alnus glutinosa (black alder)	*Frankia*	Netherlands	22.4 g N $\mathrm{tree}^{-1}\mathrm{y}^{-1}$	Akkermans and van Dijk (1976)
Alnus rubra (red alder)	*Frankia*	Washington	85	Cole, Gessel and Turner (1978)
Alnus glutinosa and *Populus*	*Frankia*	South-east Canada	130–170	Côté and Camiré (1985)
Caenothus velutinus Douglas (snowbrush)	*Frankia*	Oregon	70–108	Youngberg and Wollum (1976)
Myrica gale (sweet myrtle)	*Frankia*	Harvard University forest (Massachusetts)	37.2	Schwintzer, Berry and Disney (1982)

(Table 6.7 cont . . .)

(b) Legumes

Legume	Rhizobium	Location	Value	Reference
Lucerne (Medicago sativa)	Rhizobium meliloti	Rothamsted, UK	70	Bell and Nutman (1971)
	Rhizobium meliloti	Rothamsted, UK inoculated	235	Bell and Nutman (1971)
	Rhizobium meliloti	Rothamsted, UK, plus lime P and K fertilisers inoculation	343	Bell and Nutman (1971)
Lucerne	NA	Central Iowa	15–136	West and Wedin (1985)
Soybeans (Glycine max)	Rhizobium japonicum	Rumania	162–182	Hera (1976)
Soybeans	Rhizobium japonicum	Beltsville, Maryland	158–214	Coale, Meisinger and Wiebold (1985)
Garden pea (Pisum)	Rhizobium leguminosarum	Washington north and south facing slopes	17–69	Mahler, Bezdicek and Witters (1979)
Cowpea (Vigna simensis)	NA	Ibadan, Nigeria	354[a] 157[b]	Agboola and Fayemi (1972)
Greengrain (Phaseolus aureus)	NA		224[a] 63[b]	Agboola and Fayemi (1972)
Calopo	NA		450[a] 370[b]	Agboola and Fayemi (1972)
Subterranean clover (Trifolium subterraneum)	Rhizobium tripoli	California	58–183	Phillips and Bennett (1978)
Rose clover	Rhizobium tripoli	California (lysimeters)	56–67	Williams, Jones and Delwiche (1977)
Subterranean clover (Trifolium subterraneum)	Rhizobium tripoli	California (lysimeters)	50	Williams, Jones and Delwiche (1977)
White clover (Trifolium repens)	Rhizobium tripoli	United Kingdom upland grazing	100–150	Newbould (1982)
White clover (Trifolium repens)	Rhizobium tripoli	United Kingdom lowland grass sward	74–280	Cowling (1982)
Tree lupin (Lupinus arboreus) under pine forest (Pinus radiata)	NA	North Island, New Zealand	78	Silvester, Carter and Sprent (1979)

6.6 SUMMARY

The volume of surface soil exploited by plant roots is physically and chemically altered by them. It is possible to identify two zones: (1) the volume 'tapped' by roots during uptake of water and nutrients, and (2) the rhizosphere, which is the zone of soil immediately adjacent to root surfaces, rich in organic matter and particularly attractive to soil micro-organisms. The below-ground input of organic matter (dead roots, root litter and root exudates) is very substantial; estimated in boreal forests to be two to five times more important in returning organic matter to soil than is above-ground litterfall, including branches. Root exudates improve the contact between root surfaces and soil particles and are a particularly important supply of easily decomposable organic substrates for soil micro-organisms. The production of exudates and micigels by roots and soil organisms aids the stabilisation of soil aggregate. Roots and earthworms are also important formers of 'biopores' or macropores which conduct soil water and solutes more rapidly than the soil matrix generally. Two main types of root–organism associations are particularly important in nutrient cycling and plant nutrition: the symbiotic relationship between roots and mycorrhizal fungi; and both non-symbiotic and symbiotic dinitrogen fixation by bacteria, actinomycetes and cyanobacteria. Environmental controls on the development and efficiency of these relationships are discussed.

PART FOUR

Processes Influenced by Soil Management Practices

Widespread and standard soil management practices include ploughing and drainage, fertilising, and the use of pesticides. These practices have two main influences on soil in which a whole series of complex processes are already operating: (1) the physical alteration of existing soil materials; and (2) the addition of new materials. In examining the effects of soil management practices on existing soil processes, the aim is to outline which new processes might be introduced to the soil system and which existing soil processes might be altered. An example of a new process introduced by intensive ploughing is the disruption of soil aggregates and resultant exposure to decomposer organisms of previously unavailable substrates locked up in the centres of peds.

The alteration of soil physical properties during ploughing has subsequent effects on hydrological processes, including infiltration, hydraulic conductivity and moisture retention. Solute chemistry is also affected, through altered porewater retention times, solute equilibration and changes in rates of solute transport and leaching. The addition of agrochemicals (both fertilisers and pesticides) to soils alters both magnitude and rate of many fundamental chemical processes, including cation and anion exchange phenomena, organic matter decomposition and nutrient mobilisation.

The effects of agricultural management practices on soil processes change with time after the initial management operation. A good example of this is the initial decrease in bulk density which accompanies ploughing. Bulk densities rapidly increase again as the soil settles after ploughing. Similarly, high initial pesticide concentrations may result in initially dramatic population declines of native soil micro-organisms, with population numbers increasing again once some degree of resistance is established.

CHAPTER 7

Soil Processes Altered by Ploughing and Drainage

7.1 INTRODUCTION

Much attention has recently been focused on the effects of ploughing and soil cultivation practices on both physical and chemical properties and processes in soil. In agriculture, the Strutt Report (1970) and Greenland (1977), for example, have expressed concern that modern intensive cultivation practices commonly used to produce a seedbed, such as ploughing followed by harrowing, disrupt the soil structure and quickly lead to puddling and compaction in wet conditions. From historical times, farmers have believed, perhaps erroneously, that more intensive mixing and pulverisation of the soil would benefit their crop. In the past, farmers have apparently ploughed and harrowed their fields up to a dozen times to prepare a good 'tilth' in the seedbed.

Russell (1977) usefully summarises currently used soil preparation practices into traditional, reduced (minimal) and conservation tillage types (Table 7.1). Reduced cultivations aim to speed up cultivation practices and to prevent damage to soil structure. In the United Kingdom, the increasing use of direct drilling since the 1960s has mainly been prompted by the development of high-return, autumn-sown cereals in combination with the widespread use of synthetic herbicides. Since operations can be completed more rapidly than conventional tillage, direct drilling practices allow larger areas to be established nearer to the optimum sowing time in October. Conservation tillage practices aim to reduce soil erosion, particularly in continental and semiarid regions.

The traditional objectives of ploughing practices were to drain and aerate the soil, to break up compacted zones, to stimulate native nutrient mineralisation and to suppress competing weed vegetation. The more widespread availability and use of herbicides to reduce weed colonisation and growth has opened the door for a reduction in intensity of agricultural cultivations. Changes in soil structure and packing caused by ploughing, initially alters soil physical properties such as bulk density, moisture retention and thermal regime. Since these factors, particularly moisture and temperature, alter rates of chemical and biological processes in the soil, rates of organic matter decomposition and nutrient mobilisation are also likely to be affected by ploughing.

An additional factor in assessing cultivation effects on agricultural soils is the influence of farm traffic, such as tractors, ploughs, sprayers and harvesters, in compacting topsoil. Current research into predicting soil strength aims to direct necessary agricultural machine operations towards

Table 7.1 Tillage practices for the cultivation of agricultural land (modified from Russell, 1977)

Tillage system	General practice	Comments
Traditional tillage		
(1) Conventional ploughing	Disruption and soil mixing to at least 25 cm, usually followed by secondary operations to prepare a seedbed (discing, harrowing, rolling, rotavating)	Produces a uniform, finely divided surface of bare soil
Reduced (minimal) tillage		
(2) Tine (chisel) ploughing	Primary soil disturbance by tines to a depth of 10–25 cm	More plant debris on the soil surface than in (1)
(3) Direct-drilling (zero-tillage, no-till, slit plant)	Seed hydraulically injected into undisturbed soil on which existing vegetation is killed by herbicides or removed by burning	Once plant debris is disposed of, soil surface is same as beneath previous crop
Conservation tillage		
(4) Stubble mulching	Crop residues left on soil surface; soil may be mixed below the surface layer	Much plant debris left on soil surface
(5) Strip tillage	Ploughing by sweep or rotary equipment in rows on the slope contour; crops planted between rows	Ground cover remains between contoured rows

times of the year when soil conditions are optimum and soil damage potentially at a minimum.

7.2 ALTERATION OF SOIL PHYSICAL PROPERTIES AND PROCESSES

Ploughing alters both soil mechanical properties, such as bulk density and structure, as well as soil hydrological properties, such as infiltration, moisture retention and hydraulic conductivity. The effects on soil physical conditions of conventional ploughing, direct drilling (or zero-tillage), drainage techniques and of agricultural traffic will be assessed and, where possible, compared.

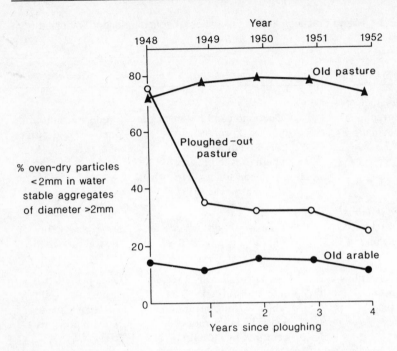

Figure 7.1 Changes in water stable aggregates after ploughing old pasture compared to remaining old pasture and old arable sites on the same soil series. (Source: Low, 1972)

7.2.1 Disruption of soil structure

Disruption of soil structure following ploughing has been widely reported (see, for example, Low, 1972) and modelled in laboratory experiments (Rovira and Graecen, 1957; Powers and Skidmore, 1984). The pattern of structural disruption for ploughed out grassland (Fig. 7.1) shows a rapid decrease in stable aggregates in the first year after ploughing, with a less dramatic decline in subsequent years. The result of several field trials, summarised in Table 7.2, indicate a consistently higher degree of aggregate stability in virgin and direct drilled soils than in ploughed and cultivated soils. The loss of aggregate strength after ploughing is mainly due to enhanced organic matter decomposition and will be discussed in Section 7.2.

Dexter and his co-workers (for example, Dexter, 1976; Ojeniyi and Dexter, 1979a; 1979b) have examined micromorphological changes in soil structure under the influence of different types and intensities of tillage. They have used statistical assessments of number, distribution and orientation of aggregates and voids in thin sections from ploughed soils to describe effects of a range of tillage types. In Dexter's (1976) initial study of soil microstructure after tillage, he found that small angles of incidence between voids and aggregates were more common in tilled soil samples

Table 7.2 Relative aggregate stabilities and organic matter content under conventional ploughing and direct drilling

Soil type	Depth	Greater aggregate stability on:		Greater organic matter content on:		Reference
		plough	direct drill	plough	direct drill	
Clay (Evesham)	surface	*		*		Douglas (1976)
Silt loam	Surface	*				Vyn *et al.* (1982)
Silt loam	0–10 cm	*				Sidiras, Henklain
	10–20 cm	*				and Derpsch (1982)
Sandy loam	0–10 cm	*		*		Tomlinson (1974)
	10–15 cm	*	*			
	15–20 cm	*	*			
Silt loam (Lethbridge)	0–2 cm				*	Carter and Rennie
	2–4 cm			*		(1982)
Clay loam	0–5 cm				*	Carter and Rennie
	5–10 cm			*		(1982)
Loam (Scott)	0–5 cm			*		Carter and Rennie
	5–10 cm			*		(1982)

than in idealised beds of spherical aggregates. This result is illustrated diagrammatically in Fig. 7.2. This result has been developed further (Dexter, 1978) to model the vertical growth of roots in ploughed soil. The model is based on the probability of a root either penetrating or being deflected by an aggregate after passing through a void. Computer-simulated structures were generated from sets of transition probabilities to represent a mouldboard plough, a tine cultivator and a rotary cultivator. Simulated root growth was found to depend on the strength of the aggregate, the angle of incidence between the root and aggregate surface (θ) and the length of the preceding void. Probability of a root penetrating an aggregate instead of being deflected by it, decreases with increasing strength of aggregate and with decreasing angle of incidence (θ).

Examination of thin sections has confirmed that smaller aggregates and pores accumulate at the base of the ploughed layer, with larger aggregates and pores at the soil surface (Ojeniyi and Dexter, 1979b). In an examination of the influence of soil moisture content on loss of pore space after ploughing, Ojeniyi and Dexter (1979a) found that maximum

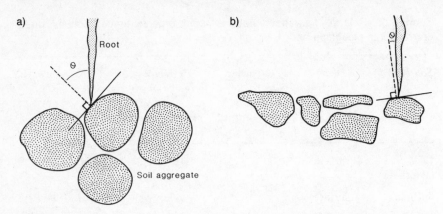

Figure 7.2 Angle of incidence (θ) (for example of a vertical root penetrating the soil) of aggregate-void surfaces. **(a)** Idealised bed of spherical aggregates (predominantly high values of (θ); **(b)** ploughed soil (predominantly low values of θ)

damage occurred at gravimetric moisture contents of around 90 per cent of the soil's plastic limit. In this condition, tillage produced the maximum number of small aggregates and the minimum number of large voids. Sidiras, Henklain and Derpsch (1982) found that proportions of stable, large aggregates (9.52–5.66 mm diameter) were highest in zero-tilled plots, while proportions of stable, small aggregates in (2.0–0.25 mm) were highest in more intensively tilled plots (Fig. 7.3). This finding is the likely precursor to the size distribution of voids and aggregates with depth in ploughed soils studied by Ojeniyi and Dexter (1979b). Dexter (1979) also used a transition probability matrix of the distribution of aggregates and voids in a soil after tillage to develop a model to predict changes in soil structure as influenced by factors such as soil moisture content, plough type and different management practices.

7.2.2 Compaction

Breakup of soil structure can initially lead to reduced bulk density, particularly in surface soils, where the material is loosened. Soane and Pidgeon (1975) found that shallow ploughing successfully reduced bulk density in the surface soil of an agricultural sandy clay loam throughout the growing season, while deep ploughing gave lower bulk density at depth. Both deep and shallow forms of conventional ploughing resulted in lower bulk densities than zero-tillage techniques (Fig. 7.4a). Soane and Pidgeon (1975) also showed that ploughing invariably resulted in large and significant reductions in soil strength, as measured by cone resistance (Fig. 7.4b). Ross and Malcolm (1982) report significantly reduced bulk density after intensive cultivation of the top 60 cm in a peaty podzolic profile for afforestation purposes (Fig. 7.5), although in this example part

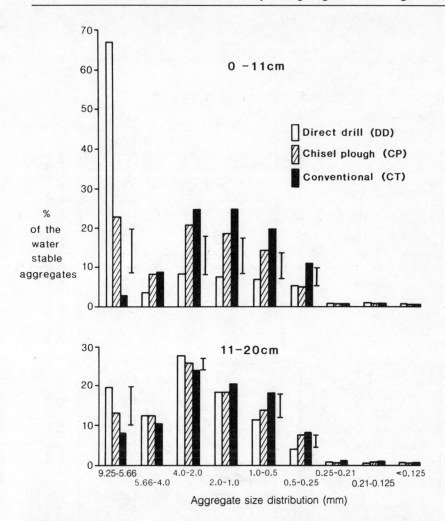

Figure 7.3 Distribution of the water stable aggregates (expressed as a percentage of an original 100 g sample of aggregates in the class 9.52–5.66 mm class range) at the soil depths 0–10 cm and 11–20 cm after 4 years of conventional tillage, chisel plough and direct drill (Londrina, Oxisol). Standard deviation bars are given. (Source: Sidiras, Henklain and Derpsch, 1982)

of the effect was due to the incorporation of surface organic matter during soil mixing.

Successive relatively shallow cultivations common in agricultural tillage often cause the formation of a smeared and compacted layer immediately below the ploughing depth. Since these root-inhibiting plough pans can usually be broken up by subsoiling using a deep tine plough or chisel plough, Greenland (1977) suggests that they may be considered as temporary rather than permanent damage to most agricultural soils.

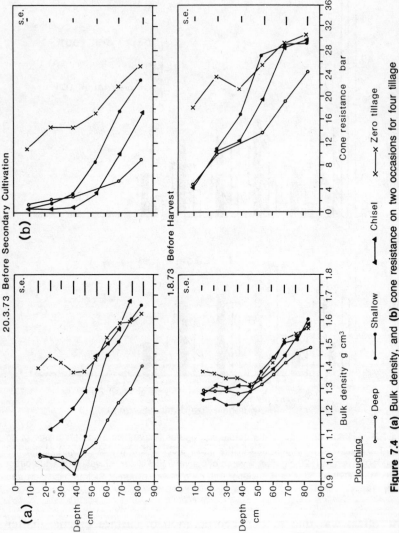

Figure 7.4 **(a)** Bulk density, and **(b)** cone resistance on two occasions for four tillage treatments applied to a loam soil growing continuous barley for 6 years. (Source: Soane and Pidgeon, 1975)

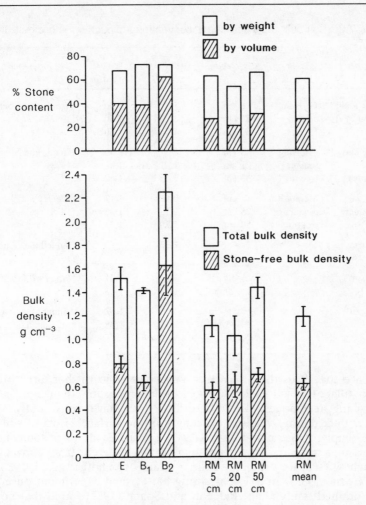

Figure 7.5 Bulk density and per cent stone content in three horizons of undisturbed soil (E, B_1, B_2) and at three depths in deeply rotavated soil (RM). (Source: Ross and Malcolm, 1982)

Voorhees (1983) has indicated that freezing and thawing in soils of the Midwest United States can help to break up compacted cultivation layers. The restriction of rooting to a shallow surface soil layer above the plough pan appears to mainly cause crop drought. In a majority of tillage studies in the sandy soils of the Atlantic Coastal Plains region of the United States, Reicosky (1983) reports that restricted root growth due to high bulk densities in the plough pan resulted in severe water stress within 3–7 days after rainfall. When plough pan compaction is relieved by disruption the resultant improvement in crop yields is a clear indication of the detrimental effects of restricted rooting (Eck and Unger, 1985).

Table 7.3 Soil bulk densities under conventional ploughing and direct drilling

Soil type	Time of sampling	Depth	Bulk density (g cm)$^{-3}$		Reference
			plough	direct drill	
Sandy loam	Growing	6 cm	1.2	1.36	Pigeon and Soane
(Macmerry)	season	21 cm	1.27	1.30	(1977)
	(mean)	33 cm	1.5	1.50	
Sandy clay	Growing	6 cm	1.3	1.45	Pigeon and Soane
loam	season	21 cm	1.41	1.42	(1978)
(Winton)	(mean)	33 cm	1.65	1.62	
Clay loam	Mean after	5 cm	0.98	1.11	Gantzer and Blake
(Le Sauer)	5 years of	25 cm	1.12	1.32	(1978)
	treatment	50 cm	1.29	1.34	
Silt loam		1.3–8.9 cm	1.33	1.50	Cannell and Finney (1974)
Silt loam		5–10 cm	1.21	1.41	Vyn et al. (1982)
		15–20 cm	1.35	1.48	
Sandy silt		0–15 cm	1.38	1.52	Tollner, Hargrove
loam		30–35 cm	1.53	1.38	and Langdale (1984)

Since the 1970s there has been particular concern that direct drilling (zero tillage) practices cause more soil compaction than conventional ploughing techniques. Directly drilled sites have consistently shown higher bulk density values, particularly in the surface 25 cm of soil (see, for example, Pidgeon and Soane, 1977; Table 7.3) although some authors report that these results are not significantly different from conventional ploughing (see, for example, Tollner, Hargrove and Langdale, 1984; Hill and Cruse, 1985). In a 7 year spring barley field experiment on a fairly well drained sandy loam, Pidgeon and Soane (1977) found that no bulk density changes occurred under continuous zero tillage after the first three years. This condition marks an equilibrium bulk density between a particular soil and imposed management practice. By assessing soils primarily according to structural stability, Wilkinson (1975), and Cannell, Davies and Pidgeon (1979) have classified UK soils into 'probability of success' classes for direct drilling. Not surprisingly, they found that the most suitable soils for successful direct drilling are those which have fairly coarse texture, good structural stability, resistance to compaction and are freely draining.

An encouraging finding from direct drilling studies has been the increased structural stability and organic matter content compared to conventional ploughing (see Table 7.2), particularly in surface soil horizons (Tomlinson, 1974). Improved aggregate stability compared to conventional ploughing results in part from increased organic matter content, but also indicates how disruptive secondary cultivations such as harrowing and rolling can be in pulverising the surface soil. Increased surface organic matter content results from crop residues being left *in situ* after harvesting to form a mulch. This is common soil conservation practice in the Midwest United States to prevent or reduce soil erosion. In the United Kingdom, the burning of straw stubble does not affect dead roots and these subsequently decompose to improve surface soil organic matter content. Soil structure and organic matter in direct drilling/ ploughing studies have been favourably compared to virgin/ploughed soils such as those reported by Abbott, Parker and Sills (1979) (see Table 7.2).

Simple penetrometer (cone) resistance values for mechanical impedance have been used as an index of the suitability of ploughed soil for root penetration. This is a reproducible, although somewhat unrealistic assessment since it is difficult to simulate root exudates that reduce friction and ease the root's passage through the soil porespace. The effect of successive tractor wheel passes on soil compaction has shown that most dramatic change occurs in the first three vehicle passes. In particular, Campbell, Dickson and Ball (1982) have shown that mechanical impedance measured by cone resistance nearly trebles with just three tractor wheel passes (Fig. 7.6). While Soane and Pidgeon (1977) found that direct drilling mainly increased mechanical impedance near the soil surface (see Fig. 7.4), at about 9–15 cm, Cassel (1982) shows that conventional ploughing mainly increases mechanical resistance below the surface soil layer, at about 20 cm. These findings have important implications for root penetration, as indicated by Ellis *et al.* (1977) who found significantly shallower rooting of spring barley in direct drilled plots compared to other forms of tillage, including conventional ploughing.

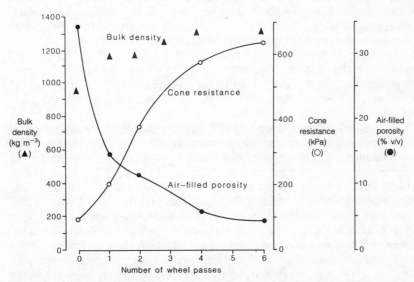

Figure 7.6 Effect of number of tractor wheel passes on soil physical properties at 30 mm depth on a sandy clay loam (Source: modified from Campbell, Dickson and Ball, 1982)

Although they attribute this result to the greater compaction measured in the surface 5–7.5 cm of direct drilled soil, it is difficult to use bulk density or mechanical impedance measures alone to suggest resistance to root penetration. The size and continuity of soil pores are probably more important determinant of root growth. Burnett and Tackett (1968) have shown that deep ploughing of compacted soil can have beneficial effects for rooting. They report significantly increased length and branching of roots in crops of sorghum and cotton grown on deeply ploughed clay soil in Texas. Several authors have indicated that total root biomass in direct drilled soils is often higher than in conventionally ploughed soils (see, for example, Drew and Saker, 1978; Ellis and Barnes, 1980), although no statistically significant increases have been reported. The work of Eck and Davies (1971) has also shown that while deep ploughing and profile mixing sometimes decreased root yields, root activity, as measured by top-root ratios, was invariably enhanced.

7.2.3   Soil moisture and aeration

The alteration of soil pore volume and pore distribution by soil cultivation techniques has an important effect on soil aeration, infiltration and soil moisture retention. While zero-tilled and direct-drilled soils are generally found to have smaller total pore space and lower porosities than ploughed soils (for example, Ellis et al., 1975; Gantzer and Blake, 1978; Ball and O'Sullivan, 1982), they also appear to have increased soil structural and hence pore channel stability. Part of this effect seems to be due to the presence in untilled and zero-tilled soil of a larger population of soil macrofauna (Abbott, Parker and Sills, 1979), particularly earthworms (Ehlers, 1976). This stable and interconnected porespace maintains well-aerated conditions in surface soils which are well suited to direct drilling. Dowdell et al. (1979) found that direct drilling of a clay soil resulted in higher oxygen concentrations at 15 cm than ploughing. During two successive wet winters, mean oxygen concentrations of 10.2 and 7.2 per cent (v/v) were obtained for direct drilled and ploughed plots respectively.

Van Ouwerkerk and Boone (1970) suggest that it is the alteration of pore size distribution during tillage, causing a loss of large pores and a predominance of small pores, that influences moisture retention after ploughing. Several authors have differentiated between (1) the capacity of tilled soils to retain moisture, and (2) their content of plant-available water. Hamblin and Tennant (1981) and Hill, Horton and Cruse (1985) report that more moisture is retained over a large suction range in zero and minimum-tilled soils than is retained in conventionally tilled soils. Although Blevins et al. (1971) and Negi, Raghavan and Taylor (1981) give the same soil moisture retention result, their crop yield and root growth studies imply that zero-tilled soils contain up to twice as much plant-available water compared to conventionally tilled soils. This result suggests a predominance of medium sized pores in zero-tilled soils.

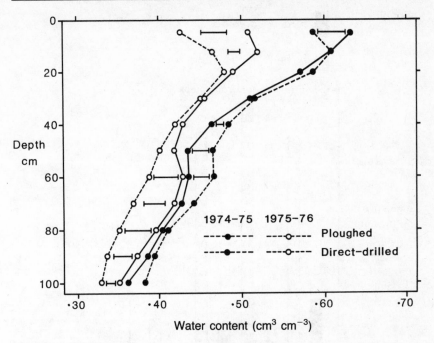

Figure 7.7 Variation in soil water content with depth in ploughed and direct-drilled soil when the greatest water content was observed after the winters of 1974–75 and 1975–76 (Denchworth series). ⊢⊣ Indicates least significant difference between treatments (P = 0.05). (Source: Goss, Howse and Harris, 1978)

Gravimetric field soil moisture results for a comparatively wet growing season (1974–75), and for a particularly dry season (1975–76) in the United Kingdom (Goss, Howse and Harris, 1978) suggest that ploughed soils can retain more soil moisture during wet seasons, while directly drilled soils can retain more profile soil moisture during dry conditions (Fig. 7.7).

Although surface porosities in direct drilled soils are lower than in ploughed soils, the maintenance of stable and interconnected 'bio-channels' in direct drilled soils is an important reason why infiltration rates can remain relatively high (Ehlers, 1976). Despite this, Steichen (1984) reports that infiltration rates for all types of ploughing are higher than for zero tillage. Commenting on unpublished seasonal soil moisture work carried out by Ehlers, Baeumer and Bakermans (1973) note that tillage systems can alter soil moisture conditions to depths in excess of 2 m. In particular, they note that during rewetting, water rapidly infiltrates into the ploughed layer of tilled plots with much slower movement in the subsoil. In untilled soil, infiltration and soil water drainage is more evenly regulated throughout the profile to the depths of earthwork reworking and root penetration.

One of the major influences of infiltration capacity, or the cumulative infiltration before runoff begins, is the nature of the soil surface.

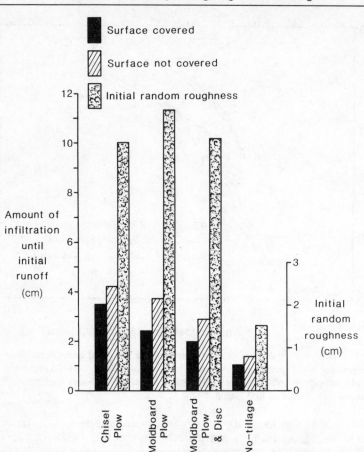

Figure 7.8 Tillage and surface cover influences on initial random roughness and cumulative infiltration before runoff from Putnam silt loam. (Reproduced with permission from Steichen, 1984, *Soil and Tillage Research*, 4, © Elsevier Scientific Publishers B.V., Amsterdam)

Steichen's (1984) results for three levels of ploughing intensity and for zero tillage, confirm that enhanced surface roughness immediately after ploughing and the presence of a surface mulch both increase infiltration capacity. Zero tillage consistently showed lowest infiltration rates (Fig. 7.8), due partly to the fact that the soil was undisturbed. Infiltration rates in ploughed soils usually decline through time due to the development of a surface soil crust by rainfall impact after aggregate disruption during intensive ploughing. This is illustrated in Fig. 7.9 by a greater decline in soil surface random roughness with increasing rainfall energy for types of ploughed soils compared to zero tillage. Direct drilled and zero-tilled soils thus show more consistent infiltration and runoff rates through time. Since it is the infiltration capacity which ultimately determines the magnitude of runoff and the likelihood of soil erosion, we can suggest that soil erosion hazard will be lowest immediately after

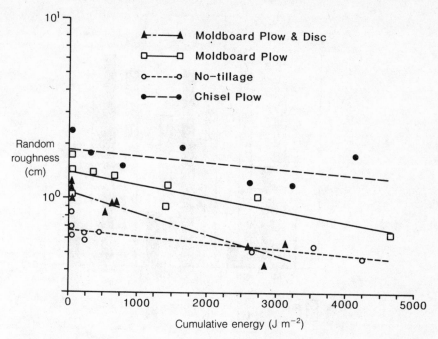

Figure 7.9 Decrease in random roughness of a Putnam silt loam as influenced by tillage treatment and cumulative kinetic energy in simulated rainfall. (Reproduced with permission from Steichen, 1984, *Soil and Tillage Research*, 4, © Elsevier Scientific Publishers B.V., Amsterdam)

ploughing, increasing through time with the development of surface crusting or compaction. Johnson and Moldenhauer (1979) showed that, irrespective of initial soil moisture condition, infiltration rates are higher in chisel ploughed soil and lower in the more intensively mixed mouldboard ploughed soil, resulting in higher runoff rates in the mouldboard ploughed soil (Fig. 7.1). This is probably because chisel ploughing creates cracks and drainage channels for preferential downward flow of infiltrating water. The higher runoff rates from mouldboard ploughed soil result in much higher rates of soil loss from these more intensively cultivated plots.

Recognising the heterogeneity of hydraulic conductivity (k) in ploughed soil, relatively few workers have attempted to measure k directly, preferring instead to use simpler descriptors such as bulk density, particle size and pore size distribution to infer hydraulic properties. Of those who have measured hydraulic properties, both Ehlers (1976; 1977) and Hamblin and Tennant (1981) used soil water diffusivities and soil moisture retention (release) curves to predict unsaturated hydraulic conductivities (k_u) over a range of volumetric moisture contents (θ). The Marshall (1958) equation of the form given in Equation **(7,1)** was used by Ehlers for calculating k values from the soil moisture retention curve.

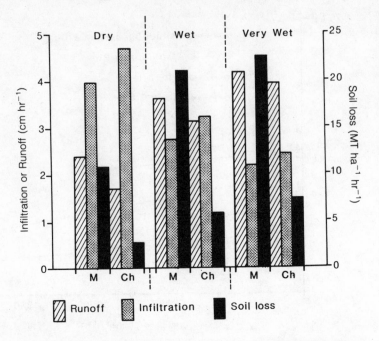

Figure 7.10 Relationship between runoff, infiltration and soil loss after chisel ploughing and mouldboard ploughing. (Source: drawn from the data of Johnson and Moldenhauer, 1977)

$$k_i = \frac{k_s}{k_{sc}} \cdot \frac{30\gamma^2}{\rho g \, \eta} \cdot \frac{\epsilon^p}{n^2} \sum_{j=1}^{m} \left[(2j+1-2i)\frac{1}{(-\psi_j)^2} \right] \quad i = 1,2,3 \ldots n$$

$$(7,1)$$

where: k_i = unsaturated hydraulic conductivity for a specific volumetric moisture content (θ)

i = last water content class on the wet end (i.e. $i = 1$ is the pore class corresponding to a saturated water content, $i = m$ is the pore class corresponding to the lowest moisture content for which conductivity is calculated)

$\dfrac{k_s}{k_{sc}}$ = matching factor (measured saturated conductivity/calculated saturated conductivity)

γ = surface tension of water (dynes cm^{-1})

ρ = density of water (g cm^{-3})

g = gravitational constant (cm sec^{-1})

η = viscosity of water (g cm^{-1} sec^{-1})

ϵ = porosity (cm^{-3} cm^{-3})

p = parameter to account for interaction of pore classes

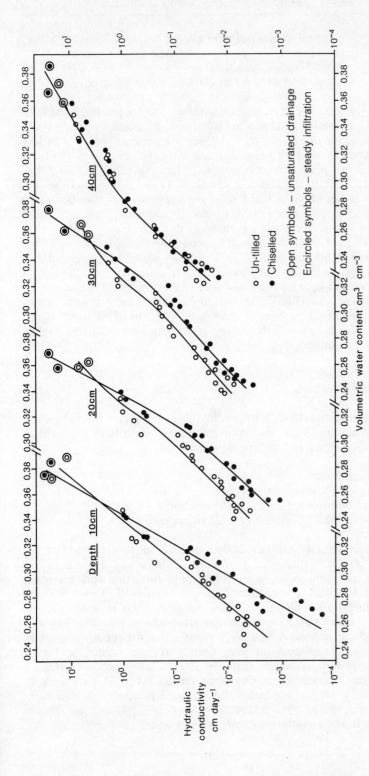

Figure 7.11 Chiselling effects on hydraulic conductivity at 10, 20, 30 and 40 cm depths in Walla Walla soil. (Source: Allmaras et al., 1977)

$$n = \text{total number of pore classes between } \theta_o \text{ and } \theta_s$$

where θ_o = zero volumetric moisture content

θ_s = saturation

$-\psi$ = tension of a given class of water-filled pores

Not surprisingly, Ehlers (1976) reports a more accurate prediction of field measured k_u for untilled soil than for tilled soil. Marshall's model gave a reasonably accurate prediction of k_u values for untilled soil, it was less good in tilled soil, and at high tensions below a depth of 60 cm. For both field measurements and calculated values, Ehlers (1976, 1977) found that at comparable depths, unsaturated conductivity in untilled soil at low tensions was higher than in tilled soil. He suggests that this is due to compaction and loss of macroporosity due to ploughing, since conductivities at high tensions are not significantly different.

Allmaras *et al.* (1977) measured k_u values in a field trial of chisel ploughing to a depth of 40 cm. Their results show lower volumetric moisture contents in chiselled soil at tensions in excess of 1.0 kPa. Chisel ploughing increased k_u at low soil moisture contents (<0.36 cm^3 cm^{-3}) in the top 30 cm of soil (Fig. 7.11). At higher moisture contents, k_u in untilled soil tended to be greater than for tilled soil. This result again reflects a loss of macroporosity during ploughing and relatively higher microporosity in untilled soil.

7.2.4 Soil temperature

In reviewing simulation models to predict tillage effects on diurnal topsoil temperatures, Cruse, Potter and Allmaras (1982) identify four groups of tillage related parameters affecting soil heat flux:

(1) soil surface albedo;
(2) surface microtopographic roughness;
(3) soil surface and subsurface bulk density; and
(4) porespace and hence amounts of soil air and soil water.

Since ploughing initially exposes darker, usually more moist soil at the soil surface, it is not surprising that Hay, Holmes and Hunter (1978) found higher per cent reflectance coefficients for drier and smoother direct drilled soil surfaces compared to ploughed soil. Enhanced soil surface roughness after ploughing (Steichen, 1984) has an effect on the transfer of heat and water vapour in the atmospheric boundary layer at the soil surface. Gausman *et al.* (1977) suggest that an equally important effect of surface roughness on soil thermal regime relates to higher radiation absorption by rough surfaces after tillage, probably due to multiple reflections occurring in the rough surface layer. The influence of soil bulk density on thermal properties is perhaps the most important control and relates to the change in volumetric proportions of soil minerals, soil organic matter, soil air and soil water.

The prediction of one-dimensional heat transfer in soil can be calculated from:

$$\frac{\delta \, (C_s T)}{\delta \, t} = \frac{\delta \left(k_s \dfrac{\delta \, T}{\delta \, z} \right)}{\delta \, z} \tag{7,2}$$

where: T = temperature (°K)
      t = time (s)
      C_s = soil volumetric heat capacity (J m^{-3} K^{-1})
      k_s = soil thermal conductivity (W m^{-1} K^{-1})
      z = depth (m)

Values for C_s and k_s in soil vary with depth and time due particularly to changes in content of soil air and soil water in the porespace. de Vries (1963) formulated an approximate equation for volumetric heat capacity as:

$$C = 1.92_{x_m} + 2.51_{x_o} + 4.18_{x_w} \text{ (MJ m}^{-3}\text{ K}^{-1}) \tag{7,3}$$

where: x_m = volumetric fraction of soil minerals
      x_o = volumetric fraction of soil organic matter
      x_w = volumetric fraction of soil water

Since variations in x_m are slow and the magnitude of x is small, C is found to vary nearly linearly with x_w; or, more simply, soil heat capacity varies directly with soil moisture content. The variations in C for air, water and for several soil types are given in Table 7.4. It is clear that ploughing and drainage operations which cause a drying out of the soil, increasing volumetric air content with a concurrent decrease in soil water content, will reduce the soil heat capacity and result in increased thermal sensitivity. van Duin (1956) showed that soil loosening which occurs

Table 7.4 Thermal diffusivity for two different cultivation treatments at 5–20 cm in a sandy loam (from Hay, Holmes and Hunter, 1978)

	Bulk density (g cm^{-3})	Mean soil moisture content (per cent dryout) (range over 25 days)	Thermal diffusivity (cm^{2}s^{-1} × 10^{-2}) (mean)
Conventional ploughing	1.16	24.8–29.3	0.63 ± 0.02
Direct drilling	1.32	24.1–27.1	1.00 ± 0.04*

* significant difference at $P < 0.001$.

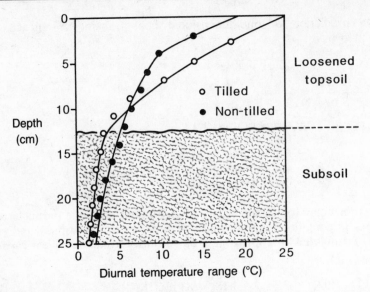

Figure 7.12 Variation in amplitude of the temperature waves with depth below the soil surface for tilled and non-tilled soil. (Source: van Duin, 1956)

during ploughing causes more heat exchange in a tilled soil to take place in the surface soil. Temperatures in tilled surface soil are thus more sensitive to atmospheric temperature fluctuations than are untilled or direct drilled soils. Tilled soils have warmer surface temperatures in summer and daytime than untilled soils, while in winter and at night the reverse is true. Thus there is a higher diurnal temperature range throughout the loosened ploughed soil layer than in the topsoils of untilled or direct drilled plots. Below the cultivation zone, the subsoil diurnal temperature range was slightly higher in untilled soil (Fig. 7.12). The result is slightly different in tilled forestry soils. Ross and Malcolm (1982) found that the diurnal temperature range at the vegetated surface of an intensively mixed peaty podzol was more than twice that at the vegetated surface of unploughed soil.

The rate at which a soil warms up and cools down is partly determined by the soil thermal diffusivity (K_T) which is given as:

$$K_T = \frac{k_t}{C_s} \tag{7,4}$$

where: K_T = soil thermal diffusivity ($m^2 \ s^{-1}$)
k_T = soil thermal conductivity ($Wm^{-1} \ k^{-1}$)

Typical values of K_T are given in Table 7.4. As with k_T, variations in soil moisture content are a major control on K_T (Fig. 7.12). In most soils beyond about 20 per cent (v/v) soil moisture, K_T begins to decline. This happens because k_T values level off (see Fig. 7.13) while values of C

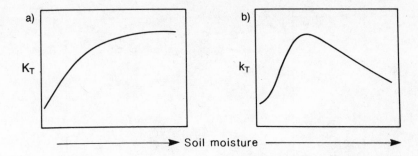

Soil moisture

Figure 7.13 Relationship between soil moisture content and **(a)** soil thermal conductivity, K_T and **(b)** soil thermal diffusivity, k_T for most soils. (Source: Oke, 1978)

continue to increase at higher moisture contents (Oke, 1978). Sandy soils have higher thermal diffusivities than other soils because quartz has a much higher thermal conductivity than clay minerals. Peats have the smallest thermal diffusivities because of the small thermal conductivity of soil organic matter. Over 25 high insolation days in East Scotland, Hay, Holmes and Hunter (1978) found that the thermal diffusivity in a sandy loam at 5–20 cm was significantly higher on a directly drilled site than on a conventionally ploughed site. They attributed this result to the higher bulk density and slightly higher moisture contents of direct drilled soil (Table 7.5). Potter, Cruse and Horton (1985) similarly reported higher thermal diffusivities in zero tilled soil and subsequently reported empirical evidence that soil thermal conductivity at one of their study sites was more than 20 per cent greater in zero-tilled soil than under conventional ploughing.

The two modelling approaches currently used for soil temperature simulation – (1) process orientated models, requiring precise and detailed initial and boundary condition inputs; and (2) semi-process or non-process orientated models which require weather station data and soil data for one depth (Cruse, Potter and Allmaras, 1982) – can both be modified for soil tillage applications. Process-related models can be subdivided into two approaches. First, Wierenga, Nielson and Hagon (1969) used a finite difference equation to approximate the one-dimensional heat flow equation:

$$\frac{[T_j^{n+1} - T_j^n]}{\Delta t} = \frac{[K_s (T_{j+1}^n - 2T_j^n + T_{j-1}^n)]}{\Delta Z^2} \tag{7,5}$$

where: j = depth interval (cm)
n = time interval (s)
K_s = thermal diffusivity ($m^2\ s^{-1}$)
T = temperature (°C)
Z = depth (cm)

Table 7.5 Thermal properties of soils and soil components (from van Wijk and de Vries, 1963)

	Density (ρ) $(kg\,m^{-3} \times 10^3)$	Specific heat (c) $(J\,kg^{-1}\,K^{-1} \times 10^3)$	Heat capacity (C) $(J\,m^{-3}\,K^{-1} \times 10^6)$	Thermal conductivity (k) $(W\,m^{-1}\,K^{-1})$	Thermal diffusivity (K) $(m^2\,s^{-1} \times 10^{-6})$
(1) Soil components					
Quartz	2.66	0.80	2.13	8.80	4.18
Clay minerals	2.65	0.90	2.39	2.92	1.22
Organic matter	1.30	1.92	2.50	0.25	1.00
Water (4°C still)	1.0	4.18	4.18	0.57	0.14
Air (10°C still)	0.0012	1.01	0.0012	0.025	20.50
(2) Soils					
Sandy soil (dry)	1.60	0.80	1.28	0.30	0.24
(40% PS)* (saturated)	2.00	1.48	2.96	2.20	0.74
Clay soil (dry)	1.60	0.89	1.42	0.25	0.18
(40% PS) (saturated)	2.00	1.55	3.10	1.58	0.51
Peat soil (dry)	0.30	1.92	0.58	0.06	0.10
(80% PS) (saturated)	1.10	3.65	4.02	0.50	0.12

* PS = pore space

Wierenga and de Wit (1970) subsequently developed a simulation based on the de Vries (1963) Equation (7,3). Neither of these methods have been adapted for predicting tillage-induced surface temperatures. Since tillage alters soil thermal conductivity and heat capacity, both important elements of the de Vries model, this approach would appear to offer potential for further development. Only one working simulation of soil temperatures after tillage has so far been published. Cruse *et al.* (1980) developed a semi-process orientated model based on soil surface radiation absorption and soil thermal inertia. The influence on soil temperatures at 5 cm depth of moisture content, soil surface roughness and decay of surface residues during conservation practices were successfully predicted within + 2°C for zero tillage and mouldboard ploughing.

7.2.5   Soil trafficability, workability and the timeliness of cultivation operations

Trafficability and workability are terms used in the study of soil deformation due to stresses caused by loading with tractors and agricultural implements. Apart from the obvious influence of such traffic on soil compaction, efficiency in ploughing is significantly reduced due to loss of drawbar pull when the tractor is operating in loose soil where considerable sinking of tyres occurs. Soil compaction caused by high wheel loading is not the only problem of agricultural trafficking. In order to generate draught, all tractors develop some wheelslip which results in soil shearing. The greater the draught, the higher will be the wheelslip for a given axle weight (Davies, Finney and Richardson, 1973). A compromise must be achieved between tractor weight and wheelslip to optimise work output.

The passage of a tractor or implement causes soil to deform until the shear force developed can withstand the applied stress. This leads to two types of failure: brittle failure and flow (compressive) failure. During ploughing, brittle failure, such as the shearing of existing structural units into smaller aggregates, is desirable since it leads to reduced soil bulk densities. Flow failure, on the other hand, caused by the plastic or granular flow of the bulk soil mass, is undesirable, since it generally leads to compaction and increased soil bulk density. Spoor (1979) defines the resistance to movement between clods and peds as the bulk shear strength and that within the clods and peds as the clod shear strength. When clod strength is greater than bulk strength and the soil deforms by brittle failure, soil structural damage during ploughing is minimal. When bulk strength is greater than clod strength, peds fragment and are crushed. All soil structure can be lost in soils of low clod strength if they are worked at high moisture contents. The optimum moisture content for ploughing operations is at the soil's lower plastic limit where bulk strength is at a maximum (Fig. 7.14). Bulk soil strength decreases with increasing moisture content to a minimum at the liquid (upper plastic)

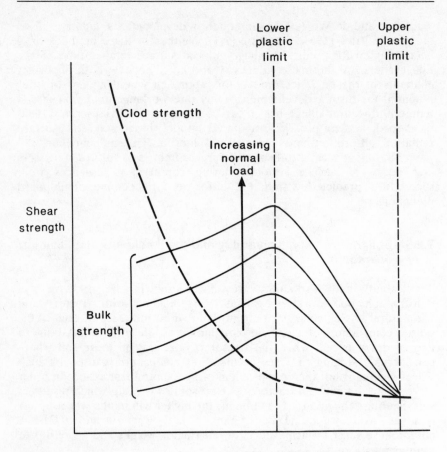

Figure 7.14 Diagrammatic relationship between shear strength and soil moisture content, indicating the influence of increased load on soil bulk strength. (Source: Spoor, 1979)

limit. Spoor (1979) illustrates how additional tractor load (ballast) can be used to artificially increase soil bulk strength at the plastic limit (Fig. 7.14). While brittle failure may be necessary for seedbed preparation, Spoor and Godwin (1979) point out that in mole drainage, the production of a stable channel by causing compaction of the soil along the walls, requires localised compressive failure.

When considering solutions to agricultural traffic compaction problems, it is perhaps useful to separate the effects of (1) tractor tyres, and (2) plough implements. Comparing the effect of both ploughing and rotavating a silt loam during wet soil conditions, Gooderham (1976) found increases in mechanical resistance for both treatments to be confined to the upper 30 cm of soil. In addition, wet ploughing was found to increase bulk density and mechanical impedence by 4 and 10 per cent respectively and to reduce air porosity by 18 per cent. Cultivating under wet soil conditions is clearly a major factor in causing compaction by the

Table 7.6   Maximum number of machinery work days for nine English counties (mid February to April) inferred from moisture deficit estimates (after C. V. Smith, 1977)

County	Maximum number of work days (2 years in 10) Soil types			Maximum number of work days (5 years in 10) Soil types		
	heavy	medium	light	heavy	medium	light
Avon	30	40	45	40	50	55
Cambridgeshire	25	45	55	50	55	60
Cheshire	15	45	55	35	50	60
Devon	25	40	45	35	45	50
Essex	35	45	50	45	55	60
Lancashire	15	40	55	30	50	55
Lincolnshire	10	20	45	30	45	55
Oxfordshire	30	45	55	40	55	60
Suffolk	25	45	55	45	55	60

loss of soil structure and in the production of a plough pan. Solutions to this workability problem are confined to choosing suitable climatic conditions in relation to the soil permeability and plasticity characteristics. While it is fairly simple to predict return to field capacity period and specified soil moisture deficit conditions from a knowledge of soil permeability and precipitation/evaporation balances (C. V. Smith, 1977), many soils remain unworkable at high moisture deficits, particularly if the topsoil has been recently wetted by rain. The *potential* number of machinery work days for a selection of English soil types is given in Table 7.6. While these generalised predictions are useful for seasonal management planning, further development of this simple approach is required to improve short-term accuracy.

Solutions to the problem of compaction by tractor tyres have focused on spreading the vehicle weight. Fig. 7.15 illustrates how the contact length of the pressure surface affects the magnitude and depth of soil compaction. A 1.55 m contact length for crawler track compared to a 0.6 m contact length for the tyre dramatically reduces the ground surface pressure and slightly reduces the depth penetration (Reaves and Cooper, 1960). Apart from using crawler tracks, an alternative way to spread the load and increase the ground surface contact area is to use cage wheels.

Shear forces from wheelslip are just as important as surface applied pressure in causing soil compaction. During wheelslip, a parallel orientation of particles occurs in the direction of the shear forces. This acts to increase cohesion and to retard water movement. Davies, Finney and Richardson (1973) found that crawler tracks caused less smearing than flexible tyres and generally concluded from their studies that

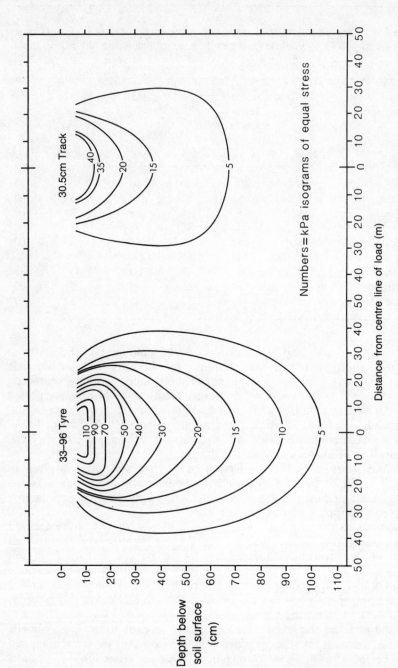

Figure 7.15 Isograms of mean normal stress under a tyre and a track. (Source: Reaves and Cooper, 1960)

excessive tractor wheelslip was more likely to cause soil damage than heavy wheel loading. In moist soil conditions, the use of ballast weighting, greater tractor speed or the use of ploughing implements with lower draught in combination with traction aids, such as tracks or double wheels, is recommended to reduce wheelslip.

7.3 ALTERATION OF SOIL CHEMICAL PROPERTIES AND PROCESSES

Although several authors recognised decades ago that agricultural soils were losing fertility under continuous intensive cultivation practices (see, for example, Jenny, 1941; Giddens, 1957), they did not differentiate between tillage effects and cropping effects. One early conclusion of their studies was that ploughing increased the rate of oxidation of native soil organic matter but no explanation for this finding was offered. A more recent survey of tilled and untilled Canadian soils has shown significant increases in organic carbon content and soil acidity over 35 years of tillage (Coote and Ramsey, 1983). This effect is almost certainly due to stubble incorporation after cereal cropping. This suggestion is corroborated by Bauer and Black (1981) in a comparison of conventional tillage and stubble mulch tillage effects over 25 years in the North Dakota region of the Great Plains. A primary determinant of soil C and N contents under stubble tillage was assessed to be the improved soil erosion control afforded by this technique (Bauer and Black, 1981).

During conventional ploughing, soil structural disruption and the altered diurnal temperature regime are considered to be important factors in increased organic matter decomposition and nutrient mineralisation. There also appears to be evidence for organic matter accumulation under direct drilling practices. These findings are assessed in the following sections.

7.3.1 Effects of soil structure disruption

During the disaggregation of soil structure which follows ploughing (see Section 7.1.1), Powers and Skidmore (1984) suggest that insoluble, probably organic bonds and bridges between individual particles and aggregates are broken by mechanical disruption. Since these bonds are responsible for aggregate stability, they found that all compressed and disturbed soils in their experimental cultivations showed poorer wet and dry aggregate stabilities than did surface soils. Rovira and Graecen (1957) were the first to show that biological degradation of organic matter in soil aggregates is just as important as physical disruption in reducing structural stability. After laboratory simulated tillage, increased oxygen uptake was attributed to the exposure to decomposer micro-organisms of organic matter which was previously inaccessible in ped centres. Greater intensities of simulated tillage, and hence disruption, increased the availability of organic matter. The resultant higher rates of oxygen uptake

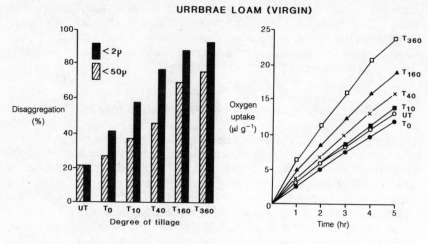

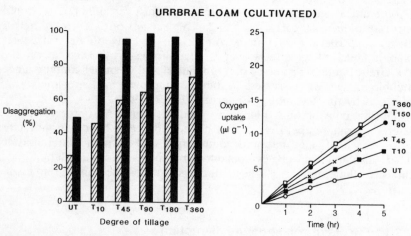

Figure 7.16 Disaggregation and microbial activity in relation to the degree of tillage; UT, untilled. Intensive tillage was simulated using a shear ring. The tillage treatment was expressed as T_Θ, where Θ was the angle of turn in degrees of the shear ring on the soil. (Reproduced with permission from Rovira and Graecen, 1957, *Australian Journal of Agricultural Research*, 8, © CSIRO Editorial and Publishing Unit)

were due to the increased activity of decomposer micro-organisms. Although this effect was shown in both virgin and previously cultivated soil, the magnitude of microbial activity following simulated tillage was higher in previously undisturbed soils (Fig. 7.16).

Tiessen and Stewart (1983) examined the carbon and nutrient content of different sized mineral and organic particles present in prairie soils which had been cultivated for many years. For virtually all samples studied, they report dramatic declines in C, N and organic P contents. Only for inorganic P was accumulation recorded in the B horizon of one of their prairie soil study sites after 65 years of cultivation (Table 7.7).

Table 7.7 Contents of C, N and P in particle size fractions of Bradwell native prairie soil and changes after cultivation (after Tiessen and Stewart, 1983)

Soil	Native nutrients				Changes after 65 years of cultivation			
	C (mg g^{-1})	N (mg g^{-1})	P_o* (µg g^{-1})	P_i (µg g^{-1})	%C	%N	%P	%P_i†
A Horizon								
Whole soil	35.0	3.25	410	374	−51	−46	−40	−22
Organic matter + sand	22.7	1.58	138	192	−67	−54	−81	n.s.
Coarse silt	34.5	3.28	367	413	−41	−47	−53	−5
Fine silt	71.4	6.61	1,006	498	−21	−19	−22	−26
Coarse clay	69.6	7.09	1,336	589	−31	−24	−23	−28
Fine clay	54.3	6.36	979	672	−45	−37	−42	−31
B Horizon								
Whole soil	13.0	1.26	238	323	−5	−6	−18	+16
Organic matter + sand	3.6	0.19	32	196	−56	−26	−40	+23
Coarse silt	17.4	1.53	240	408	−37	−35	−44	+20
Fine silt	33.5	3.13	683	397	−19	−9	−34	+10
Coarse clay	29.0	3.02	764	427	−16	−10	−27	n.s.
Fine clay	22.2	2.57	442	443	−27	−19	−31	−12

* organic phosphorus
† inorganic phosphorus
n.s. non-significant difference
Differences significant if >5% for C, N and P_o
>8% for P_i

This finding can be explained as the immobilisation of the mineralised P after cultivation, probably in non-labile calcium phosphates. Perhaps the most interesting and significant result of Tiessen and Stewart's (1983) study was the relative abundance in 60 year tilled soil of biologically resistant and well-humified organic matter in combination with fine silt (2–5 μm) and coarse clay (0.2–2 μm) mineral fractions. This result confirms the hypothesis that in old ploughed soils the remaining organic matter has a reduced ability to supply nutrients by mineralisation. Examination of the *rates* of organic matter losses led Tiessen and Stewart to qualify their original thesis and conclude that losses of organic matter from old ploughed prairie soils do not level off, but continue after 60 and 90 years at very much slower rates.

7.3.2 Effects of soil mixing

Several authors have suggested that in soil profiles containing distinctive organic and mineral horizons, the mixing of these materials during ploughing enhances organic matter decomposition and nutrient mineralisation (for example, Salonius, 1983; Ross and Malcolm, 1988). This idea is particularly important in forestry site preparation and in the reclamation of heathland soils where a prime aim of cultivation is to mobilise the native soil nutrients prior to planting (Ross, 1984). A major paradox in commercial silviculture is that while an organic horizon develops under early to mid-rotation conifers, the trees often suffer nutrient, particularly nitrogen, deficiency. Miller (1969) and Lamb (1975) suggest that since the development of a thick organic horizon under plantation conifers removes nutrients from active cycling, forest productivity may be closely related to the rate at which these accumulations can be mineralised after site preparation and throughout the rotation. One study has looked at this problem in detail for a site in South-east Scotland. In a peaty podzol to which different intensities of cultivation had been applied in an afforestation trial, Ross and Malcolm (1988) found that increased cultivation intensity resulted in smaller sized pieces of the original surface organic matter, mixed into the main mineral soil matrix. The main effect of increased mixing intensity was to increase the surface area of peat exposed to the soil decomposer population. By modelling this effect in the laboratory, they found a direct relationship between increased surface area of peat and the rates of carbon and nutrient mineralisation (Fig. 7.17). It is possible from these studies to calculate crudely

(1) the period required after cultivation before all native soil organic matter is mineralised; and
(2) the ability of nutrient mineralisation from the decomposition of native soil organic matter to supply the nutrient uptake requirements of the planted trees (Table 7.8).

Table 7.8 (a) Years till all organic matter reserves of elements are mobilised with different levels of peat-mineral soil mixing intensity (from Ross, 1979). (b) Mobilised nutrients in different intensities of mixed peat-mineral soil expressed as a percentage of conifer nutrient uptake requirements (after Ross, 1984)

(a) *Mixing treatment*	*Years till exhaustion*[*][1]				
	N^2	*P*	*K*	*Ca*	*Mg*
Least intensive mixture	550	59	12	30	50
Most intensive mixture	100	50	6.5	25	22
(b) *Mixing treatment*	*%N*	*%P*	*%K*	*%Ca*	*%Mg*
Least intensive mixture	8–10	50–all	60–all	20–45	30–60
Most intensive mixture	18–40	65–all	excess	23–50	60–all

* the calculation assumes:
[1] mean 10°C temperature for field conditions
[2] that only half of the soil N is ever mobilised

Clearly the choice of ploughing techniques for forestry and heathland applications must attempt to optimise nutrient provision to offset the need for costly fertilising, while minimising mineralisation in excess of crop requirements.

7.3.3 Effects of altered temperature regime

The larger temperature range characteristic of cultivated topsoils coupled with mixing of L (litter), F (fermentation), H (humic) and mineral horizons during ploughing act as stimuli to enhance soil microbial respiration and organic matter decomposition. Salonius (1983), examining these two effects separately and in combination, concluded that there were different reasons for the respiration stimulus at different temperatures. At temperatures <10°C, the main stimulus to decomposition is the proximity of soil organisms to new organic substrates. At temperatures around 10°C, the intermixing of L, F and H layers alone allows a more diverse population of organisms to attack the more diverse organic food substrate. At temperatures in excess of 20°C, decomposition and respiration rates are thermally driven. These experiments at constant, elevated temperatures indicate that microbial respiration and organic matter decomposition are stimulated by soil mixing only at temperatures ⩽10°C (Salonius, 1983). Ross (1985) has shown that diurnal field temperatures consistently produced lower soil nutrient mineralisation rates than constant incubation temperatures of 20°C. The suggestion that elevated temperature ranges in cultivated soil enhance decomposition and nutrient mineralisation rates thus requires some qualification. In particular, it is important to examine (1) the magnitude of temperature

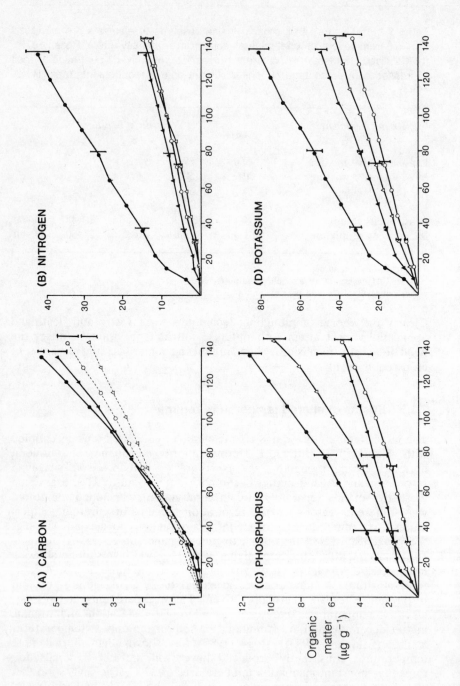

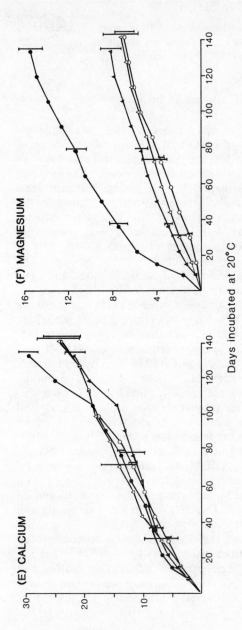

Figure 7.17 (A-F) Accumulated mobilised C, NH_4^+–N, PO_4^{3-}, K^+, Ca^{2+}, Mg^{2+} for 140 day incubation period in $\mu g\,g^{-1}$ organic matter. Standard error bars are indicated. ● Most intensive mixture (largest organic matter surface area); ▲ second most intensive mixture (second largest organic matter surface area); ○ second lowest intensity of mixing (second smallest organic matter surface area); △ least intensive mixture (smaller organic matter surface area). (Source: modified from Ross and Malcolm, 1988)

ranges in soil after different intensities of ploughing; and (2) determine whether these differences are large enough to result in significantly different rates of decomposition and nutrient mineralisation.

7.3.4   Effects of zero tillage and cultivation intensity

Just as deteriorating soil structure has influenced the move towards reduced cultivations to conserve soil physical conditions, so losses of organic matter and declining soil fertility have promoted the study of reduced cultivation effects on soil chemical processes. Keeney and Bremner (1964) showed comprehensively that as well as reducing total soil N content, conventional ploughing caused a marked change in the availability of different soil N fractions. They found a significant increase in non-exchangeable ammonium in cultivated soils and concurrent significant decreases in all extractable forms of nitrogen. The largest decline occurred in amino acid nitrogen which is likely to be one of the more easily decomposed forms of organic N in well-cultivated soil. The magnitude of these changes appears to be related to the intensity of soil mixing afforded by the tillage technique. Lynch (1984) for example suggests that aeration and soil mixing by conventional ploughing promotes aerobic biochemical processes, including nitrogen mineralisation and nitrification. This results in increased levels of NH_4^+-N and NO_3^--N in conventionally ploughed soils compared to directly drilled soils (Arnott and Clements, 1966; Dowdell and Cannell, 1975). The uptake and leaching of mobilised forms of nutrients in ploughed soils undoubtedly leads in the longterm to their relatively poor organic matter and nutrient status. This suggestion is confirmed in longterm studies where direct drilling practices have led to organic matter accumulation (Powlson and Jenkinson, 1980) and increases in organic C and N (Dick, 1983).

Very little field evidence is yet available to confirm the experimental finding that increased mixing intensity results in increased carbon and nutrient mineralisation. In one study, reporting the results of 18 and 19 year continuous tillage experiments, Dick (1983) found a declining soil organic C, N and P sequence in the order: zero tillage > minimum tillage > conventional tillage. While these results appear to confirm the hypothesis, clearly caution is required when applying such generalisation to a wide variety of soil types. The problem of organic matter and nutrient losses after ploughing appears to be most severe on coarse textured soils, particularly those cultivated continuously using conventional ploughing practices.

7.4   SUMMARY

Several reports in the 1970s indicated that highly intensive ploughing and harrowing methods severely disrupt soil structure. The most important results of aggregate disruption is soil compaction, with adverse effects on infiltration, drainage and aeration. Increased soil bulk density as a result of compaction causes mechanical impedance for plant roots, particularly young seedlings. In addition to intensive ploughing, the passage of agricultural traffic at each stage of crop growth and development (seeding, fertilising, application of pesticides, harvesting), causes compaction and soil damage. Frequently traffic damage can be minimised by (1) waiting until soil dries out sufficiently to increase load bearing capacity; and (2) spreading the load of the machinery over a larger tyre/soil contact surface area. The use of direct drilling techniques which minimise soil disturbance improves soil infiltration, aeration, organic matter content and soil organism numbers. Increased numbers of earthworms in directly drilled soil produce 'biopores' which aid infiltration, drainage and aeration. Improved aeration and temperatures in tilled soils, combined with disaggregation and mixing of mineral subsoils and organic surface soils increases the mineralisation rate of soil nutrients. This can be important in nutrient-deficient forestry soils, where regular fertilising is required for establishment and maintenance of longterm tree crops.

CHAPTER 8

Fate of Fertilisers Applied to Soil

8.1 INTRODUCTION

Two main issues are currently attracting attention in fertiliser research:

(1) fertiliser efficiency and percentage recovery of nutrients in harvested crops; and
(2) problems of fertiliser leaching to drainage waters to cause groundwater and watercourse pollution.

Although seven macronutrients and at least six micronutrients were listed in Chapter 5 as being important for plant growth, only three major nutrients – nitrogen, phosphorus, and to a lesser extent, potassium – are widely applied to soils as inorganic fertilisers. While the use of nitrogen fertilisers in England and Wales increased substantially during the period 1950–80, particularly for permanent and temporary grassland (23 and 42 per cent respectively), the use of phosphorus and potassium fertilisers for a range of crops including grassland has, on average, only doubled (Cooke, 1980). This heavy reliance on high inputs of nitrogen for the maintenance of crop productivity rates in western agriculture has directed most attention towards the dynamics of nitrogen fertilisers, nitrogen transformations and the transport of the inorganic nitrogen species: NH_4^+ and NO_3^-. Increasing interest in highly weathered tropical soils exhibiting variable charge properties is currently focusing interest on phosphorus budgeting and the use of phosphate fertilisers in both tropical and temperate agriculture. As far as potassium fertilising is concerned, current research on the thermodynamics of potassium exchange on soil colloid exchange complexes is being used to examine the efficiency of K fertiliser retention and plant-availability in the rhizosphere.

The efficiency and budgeting (recovery versus loss) of inorganic fertilisers applied to soils are different for N, P and K. Reasons for these differences lie in the form of fertiliser applied, ionic transformations in relation to soil environmental conditions and differences in processes of ionic adsorption/exchange, transport and loss from the soil profile. Some of the processes controlling the transformation and retention of applied N, P and K in soil will be examined in the following sections to explain differential nutrient losses and efficiencies in relation to soil environmental conditions.

8.2   FERTILISER FORMS AND USE

The form in which fertiliser is added to soil determines how long it may take for a plant to respond to additional nutrition. Most fertilisers are complex inorganic minerals or organic compounds which must be broken down into soluble forms before a plant can absorb them. These types of fertilisers must be added well in advance of the plant's actual needs. Insoluble and slow-release compounds are valuable additives to wet and waterlogged soils, conditions that would wash away soluble fertilisers. A second important factor in determining how quickly a plant can benefit from a fertiliser application is the degree of fragmentation of the compound. Finely powdered materials become readily available while gravel-sized chips of pelleted materials may take some time to become available for uptake. Commercial inorganic fertilisers are granulated to increase their lifespan in the soil.

The timing and method of fertiliser application may also be quite critical in determining the magnitude of nutrient losses from soil. Both P and K can be strongly adsorbed onto soil colloids (see Sections 8.5.2 and 8.6), so the timing of application of these nutrients is not as critical as for very soluble fertilisers such as nitrates which are easily leached. Nitrate fertilisers must be applied just before they are required, in quantities small enough to be immediately used by the crop. Fertilisers can be broadcast on the soil surface or can be mechanically placed in a drill at a depth adjacent to the roots of the crop. Fertiliser placement techniques are designed to minimise losses due to surface erosion or runoff and losses to the atmosphere by volatilisation.

Some of the most commonly used forms of N, P and K fertilisers are listed in Table 8.1, together with their percentage element compositions. The majority of compounds contain much less than 50 per cent of the nutrient element in question. This has led to the use of ammonia and nitrogen liquids which contain up to 80 per cent N. These materials are expensive and suffer from problems of gaseous loss to the atmosphere. They tend to be used for large-scale fertilising operations only and must be placed at depth in the soil to maintain efficiency. A second direction in nitrogen fertiliser technology is the development of slow-release materials for use in wet and flooded soils such as rice paddies. These compounds are designed to overcome the problems of excess solubility while still supplying N in the form of NO_3^-. Two approaches have been used. First, organic materials such as formaldehyde urea and isobutylidene diurea (IBDU) must be decomposed to release simple inorganic forms of N for plant uptake. They contain 38 and 28 per cent total N respectively.

Second, fertiliser granules or particles should be coated with semipermeable membranes, waxes, resins or elemental sulphur to form a layer on the fertiliser surface which must be negotiated before nutrients can be released by solution or microbial attack. An alternative method for slowing down the release of inorganic N from nitrogen fertilisers is to add materials that inhibit nitrification. Nitropyrin, or N-Serve (2-chloro-6-

Table 8.1 Composition of some commonly used fertilisers. Data tabulated from Cooke (1980)

(a) *Nitrogen*				Per cent N composition
solids	Ammonium sulphate	$(NH_4)_2SO_4$		21
	Potassium nitrate	$K\,NO_3$		13.8
	Ammonium chloride	$NH_4\,Cl$		26
	Ammonium nitrate	$NH_4\,NO_3$		35
	(Nitro Chalk, if mixed with $CaCO_3$)			
	Urea (Carbamide)	$CO(NH_2)_2$		46
	Calcium cyanamide	$Ca\,CN_2$		21–22
	Gold-N (sulphur-coated, slow-release fertiliser)			32
liquids	Ammonia gas liquors (dilute NH_3 solutions)			1–4
	Aqueous ammonia gas	(NH_3)		21–29
	Anhydrous ammonia gas	(NH_3)		82

(b) *Phosphorus*			Per cent P composition
	Single superphosphate	$Ca(H_2PO_4)_2$; $Ca\,SO_4$	8–9.5
	Triple superphosphate	$Ca(H_2PO_4)_2$	20
	Monoammonium phosphate	$(NH_4)H_2PO_4$	20–30
	Diammonium phosphate	$(NH_4)_2\,HPO_4$	21
	Basic slag		5–10
	Ground mineral phosphate (GMP)/crushed rock phosphate (CRP)		12.5

(c) *Potassium*				Per cent K composition
	Muriate of potash	(potassium chloride)	KCl	50
	Sulphate of potash	(potassium sulphate)	K_2SO_4	40–42
	Saltpetre	(potassium nitrate)	KNO_3	36

trichloromethylpyridine) added to ammonium fertilisers will inhibit the activity of the bacterium *Nitrosomonas* for a period of four to eight weeks if maintained to a concentration of 20 mg g^{-1} in the soil (Fink, 1982).

The nitrogen, phosphorus and potassium (NPK) content of combined fertilisers is expressed as the ratio of [total N : phosphoric acid (P_2O_5) : potash (K_2O)]. An example is the Norsk Hydro combined fertiliser '52 Regular' which has the ratio 20 : 10 : 10 and is recommended mainly for less intensive grasslands (hay and silage) and for spring cereals. Since the P, K, Ca and Mg contents of commercial fertilisers are expressed as oxides, their total nutrient contents are calculated using the relationships in Table 8.2.

Table 8.2   Fertiliser conversion factors

$P_2O_5 \rightarrow P$	multiply by 0.436
$K_2O \rightarrow K$	multiply by 0.830
$CaO \rightarrow Ca$	multiply by 0.715
$MgO \rightarrow Mg$	multiply by 0.603

8.3   FERTILISER EFFICIENCY

There are five possible 'fates' of nutrients applied to soil as fertilisers:

(1) *Uptake* by plants and animals (immobilisation); this represents a loss from the soil when crops are harvested.
(2) *Fixation* (adsorption/exchange) in the soil.
(3) Leaching and *loss in soluble form* to drainage waters.
(4) Volatilisation and *loss in gaseous form* to the atmosphere.
(5) Surface *loss in solid form* by runoff and erosion.

Assessment of the importance of each of these for N, P and K fertilisers will be made in the following sections.

The overall efficiency of a fertiliser application is usually calculated as the percentage recovery in the crop. Assessed in this way, 'efficiency' is obviously dependent on the form and amount of fertiliser applied, the deficiency status of the soils, the growth stage of the plant and roots in relation to nutrient uptake requirements and the time period over which 'recovery' is being assessed. Commonly, recovery rates are calculated over one year on deficient soils and are crop specific. Despite the problems listed above, results from such studies do reveal important trends.

Although the vast majority of early N-recovery studies using different fertiliser sources and different crops reported recovery rates ranging from 50 to 100 per cent (reviewed by Allison, 1966), it has been suggested more recently that for mineral fertiliser sources, maximum field recovery rates of 50–60 per cent in the first year are probably more realistic (Fink, 1982). A whole series of nitrate-N fertiliser studies carried out at the Letcombe Laboratory in southern England indicate crop N recoveries of 39–63 per cent for grassland, spring barley and winter wheat in lysimeter studies and recoveries of 60–72 per cent for winter wheat in field trials (Table 8.3). These fairly high uptake rates can be achieved because NO_3^--N is carried to plant roots in solution by mass flow. Where nitrogen is applied to crops in the form of organic manures, Fink (1982) suggests that only 20–30 per cent of applied N can be recovered by crops in the first year. These lower recoveries reflect the delay in availability to plants of simple forms of inorganic N by mineralisation during organic matter decomposition.

Table 8.3 Topsoil NO_3-N and N content of spring barley crop after application of four different N fertilisers to a flinty clay loam (from Addiscott and Cox, 1976)

| Fertiliser | Application rate | Soil NO_3-N(mg ha^{-1}) | | Nitrogen in barley (kg N ha^{-1}) | Apparent nitrogen recovery (kg N ha^{-1}) |
		0–26 cm	26–52 cm		
Ca(NO$_3$)$_2$	100 kg N ha^{-1}	5.1	7.3	101	28
(NH$_4$)$_2$SO$_4$	100 kg N ha^{-1}	5.8	7.7	108	35
Urea	100 kg N ha^{-1}	5.6	7.7	93	20
Sulphur-coated urea	100 kg N ha^{-1}	7.3	7.1	95	22
Control	–	5.8	6.1	73	–

Much lower amounts of fertiliser P (5–15 per cent) are commonly recovered by annual crops in the short term (Russell, 1973; Fink, 1982). Recovery can be increased to 50 per cent if fertiliser is placed at a depth in soil adjacent to seedling roots (Cooke, 1966). Applied phosphates are quickly transformed to sparingly water-soluble forms in soil, and low plant uptake rates reflect the slow diffusion of the remaining soluble P over short distances close to root surfaces (Nye and Tinker, 1977). Long-term field experiments indicate that larger amounts of applied fertiliser P (as much as 50–100 per cent, over 10–40 years) can eventually be recovered by crops (Widdowson and Penny, 1973; Fink, 1982). Unlike nitrogen, Fink (1982) suggests that a larger amount of P (up to 30 per cent) may be recovered by annual crops from organic P manures in the short term than from inorganic P fertilisers. It is now clear from combined fertiliser studies that nitrogen addition increases P recovery rates from P fertilisers. In a study of spring wheat, Halvorson and Black (1985) found that fertiliser P recovery in the grain for 22, 45, 90 and 180 kg ha^{-1} P treatments averaged 32, 25, 23 and 13 per cent respectively without N fertiliser and 45, 38, 37 and 24 per cent respectively with 45 kg N ha^{-1} as NH$_4$NO$_3$.

Fairly high K recovery rates of 75–90 per cent are quoted for grassland and for some cereal crops (Widdowson and Penny, 1973; Fink, 1982). Even for most arable crops, such as potatoes, kale and some cereals, the uptake efficiency is usually around 50 per cent (Russell, 1973). Recovery of applied K is mainly determined by the ability of the soil to retain potassium in a readily available form, usually through cation exchange, within the root zone of the plant. Uptake rates are then determined by rates of cation exchange at colloid surfaces and rates of diffusion to the root surface (see Section 8.6).

8.4   FATE OF NITROGEN FERTILISERS APPLIED TO SOIL

All five of the 'fates' associated with fertiliser economy in Section 8.3 are applicable to N fertiliser budgeting. Since the early reviews of Allison (1955 and 1966) in which levels of 'unaccounted-for-N' in N fertiliser budgets were as high as 50 per cent, many attempts have been made to understand the nature of these 'invisible' losses. An understanding of the transformations of organic and inorganic forms of native soil nitrogen can be used to identify and explain the processes associated with fertiliser N transformations and transport in soil. Figure 8.1 illustrates the seven processes associated with the transformation and cycling of nitrogen in soil. Of these, chemical fixation, the transformation processes (ammonification, nitrification and denitrification), and the transport processes (NO_3 leaching (not illustrated) and volatilisation) are of main concern in fertiliser budgeting (Table 8.4).

Table 8.4 Processes of the nitrogen cycle, relevant to nitrogen fertiliser budgeting

Type of process	Process description
Storage	*Chemical fixation* of NH_4^+ by cation exchange on negatively charged soil colloids
Transformation	(1) *Ammonification.* Mineralisation of organic N manures by a wide range of decomposer organisms
	(2) *Nitrification.* Bacterial transformation of NH_4-N firstly to NO_2-N by *Nitrosomonas* then to NO_3-N by *Nitrobacter*
	(3) *Denitrification.* Reduction of NO_3 to NO_2 followed by further reduction to the gases N_2O, NO, N_2 by *Thiobacillus denitrificans* and other bacteria
Transport	(1) NO_3 *leaching.* Nitrate is particularly soluble and subject to mass flow and drainage out of soil profiles during storm events
	(2) *Volatilisation* (gaseous loss to the atmosphere)
	(a) N_2O, NO, N_2 after denitrification has taken place
	(b) NH_3 gaseous losses from NH_4 fertilisers applied to calcareous soils and through urea hydrolysis

8.4.1   Fixation of ammonium nitrogen

Ammonium ions are subject to cation exchange onto negatively charged soil colloids (see Chapter 5). Only in soils with very low cation exchange capacities is NH_4^+-N leached in significant quantities. MacKowen and

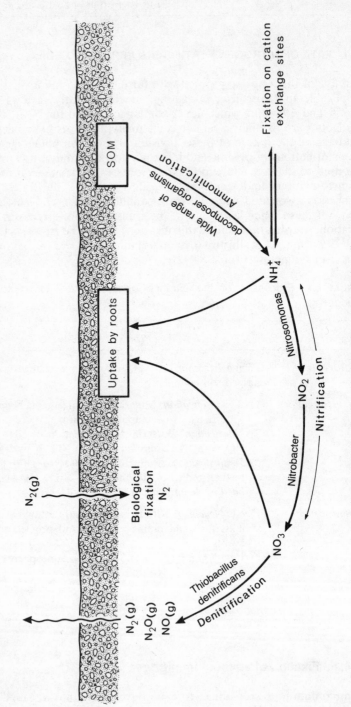

Figure 8.1 The processes of the nitrogen cycle in soil

Tucker (1985) showed that leaching losses of ammonium from coarse textured sandy soils could be very significantly reduced (84 per cent reduced to 6 per cent) by mixing a high CEC zeolite with the soil. By reviewing the 'rules' of cation exchange processes (Sections 5.2.2 and 5.2.3), we can see that cations are exchanged onto negatively charged soil colloids according to the lyotropic series. Small, univalent cations can approach charged surfaces more closely and are thus held more strongly. For the major nutrient cations in soil, the order of replaceability is:

$$Na^+ > K^+ \approx NH_4^+ > Mg^{2+} > Ca^{2+}$$

As this sequence suggests, fixation of K^+ and NH_4^+ in soil is considered to be analogous. Their similarity in exchange behaviour is due to the fact that both cations are monovalent and have similar dehydrated ionic radii ($K^+ = 0.133$ nm, $NH_4^+ = 0.143$ nm). K^+ and NH_4^+ ions are fixed primarily by $2:1$ type clay minerals. Where substantial isomorphic substitution produces a high charge in both Si-O tetrahedral and Al-O-OH octahedral sheets, strong interlayer bonding develops, as in the case of micas, and exchanging cations cannot penetrate between lattice layers. In minerals such as montmorillonite, the much weaker interlayer bonding allows water molecules and cations to penetrate between layers, causing the mineral to expand. Lying between these two extreme conditions are the vermiculites. According to Nommik and Vahtras (1982), K^+ and NH_4^+ ions can snugly fit into 'holes' between lattice layers in vermiculite and degraded illite. The small size of these ions allows the clay lattice layers to approach each other closely and to become bound, preventing rehydration or re-expansion of the crystal (Fig. 8.2). This discovery led to the suggestion that NH_4^+, like K^+, is held on soil colloids with two different degrees of strength, depending on whether it is fixed in the internal lattice, or exchanged onto colloid surfaces (see Fig. 8.2, stage 3). Nommik and Vahtras (1982) suggest that NH_4^+ availability to plant roots follows the equilibrium:

$$\begin{array}{c} NH_4^+ \\ \text{uptake by} \\ \text{roots} \end{array} \rightleftharpoons \begin{array}{c} \text{Soluble} \\ NH_4^+ \end{array} \rightleftharpoons \begin{array}{c} \text{Exchangeable} \\ NH_4^+ \end{array} \rightleftharpoons \begin{array}{c} \text{Fixed} \\ NH_4^+ \end{array} \qquad \textbf{(8,1)}$$

Since K^+ and NH_4^+ cause bonding of lattice layers in some clay minerals, it is not surprising to discover that soil CEC_t can be measurably reduced by this process (Page, Burge and Ganje, 1967).

Total soil ammonium adsorption capacity has been predicted with some degree of success using both the Langmuir and Freundlich adsorption models discussed in Section 5.3.3 (Hunt and Adamsen, 1985). Kowalenko and Cameron (1976) differentiated ammonium adsorption into fixed and exchangeable forms as outlined in Equation **(8,1)**. They found that a Langmuir kinetic model best described the relationship between exchangeable and fixed NH_4^+ while a non-linear Freundlich equilibrium model best described the relationship between soluble and

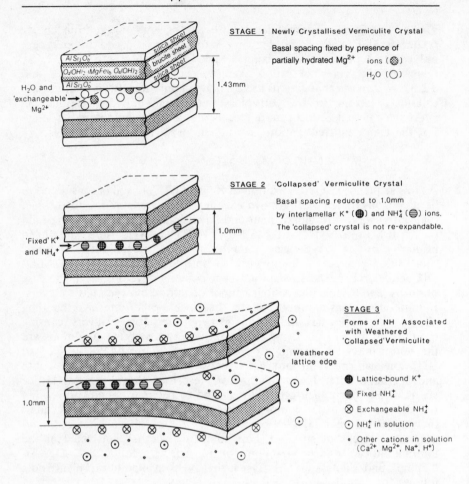

Figure 8.2 Fixation of the ammonium and potassium ions in the interlayer of vermiculite

exchangeable NH_4^+. The good fit of the Freundlich model for the exchangeable NH_4^+ : soluble NH_4^+ equilibrium agrees with the results of many authors, summarised in Chapter 5, Section 5.3.3, who found that the basic Freundlich equation gave the best fit for cation exchange generally.

The fixation of applied NH_4^+ fertiliser appears to be rapid. In both field and laboratory studies, between 40 and 50 per cent of applied NH_4^+-N has been shown to be fixed within two days of application (Sowden, MacLean and Ross, 1978; Kowalenko and Cameron, 1976). 'Recently fixed NH_4^+' is clearly available for exchange and uptake, as shown by Kowalenko and Cameron (1978) who found that as much as 70–90 per cent is re-released over the growing season. Dalal (1975) has shown that the amount of NH_4^+ fixation depends on the associated anion in the ammonium salts which are applied to soil. The NH_4^+ adsorption maxima obtained from $(NH_4)_2SO_4$

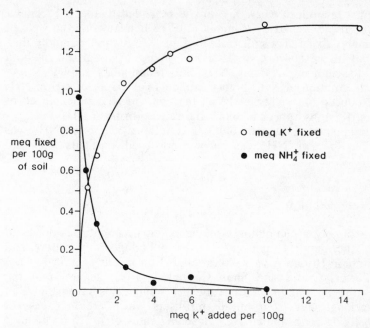

Figure 8.3 Reduced NH_4^+ fixation in a silty clay loam as influenced by previous fixation of K^+. (Reproduced from Stanford and Pierre, 1947, *Soil Science Society of America Proceedings*, Vol 11, 155–60. By permission of the Soil Science Society of America Inc.)

and NH_4NO_3 were 63 and 45 per cent respectively of that from $NH_4H_2PO_4$.

The parallel fixation of K^+ and NH_4^+ in soil results in a depression of NH_4^+ fixation in soils high in K, and to a lesser extent, those high in Cs and Rb. The early study by Stanford and Pierre (1947) illustrates the effect of previously fixed K^+ on the fixation of NH_4^+ in a silty clay loam (Fig. 8.3). The presence of other cations including H^+ have been sown to influence NH_4^+ fixation by soil. Nommik and Vahtras (1982) suggest that the addition of cations with high replacing power leads to a decrease in exchangeable NH_4^+. Divalent and trivalent cations depress the fixation more than monovalent cations. Several studies have also indicated that H^+ ions depress NH_4^+ fixation. This has been shown by higher NH_4^+ fixation with increasing pH (see, for example, Nommik, 1957)

Critical in assessing the importance of NH_4^+ fixation in fertiliser budgeting are the degree to which fixed NH_4^+ is available (a) to be converted to NO_3^- by nitrifying bacteria, and (b) for uptake by plants. The discovery that K^+ ions interfere with the defixation of NH_4^+ ions has helped to explain previously variable and usually low rates at which fixed NH_4^+ is nitrified to NO_3^- in soil. Welch and Scott (1960) found that the addition of 100, 200 and 300 $\mu g\ g^{-1}$ of K to soils reduced the nitrification of fixed NH_4^+ by 30, 80 and 90 per cent respectively. Nommik (1957) found a similar reduction in the immobilisation of fixed NH_4^+ by heterotrophic microflora when K^+ ions were added to the soil. Similar results were also obtained by Nommik (1957) when studying the yield of oats in a highly fixing clay soil. Oats took up NO_3^--N more rapidly than NH_4^+-N, but in the absence of K^+ ions, yields of oats grown on both $Ca(NO_3)_2$ (calcium nitrate) and $(NH_4)_2SO_4$ (ammonium sulphate) were similar. The addition of K to both treatments caused no effect on NO_3 plots but reduced oat yields by 10–20 per cent on NH_4 plots. In all of these cases, K^+ ions appear to block the release of fixed NH_4^+ ions.

When organic manures and solutions containing ammonia (NH_3) gas are added to the soil, NH_4^+ can become adsorbed, particularly by soil organic matter.

8.4.2 Nitrate leaching

There is a rapidly accumulating volume of environmental evidence to show that the major source of nitrates polluting streamwaters and groundwaters in Britain is agricultural land. In a recent overview of the subject, Foster, Cripps and Smith-Carington (1982) suggest that the modern British farming practices of the 1970s and 1980s have resulted in annual leaching rates from nitrate fertilisers which are in excess of 80 kg NO_3 ha^{-1}. High annual NO_3^- leaching rates of up to 55 kg ha^{-1} have also been reported from agricultural fields in North Carolina (Jacobs and Gilliam, 1985). Apart from the loss of fertiliser efficiency caused by such losses, NO_3^- polluted groundwaters can result in eutrophication of watercourses and the growth of algal blooms. Waters high in NO_3^- are also a health hazard, particularly to young babies who can develop methaemoglobinaemia ('blue baby' syndrome). While the European Economic Community recommend that levels of NO_3^- in drinking water should be $\leqslant 50$ µg NO_3^- ml^{-1}, levels up to 100 µg NO_3^- ml^{-1} are thought to be acceptable (EEC, 1980). It is important here to point out the difference in reporting NO_3^- concentrations and NO_3^--N concentrations in water. Since only 14/62 or 22.58 per cent of the NO_3^- ion is made up of N (atomic weight of N/molecular weight of NO_3^-), a concentration reported as 100 µg NO_3^- ml^{-1} is equivalent to a concentration reported as 22.58 µg NO_3^--N ml^{-1}. Confusingly, nitrate levels in soils and sediments are usually reported in µg NO_3^- ml^{-1}, while stream and groundwater quality are usually reported in µg NO_3^--N ml^{-1}.

In Table 8.5 the concentrations of NO_3^- in soils, rocks and streamwaters are compared. The amounts of leached NO_3 measured in field plots and lysimeters depends on the rate of fertiliser application. Although Dowdell, Webster and Mercer (1984) found little difference in amounts of NO_3 leached at the low fertiliser application rates outlined in Table 8.5, the higher application rates used by Barraclough, Geens and Maggs (1984) yielded increasing NO_3^- losses in both field and lysimeter experiments. In studies carried out by Imperial Chemical Industries (ICI) at their field experimental station at Jealott's Hill, NO_3^- leaching from NO_3^- fertilised grasslands rarely exceeded EEC recommended levels with rates of 250 kg N ha^{-1} (three times the average national application rate). Only at application rates of 750 kg N ha^{-1} was leachate NO_3^- concentration unacceptably high (Hood, 1976). Colbourn (1985) found that losses of NO_3^- from ploughed and fertilised Denchworth clay soil was slightly higher (mean of 17.7–132.9 µg NO_3^- ml^{-1} in drainage waters over the period 1980–84) than from directly drilled soil (mean of 39.9–101.9 µg NO_3^- ml^{-1}). The very wide range of lysimeter studies surveyed by Wild and Cameron (1980) yield NO_3^- leaching concentrations in the United Kingdom, West Germany and the United States of 3.5–194.4 µg NO_3^- ml^{-1} for zero fertiliser applications to 4.4–1656.3 µg NO_3^- ml^{-1} from soils receiving 80–500 kg N ha^{-1} y^{-1}. It is clear that some soil leachates, chalk porewaters and some river waters exceed EEC recommended levels for drinking waters. Foster, Cripps and Smith-Carington (1982) show that concentrations of NO_3^- in the porewater of chalk underlying longterm arable land in Yorkshire and Cambridgeshire greatly exceed EEC acceptable levels, while Walling and Webb's (1981) data suggests that NO_3^- levels, even in unpolluted streamwaters in some parts of South-east England, exceed EEC recommendations (see Table 8.5).

Quantification of nitrate leaching losses and calculation of fertiliser budgets have been facilitated by studying transport of ^{15}N-labelled fertilisers in field lysimeters or in soil monoliths built into lysimeters in the laboratory. For N budgeting, a totally enclosed lysimeter design is necessary. Systems such as that illustrated in Fig. 8.4 have been used in agricultural, forestry and peat soils (see, for example, Overrein, 1968; Pratt et al., 1967; Dowdell and Webster, 1980; Malcolm and Cuttle, 1983; Barraclough et al., 1984). Apart from plant uptake, leaching of NO_3^--N and volatilisation of NH_3 and N_2O are the main losses of N from artificial fertilisers. In the lysimeter depicted in Fig. 8.5, leachates are collected by drainage under gravity of artificially applied rainfall or irrigation water. Since no suction is applied, this technique is called zero tension lysimetry. Volatilised gases are sampled in the airspace above the soil. The main disadvantage of the use of lysimeters in leaching studies is that in most cases the soil must be repacked and does not retain initial structure, aeration and drainage.

An alternative approach to studying nutrient losses by leaching is to monitor the composition of effluent from field tile drains. This overcomes the problems of soil disturbance during lysimeter installation. The value of these techniques have, however, been queried for two main reasons:

(1) the total volume of soil drained per tile depends on soil type and cannot be calculated very accurately; and
(2) the spatial variability of macropore flow and seepage in field soil must be overcome using replication.

Three main aspects of NO_3 leaching from N fertilisers require attention:

Table 8.5 Nitrate levels in (a) soil, (b) groundwaters and (c) streamwaters in relation to EEC drinking water recommendations (all concentrations reported in μg NO_3 ml^{-1} in solution)

EEC drinking water	Recommended 0–50	Acceptable 50–100	Not recommended >100
(a) *N Fertilised soils*	Fertiliser rate	Leachate NO_3 concentration	
		Field drainage	Lysimeter
Barraclough et al. (1984)	250 kg N ha^{-1}	0.3– 3.9	0.2– 2.0
(sandy loam over three years,	500 kg N ha^{-1}	5.1–22.9	5.8– 34.0
grassland. fertiliser = NH_4NO_3)	900 kg N ha^{-1}	66.0–98.2	81.5–117.2
Dowdell, Webster and Mercer (1984)	0 kg N ha^{-1}	–	22.1–160.0
(silt loam over four years,	80 kg N ha^{-1}	–	31.0–132.9
ryegrass. fertiliser = $Ca(NO_3)_2$)	120 kg N ha^{-1}	–	44.2–132.9
(b) *Chalk porewaters*	Location	Porewater NO_3 concentration	
Foster, Cripps and Smith-Carington (1982)	West Norfolk	22.1–177.1	
(5–15 m depth beneath longstanding arable land)			
	East Yorkshire	22.1–199.3	
Triassic sandstone porewaters			
Young and Gray (1978)	Greater Manchester	27.0– 48.3	
(20–40 m depth beneath grassland)			

(Table 8.5 cont...)

(c) Streamwaters

Walling and Webb (1981)
(averaged data)

Location	Streamwater NO_3 concentration			
	mean NO_3 concentration		range of maximum NO_3 concentrations	
	(1953–67)	(1976–77)	(1956–67)	(1976–77)
Upland Britain 0.4–17.7				
Lowland Britain 17.7–66.4				
Thomlinson (1970) (data for 1953–67)				
River Severn (Tewkesbury)	13.73	–	14.2–34.1	–
River Stow (Langham)	27.46	8.0–99.6	34.5–96.5	10.2–145.7
River Tyne (Wylan)	1.77	–	2.7– 8.9	–
River Great Ouse (Bedford)	18.60	19.9–89.5	21.3–72.6	31.0–106.7
Greene (1978) (data for 1976–77)				
River Tees (Darlington)	2.21	–	3.5–11.1	–
River Dee (Chester)	4.87	–	11.1–19.9	–

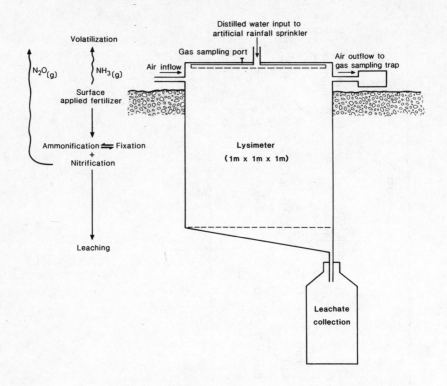

Figure 8.4 Lysimeter design for studying N–fertiliser losses in the field

(1) the form of nitrogen fertiliser applied;
(2) the seasonality of leaching in response to rainfall (which controls rates of transport processes) and temperature (which controls rates of the N transformation processes: mineralisation and nitrification) and;
(3) the influence of soil profile characteristics.

Form of nitrogen fertiliser applied

Three main forms of nitrogen are applied to soil as fertilisers: nitrates, in the form of KNO_3, NH_4NO_3 or $Ca(NO_3)_2$, ammonium salts, mainly $(NH_4)_2SO_4$ or NH_4Cl, and organic forms such as urea. It is not surprising that studies of comparative leaching losses in both forestry and agricultural soils indicate rapid and largest leaching losses from nitrate fertilisers with slower rates of leaching from ammonium salts and urea (Fig. 8.5). A similar sequence of loss during the growing season was recorded by Addiscott and Cox (1976) for spring barley grown on a flinty clay loam at Rothamsted. While NO_3^--N content of topsoils in the spring after application of four fertilisers are not significantly different (see Table 8.3), the apparent recovery of N by the barley crop is significantly

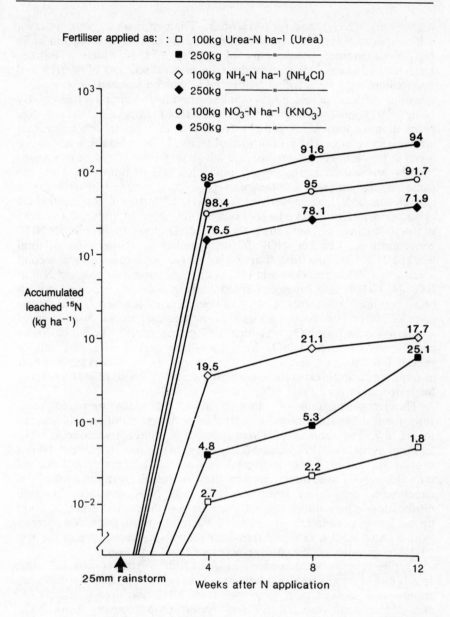

Figure 8.5 Accumulated leaching losses of tracer ^{15}N, $^{15}NH_4$ and $K^{15}NO_3$ (figures on graph represent ^{15}N leached as a per cent of total leached N). (*NB*: logarithmic scale on y-axis). (Source: Overrein, 1968)

higher for $Ca(NO_3)_2$ and for $(NH_4)_2SO_4$. The two main reasons given for differences in N leaching losses are, first, that the ability of the soil to fix NH_4-N determines the amount of applied NH_4-N which is initially leached; and second, that urea must be mineralised and both urea and ammonium must be nitrified before NO_3^- can be leached. These issues question the form in which N is being leached from applied fertilisers. By using ^{15}N tagged fertilisers, Overrein (1968) calculated the relative proportions of fertiliser N and non-fertiliser N in lysimeter leachates from a brown forest soil. The proportions of leached ^{15}N to total leached N are given in Fig. 8.5. His results indicate a high degree of retention of applied urea-N, particularly at the lowest application rate of 100 kg ha^{-1}. At the same application rate, substantial NH_4^+ fixation resulted in only 20 per cent of leached N being derived from NH_4 Cl fertiliser. At the highest application rate of 250 kg ha^{-1}, leached NH_4^+ rose to about 75 per cent of total leached N, indicating reduced NH_4^+ fixation at higher NH_4^+ concentrations. Leached NO_3^- fertiliser makes up 98 per cent of total leached N in just the first four weeks of the experiment. In a second series of experiments, Overrein (1971) differentiated total leached N into NO_3^--N, NH_4-N and organic N fractions (Fig. 8.6). As expected, for all NO_3^- fertiliser treatments, nearly 100 per cent of leached N was in the form of NO_3^-. For the three highest urea treatments, the proportion of nitrogen leached as NO_3 was high (50–75 per cent). The much lower proportions of leached NO_3 from the three highest NH_4^+ treatments reflects NH_4^+ fixation and a lag in nitrification. These results suggest that in fertilised, acid forest soils, mineralised urea-N is more readily available for nitrification than fixed NH_4^+.

The transformations of nitrogen in an artificially prepared sand/sand–loam substrate continuously leached with urea solution is illustrated in Fig. 8.7. The sand texture was specifically chosen to minimise NH_4^+ fixation. Ardakani, Volz and McLaren (1975) found that about 80 per cent of applied urea was recovered at a depth of 50 cm on the second day. While NH_4^+ was mineralised by urea hydrolysis from the start of the experiment, only after seventeen days was NO_3^- produced through nitrification. The authors suggest that this was the period taken to build up an active population of nitrifying bacteria, particularly *Nitrobacter*. Within three weeks, nitrogen transformations were completed in the top 10–15 cm because of the greater activity of soil organisms.

Fertiliser recovery studies for a range of soils in Alberta show the same trends as Overrein's results. Malhi and Nyborg's (1983) fertiliser recovery results show highest crop recoveries from NH_4^+ and urea fertilisers (97 and 71 per cent respectively) and lowest crop recovery from NO_3^- fertiliser (60 per cent). Their data show, however, that fertiliser losses are due to denitrification, not leaching. Nitrification inhibitors have been used in an attempt to delay fertiliser N losses through NO_3^- leaching and volatilisation. One of these inhibitors, nitropyrin, has been found to delay both nitrification and the release of fixed NH_4^+-N (Aulakh and Rennie, 1984).

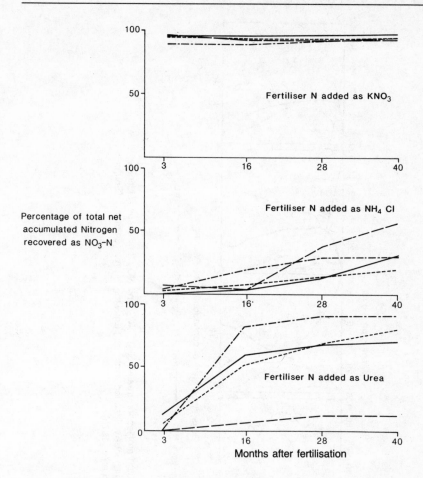

Figure 8.6 Percentage of total net accumulated N recovered as NO_3-N for three fertilisers: KNO_3, NH_4Cl and urea-N at four application rates: 100 kg ha^{-1} (– – –), 250 kg ha^{-1} (–·–·), 500 kg ha^{-1} (---) and 1000 kg ha^{-1} (——). (Source: drawn from the data of Overrein, 1971)

Influences of soil profile characteristics

The influences of soil profile characteristics on nitrate leaching can be resolved into two main topics. First, the effect of soil texture, particularly percentage of clay content, on water transmission rates and on the development of anaerobic conditions which promote denitrification. Second, the influence of macropore or bypassing flow on NO_3 concentrations in drainage waters.

Experiments comparing NO_3^- leaching losses from different soil types have confirmed that losses are generally higher from coarse textured sandy soils than from fine textured clay soils (see, for example, Pratt,

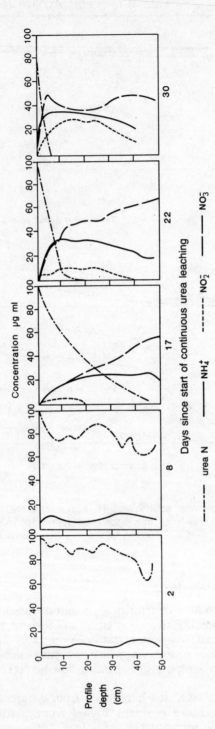

Figure 8.7 Concentration profiles for urea-N, NH_4^+, NO_2^- and NO_3^- in the soil solution of an artificial sand, continuously leached with 100 μg N ml^{-1} urea solution. (Source: Ardakani, Volz and McLaren, 1975)

Lund and Warneke, 1980; Kolenbrander, 1969). Lower hydraulic conductivity in soil clay layers was found by Devitt *et al*. (1976) to cause reducing conditions and denitrification, so that overall N losses are sometimes higher from clay soils than from sandy soils.

Soil structure as well as texture determines NO_3^- behaviour in soil. Nitrate concentration during macropore or bypass leaching appears to depend on the time elapsed since NO_3^- fertiliser application. For newly applied NO_3^- fertiliser, Smettem, Trudgill and Pickles (1983) and Barraclough, Hyden and Davies (1983) report rapid NO_3^- transmission, resulting in higher NO_3^- concentrations in leachate than would occur through miscible displacement. Smettem, Trudgill and Pickles (1983) and Wild (1972) suggested that over time after fertilisation, NO_3^- relocated within soil aggregates is protected from leaching when bypass flow occurs. This results in much lower NO_3^- concentrations in leachate than would occur through miscible displacement. The rate at which NO_3^- held in soil peds equilibrates with the soil solution is controlled both by the concentration of NO_3^- within the ped and the ped surface area available for exchange. For an intra-aggregate NO_3^- concentration of $14 \ \mu g \ N \ h^{-1}$, Darrah, Nye and White (1983) calculated that nitrate would diffuse to a macropore of zero NO_3^- concentration at a rate of $2.3 \ \mu g \ h^{-1} \ N \ cm^{-2}$ of ped surface in contact with the soil solution. This means that soils with fine crumb structures and large ped contact surfaces will show the most efficient NO_3^- diffusion from aggregates into macropores. Haigh and White (1986) suggest that the fine crumb structure of A horizons and agricultural topsoils allows thorough mixing of percolating waters with soil and the efficient diffusion of NO_3^- into the soil solution. This water then drains quickly through B horizons via macropores and cracks. Drainage water interacts with a very small contact surface in the B horizon and so its NO_3^- concentration remains high.

Seasonal influences on nitrate leaching

It is doubtful whether NO_3^- leaching or denitrification contributes most to the loss of N fertilisers from soils. There is general agreement that both processes are greatest in the period immediately after rainstorm events in autumn and winter. Nitrogen immobilisation and accumulation on the other hand occur during dry conditions in the summer.

Perhaps some of the clearest illustrations of nitrate leaching in relation to rainstorms and soil water drainage are presented by Cameron, Kowalenko and Ivanson (1978) who examined the migration of NO_3^--N and NH_4^+-N from an NH_4NO_3 fertiliser applied to a sandy podzol in the Canadian Great Plains (Fig. 8.8). In two separate study years, fertiliser NH_4-N was quickly lost from the soil; control plots indicating that it had been nitrified. The pattern of NO_3^- behaviour in the two study years showed high nitrification rates (0.65–$1.25 \ kg \ N \ ha^{-1} \ d^{-1}$) in the summer, which masked any small NO_3^- leaching losses, with the main NO_3^- leaching losses occurring in late autumn and early spring. Most of the nitrate leaching is related to rainfall events. In Fig. 8.8 the two largest accumulated rainfall periods, 28 May/14 June and 12 July/3 Aug., show the most dramatic losses in the NO_3^--N profile. Nitrate moves down the profile as a diffuse bulge rather than the sharp plug which might have been expected in the sandy soil. Cameron, Kowalenko and Ivarson (1978) suggest that more research is required to examine field variability in dispersion processes.

Nitrate leaching models

As indicated in Section 5.5.2, a large number of solute particularly nitrate, leaching models have been developed for a range of soil and cropping conditions (see, for example, Burns, 1974; Jury et al., 1976; Addiscott, 1977; Saxton, Schuman and Burwell, 1977). Since NO_3^- is not significantly adsorbed in soil, most NO_3^- leaching models are simple modifications of water transport models. In the model of Burns (1974), leaching under the influence of rainfall and evaporative rise is calculated for a series of consecutive 3.75 cm thick soil layers. The model assumes zero anion adsorption and complete convective mixing in each layer. Although the model was successfully field verified for NO_3^- transmission in a sandy loam, its adaptability may be limited by the need for fairly shallow soil layers and small range of rainfall/evaporation ratios. The model has since been modified to simulate fertiliser placement effects (Burns, 1976) and the influence of NO_3 leaching on crop yields (Burns, 1980).

In an attempt to account for the differential retention of soil water and solutes in field soils, Addiscott (1977) developed a leaching model for structured soil in which water and solutes are partitioned into mobile (pF 1.7–3.3) and immobile (> pF 3.3; halfway between the wilting point (pF 4.2) and oven dryness) phases. The mobile phase is displaced during leaching and it is assumed that equilibrium is attained between mobile and immobile solutes during periods when there is no water movement. Addiscott achieved good correlations between predicted and actual winter NO_3 leaching in a stony clay loam. Allocating values to the mobile and immobile water phases is identified as a major problem, particularly if a comparison is required between wet, winter conditions and dry periods when soils dry out and salts accumulate at the surface. A second problem with this and other simple leaching models is the failure to account for other N transformation processes such as the mineralisation of soil organic N. The simplest solution to this problem is the inclusion of subroutines in the leaching simulation model which handle the relevant N transformation processes: mineralisation, nitrification, NH_4^+ fixation, NH_3 volatilisation and denitrification. This is the approach adopted by Frissel, van Veen and Kolenbrander (1980). In freely draining sandy textured soils, it may be relevant to consider only mineralisation and nitrification, whereas in waterlogged clay soils it may be necessary to include all five transformation subroutines. When mineralisation and nitrification were simulated during NO_3 leaching in a loamy sand, Jury et

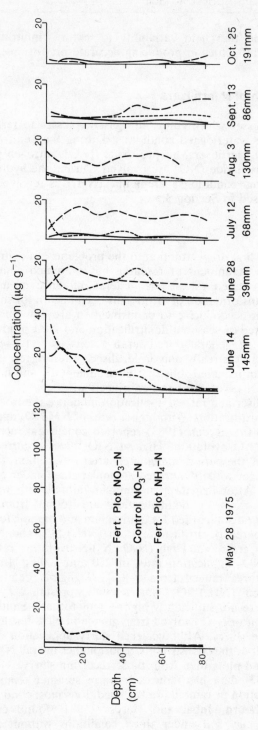

Figure 8.8 Leaching of NH$_4$-N and NO$_3$-N after a May application of 255 kg ha^{-1} as NH$_4$NO$_3$ on a sandy podzol. Rainfall amounts between sampling periods are given at the base of each graph. (Source: modified from Cameron, Kowalenko and Ivarson, 1978)

al. (1976) still found that field variability in texture, infiltration and percolation caused difficulties in producing accurate predictions.

8.4.3　Volatilisation of N fertilisers

The two main processes which contribute gaseous losses to fertiliser N budgets occur under waterlogged conditions. During the denitrification, or reduction, of NO_3^- fertilisers, nitrous oxide (N_2O), nitrogen gas (N_2) and occasionally nitric oxide (NO) are evolved. During the hydrolysis of urea under reducing conditions, ammonia (NH_3) is evolved. Both processes are discussed in Section 3.5.

Denitrification

Since Allison (1955) first drew attention to the problems of denitrification of N fertilisers, a vast amount of research has examined contributing factors and has quantified rates. The general influence of anaerobic conditions on the transformation of nitrate has been known for some time and the scenarios most conducive to denitrification are now quite clear. Although some degree of seasonal denitrification and N_2O production is frequently reported in generally well-aerated soils (see, for example, Dowdell *et al.*, 1972), it is really only in fertilised, waterlogged soils such as rice paddies and in irrigated soils that environmentally and financially significant losses of N occur through denitrification.

The form of fertiliser and rate of application influence the rate of N_2O evolution during denitrification. With higher rates of NH_4NO_3 application to a cut ryegrass sward, Ryden (1983) reported concurrent increases in denitrification and N_2O evolution. Highest N_2O fluxes occurred under wet conditions during the period immediately after fertilisation. Different forms of nitrate appear to be denitrified at similar rates under grassland swards (Table 8.6). After irrigation or summer rainfall with warm soil conditions, highest rates of denitrification are found from nitrate fertilisers with lower rates recorded for ammonium and organic fertilisers. Ryden (1981) compared denitrification from $(NH_4)_2SO_4$ and NH_4NO_3 fertilisers applied to grassland. High (N_2O+N_2)-N evolution rates were measured for NH_4NO_3 application rates of 250 and 500 kg N ha^{-1} y^{-1} when soil temperatures ranged from 5–8°C. Negligible denitrification was measured when $(NH_4)_2SO_4$ was applied, presumably because nitrification rates were not sufficiently high. Eggington and Smith (1986) compared the amount of N_2O evolved from grassland plots fertilised with $Ca(NO_3)_2$ and cattle slurry. Although each slurry application caused a pulse of N_2O evolution, they consistently report higher overall N_2O fluxes from nitrate amended plots than from those receiving slurry.

As Ryden's (1983) data has indicated, warm summer temperatures combined with irrigation or rainfall are the conditions most conducive for denitrification. Colbourn, Iqbal and Harper (1984) found enhanced denitrification in a clay soil under these conditions without fertiliser

Table 8.6 Influences of different N fertilisers on gaseous N flux from field soils

Agricultural system	Fertiliser application	(N_2O+N_2) Flux		Reference
		daily	annual	
Ryegrass	300 kg N ha^{-1} (KNO$_3$)	1.42 kg N ha^{-1}d^{-1}	–	Ryden (1981)
Ryegrass	285 kg N ha^{-1} (KNO$_3$)	0.2–2.0 kg N ha^{-1}d^{-1}	–	Rolston et al. (1982)
Ryegrass	63 kg N ha^{-1} (NH$_4$NO$_3$)	1.2 kg N ha^{-1}d^{-1}	–	Ryden (1981)
Ryegrass	250 kg N ha^{-1} (NH$_4$NO$_3$)	–	11.1 kg N ha^{-1}y^{-1}	Ryden (1983)
Ryegrass	500 kg N ha^{-1} (NH$_4$NO$_3$)	–	29.1 kg N ha^{-1}y^{-1}	Ryden (1983)
Permanent grass	70 kg N ha^{-1} (NH$_4$NO$_3$)	0.05 ± 0.01 kg N ha^{-1}d^{-1}	–	Colbourn, Iqbal and Harper (1984)
Permanent grass	100 kg N ha^{-1} Ca(NO$_3$)$_2$	1–2 kg N ha^{-1}d^{-1}	–	Webster and Dowdell (1982)
Ryegrass	100 kg N ha^{-1} Ca(NO$_3$)$_2$	–	3.31 kg N ha^{-1}y^{-1}	Lippold et al. (1981)
Ryegrass	100 kg N ha^{-1} Ca(NO$_3$)$_2$	–	2.6–11.2 kg N ha^{-1}y^{-1}	Eggington and Smith (1986)
Winter wheat	100 kg N ha^{-1} Ca(NO$_3$)$_2$	0.3 ± 0.07 kg N ha^{-1}d^{-1}	–	Colbourn, Harper and Iqbal (1984)
Ryegrass	1200 kg N ha^{-1} cattle slurry		1.2–5.3 kg N ha^{-1}y^{-1}	Eggington and Smith (1986)

application. By examining the drying and wetting associated with different frequencies of irrigation, Rolston *et al.* (1982) found that frequent, small irrigations maintained highest levels of NO_3^- in the upper levels of the soil profile, in the zone accessible to plant roots. They recorded higher denitrification rates after the first irrigation and subsequently with higher frequency irrigations. Their soil management recommendations favour frequent small irrigations which may slightly increase denitrification losses of applied N fertiliser, but reduces leaching losses and improves plant uptake efficiency.

In laboratory studies of denitrification, Colbourn, Iqbal and Harper (1984) measured substantially higher denitrification losses $(30-40 \text{ kg N ha}^{-1}\text{d}^{-1})$ compared to rates measured in the field $(< 0.1 \text{ kg N ha}^{-1}\text{d}^{-1})$. They found that nitrification was frequently the rate-limiting step in gaseous N loss from incubated soil. Only when heavy rain followed fertilisation did O_2 concentrations in field soil drop as low as laboratory conditions, indicating that nitrification was probably the main rate limiting step in gaseous N loss from fertilised field soils.

Several factors, including N fertilisation, have been shown to influence the ratio of N_2 to N_2O evolved during denitrification (see Section 3.5). According to Colbourn, Iqbal and Harper (1984), the $N_2 : N_2O$ ratio (or the mole fraction of N_2O) from fertilised soil was higher than the mole fraction from unfertilised soil (0.60 compared to 0.17). Like Cho and Sakdinan (1978), they suggest that the presence of excess NO_3^- inhibits the reduction of N_2O to N_2, particularly in NO_3^- fertilised soil and soils with high nitrification rates.

Comparative N losses due to leaching and denitrification

Budgets for the estimation of N fertiliser efficiency have been calculated since Allison's (1955) indication that unaccounted N losses ranged from 0 to 50 per cent of applied N fertiliser. The N losses of these early N budgets were assumed to represent the combined effects of leaching and denitrification. Studies of N fertiliser leaching and denitrification (Table 8.7a and b) indicate that substantial N losses can be incurred with high fertiliser application rates and when soil environmental conditions are optimal for these processes. The 70 per cent denitrification loss reported by Rolston, Hoffman and Toy (1978), for example, was recorded when the soil was warm and near saturation (23°C and 1.5 kPa). Similarly, the high leaching losses reported by Chichester and Smith (1978) reflect well-drained soil conditions in a humid climate in which 45 per cent of the annual rainfall percolated through the soil. The 32–34 per cent N leaching losses they report mainly occurred between November and May during the period of peak rainfall. In temperate conditions with average fertiliser application rates of 100–400 kg $N \text{ ha}^{-1}\text{y}^{-1}$, leaching and denitrification losses range from 1.5 to 12.4 per cent (Table 8.7b). Denitrification losses exceed leaching losses with high N fertiliser application rates, in heavy clay soils and after irrigation or rainfall, while leaching losses are only likely to exceed denitrification

Table 8.7 Comparative nitrogen losses after nitrogen fertiliser applications, attributable to (a) leaching and (b) denitrification (all values expressed as percentages of applied fertiliser nitrogen)

Agricultural system	Fertiliser application (y^{-1})	Per cent loss by leaching	Reference
(a) *Nitrogen-leaching losses*			
Grassland	250 kg N ha^{-1} (NH_4NO_3)	0.14–1.5	Barraclough *et al.* (1983, 1984)
(over three years)	500 kg N ha^{-1} (NH_4NO_3)	3.1– 5.4	
	900 kg N ha^{-1} (NH_4NO_3)	16.7–18.1	
Winter wheat (sandy loam)	125 kg N ha^{-1} (NH_4NO_3)	27.0–31.3	Cannell *et al.* (1977)
(clay)	95 kg N ha^{-1} (NH_4NO_3)	19.7–31.2	
Ryegrass	100 kg N ha^{-1} ($Ca(NO_3)_2$)	2.2– 5.4	Webster and Dowdell (1985)
Ryegrass (silty clay)	100 kg N ha^{-1} ($Ca(NO_3)_2$)	7– 8	Dowdell and Webster (1981)
(silty clay loam)	100 kg N ha^{-1} ($Ca(NO_3)_2$)	4–14	
Winter wheat	100 kg N ha^{-1} ($Ca(NO_3)_2$)	2– 7	Jaakkola (1984)
Corn	336 kg N ha^{-1} ($Ca(NO_3)_2$)	32–34	Chichester and Smith (1978)
(b) *Nitrogen-denitrification losses*			
Ryegrass	250 kg N ha^{-1} (NH_4NO_3)	4.4	Ryden (1981)
	500 kg N ha^{-1} (NH_4NO_3)	5.8	
Winter wheat	100 kg N ha^{-1} ($Ca(NO_3)_2$)	9.5	Colbourn, Iqbal and Harper (1984); Colbourn, Harper and Iqbal (1984)
Ryegrass (23°C, 1.5 kPa)	300 kg N ha^{-1} (KNO_3)	70	Rolston, Hoffman and Toy (1978)
Ryegrass	285 kg N ha^{-1} (KNO_3)	0.7– 6.4	Rolston *et al.* (1982)
Celery	336 ⎱ kg N ha^{-1}	6–12.4	Ryden and Lund (1980b)
Cauliflower	528 ⎰ (as $(NH_4)_2SO_4$)	5– 5.5	
Artichokes	176 plus anhydrous NH_3	11.1–15.3	
Irrigated arable plots	290 ⎱ kg N ha^{-1}	10–12.1	Ryden and Lund (1980a)
	475 ⎰ (as $(NH_4)_2SO_4$)	13.0–29.4	
	665 plus urea	20.3–30.7	
Ryegrass in enclosed microplots	420 kg N ha^{-1} (as NH_4NO_3)	6.1	Bristow, Ryden and Whitehead (1987)

losses in coarse textured, freely drained, sandy soils. Both processes are highly seasonal. Peak leaching losses occur in temperate climates during winter rainfall, particularly if nitrate fertilisers have been applied to winter cereal crops. Peak denitrification losses occur after late spring fertilisation when soils are warming rapidly and once irrigation is necessary.

An unfortunate aspect of many fertiliser N budgets is that rarely are all losses quantitatively assessed. A typical approach is to examine one loss in detail, then to sum other losses and estimate them by difference. Typical examples of such studies are given in Table 8.8. They clearly indicate that volatile losses of N can range from 2 to 59 per cent of applied N fertiliser, depending on the soil type, cropping system and amount of fertiliser applied.

Fate of N fertilisers in flooded rice paddies

The inefficiency of conventional N fertilisers in waterlogged rice soils has been recognised for some time. Losses of fertiliser nitrogen in floodwater, runoff and denitrification are usually quoted as the main causes of poor crop recovery (see, for example, Savant and DeDatta, 1982). Krishnoppa and Shinde (1978) found that 24.3 per cent of applied N-urea was recovered in the rice crop, 9.7 per cent was lost by NH_3 volatilisaton and 7.5 per cent was lost in leaching. Unaccounted nitrogen in their budget (presumed to be denitrification) amounted to 31.1 per cent of applied urea-N. Several authors believe that the main reason for poor crop recoveries is the speedy transformation of fertiliser N within 30–40 days after application (see, for example, Patrick and Reddy, 1976b; Manguiat and Broadbent, 1977). The fate of applied N in a fertilised paddy field is illustrated in Fig. 8.9. Deep placement of fertiliser reduced the proportions of unaccounted N and evolved NH_3 by more than 10 per cent.

Rice and other wetland and aquatic plants are able to respire in waterlogged soil by transporting O_2 from aerial parts through aerenchyma tissues to the roots. In the rhizosphere O_2, surplus to respiratory requirements, diffuses into the soil to produce an oxygenated root zone, surrounded by anaerobic soil. Reddy and Patrick (1986) have shown that N losses through denitrification in the rhizosphere of flooded rice plants is enhanced by these conditions (Table 8.9). The main control on N loss from the rhizosphere is the competition for NO_3^- uptake between rice roots and denitrifying bacteria (Reddy and Patrick, 1986).

To overcome the problems of rapid fertiliser–N transformation under waterlogged conditions, a large number of slow release N fertilisers and nitrification inhibitors or retarders have been developed. Slow release N fertilisers are mainly produced by incorporating nitrogen into organic compounds which are decomposed very slowly in soil. Condensation products of urea, such as IBDU (isobutylidene-diurea), and aldehydes, such as UFA (ureaform-acetaldehyde), have most commonly been employed. Taslim and Verstraeten (1977) tested the rates of NH_4^+-N release from a range of urea and aldehyde condensation products. On the

Table 8.8 Comparative fertiliser nitrogen budgets, designed to quantify nitrogen losses

Reference	Agricultural system	Fertiliser application	Nitrogen budget Per cent			
			crop	soil	drainage	unaccounted
Webster and Dowdell (1985)	Ryegrass (lysimetres) (over 4 years)	400 kg N ha^{-1} Ca(NO$_3$)$_2$	57–61	31–35	2–5	2–7
						Per cent denitrified (by difference)
Webster and Dowdell (1984)	Ryegrass silty clay Ryegrass silty clay loam (over 4 years)	400 kg N ha^{-1} Ca(NO$_3$)$_2$	52–54 48–63	21–25 19–25	7–18 4–18	14–18 9–13
			crop	soil	Per cent leached and runoff	lost
Chichester and Smith (1978) Catchpool (1975)	Corn (lysimeters) Rhodesgrass pasture	336 kg N ha^{-1} Ca(NO$_3$)$_2$ 150 kg N ha^{-1} (NH$_4$NO$_3$)	25–32 13	14–32 16	33–36 12	5–23 59
			crop	soil	Per cent drainage	unaccounted
Hood (1976)	Grassland	377 kg N ha^{-1} 934 kg N ha^{-1}	63 39	– –	6 9	31 52
			crop	soil	Per cent leachate	unaccounted
Overrein (1972)	Pine forest soil (lysimeters)	100 kg N ha^{-1} (KNO$_3$) 250 kg N ha^{-1} (KNO$_3$) 100 kg N ha^{-1} (NH$_4$)$_2$SO$_4$ 250 kg N ha^{-1} (NH$_4$)$_2$SO$_4$ 100 kg N ha^{-1} (urea) 250 kg N ha^{-1} (urea)	2 1 5 16 13 9	15 12 95 61 73 64	82 90 3 17 <1 7	1 +3 +3 6 14 20

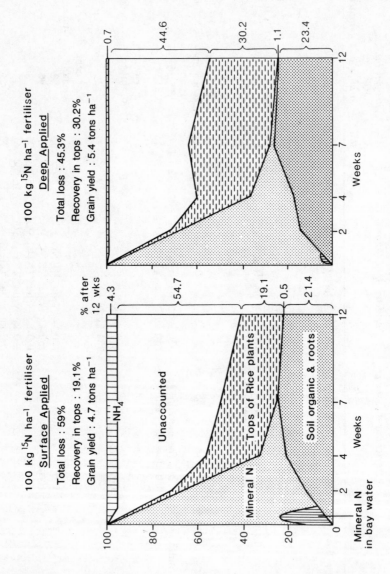

Figure 8.9 Fate of fertiliser nitrogen applied to wetland paddy. (Source: drawn by Savant and DeDatta (1982) from the data of Wetselaar, 1975)

Table 8.9 Mass balance of added $^{15}NH_4$-N in a flooded Crowley silt loam, comparing rice planted, with unplanted, soil cores (from Reddy and Patrick, 1986)

Nitrogen-fraction	Per cent of added ^{15}N	
	Rice planted	unplanted
Soil		
Inorganic nitrogen	1.8	73.9
Organic nitrogen	15.1	21.6
Total soil nitrogen	16.9	95.5
Plant		
Total plant nitrogen	65.4	–
Total nitrogen recovered	82.2	95.5
Unaccounted nitrogen	17.8	4.5

basis of longevity and per cent N availability to the crop, only IBDU and DMU (dimethylolurea) were found to be suitable slow release N compounds (Fig. 8.10).

Nitrification inhibitors such as Nitropyrin have shown promising results in both laboratory and field studies (Prasad, Rajale and Lakhdive, 1971). Several aspects of their use have been reported. Aulakh and Rennie (1984) found that Nitropyrin added at the time of urea application decreased nitrification, thus reducing denitrification losses and down-profile N transport. Significantly higher crop yields have been reported for fertilised soils treated with Nitropyrin (see, for example, Nelson, Huber and Warren, 1980), particularly where the leaching of nitrate from sandy soil has been delayed (for example, Gasser and Hamlyn, 1968). Despite the apparent advantages of slow release N fertilisers and nitrification inhibitors, these materials are expensive and their use is still less widespread than we might wish.

Ammonia volatilisation from N fertilisers

Perhaps the most obvious loss of N as ammonia (NH_3) gas occurs when anhydrous or aqueous ammonia solutions are used as fertilisers, but reduced losses can be achieved by deep injection of the solutions. The increased NH_4^+ diffusion path between soil and the atmosphere delays N loss and allows more efficient crop uptake. Ammonia volatilisation can also occur when urea fertilisers are hydrolysed in soil or when ammonium salts are applied to calcareous soils.

NH_3 is formed in the soil solution according to:

$$NH_4^+ \rightleftharpoons NH_{3(aq)} + H^+ \tag{8,2}$$

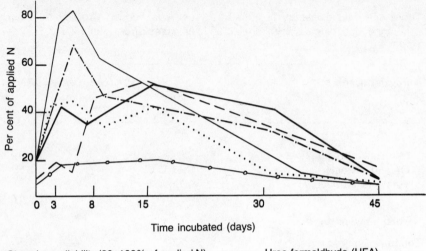

Class I   availability (80–100% of applied N)　---------　Urea formaldhyde (UFA)
—————— Urea　　　　　　　　　　　　　　　—————— Dimethylolurea (DMU)
— · — · · Isobutylidenediurea (IBDU)　　　　Class III availability (0–50% of applied N)
Class II   availability (50–80% of applied N)　—○——○— Ureaform (UF)
— —   Hexamethylene tetramide (HMT)

Figure 8.10 NH_4^+-N released, as a percentage of applied N, in six different urea forms after different incubation periods under water-logged conditions. (Source: Taslim and Verstraeten, 1977)

Thus, any NH_3-forming N fertiliser may have an associated NH_4 loss. The most commonly affected compounds are $NH_4 NO_3$, $(NH_4)_2 SO_4$, $NH_4 Cl$ and urea. While ammonium salts immediately produce NH_3 in solution, urea must firstly undergo hydrolysis in the presence of the enzyme, urease:

$$\underset{\substack{\text{urea}\\ \text{(carbamide)}}}{CO(NH_2)_2} + 3H_2O \underset{H_2O}{\overset{\text{urease}}{\longrightarrow}} \underset{\substack{\text{ammonium}\\ \text{carbonate}}}{(NH_4)_2CO_3} + H_2O \tag{8,3}$$

Ammonium carbonate then dissociates in water:

$$(NH_4)_2CO_3 + H_2O \rightleftharpoons 2NH_4^+ + HCO_3 + OH^+ \rightleftharpoons 2NH_3\uparrow + CO_2\uparrow + 2H_2O$$
$$\underset{2NH_4OH}{\updownarrow}$$

$$\tag{8,4}$$

The amount of NH_3 lost from urea depends on the rate of fertiliser applied. Hargrove and Kissel (1979) found that 31, 26, 19 and 13 per cent of the applied urea-N at 56, 112, 224 and 448 kg N ha^{-1} respectively was evolved as NH_3 after two weeks of incubation (Fig. 8.11). Losses of NH_3 from urea also increase dramatically if fertiliser is applied when the soil is wet (Volk, 1966). Large losses occur when NH_4^+ and urea fertilisers are applied to flooded soils, particularly if they are also limed. Under these

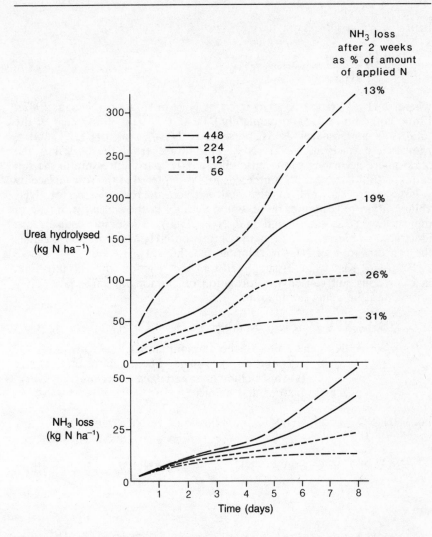

Figure 8.11 Amount of urea hydrolysed and subsequent NH_3 losses with time for four application rates of prilled urea in the laboratory. (Source: Hargrove and Kissel, 1979)

conditions, most NH_3 losses occur during the first week after N application (Ventura and Yoshida, 1977)

Given Equation **(8,2)**, it is not surprising that ammonia loss from soil depends on pH. The concentration of H^+ ions in solution affects the $NH_4 \rightleftharpoons NH_{3(aq)}$ equilibrium. The transformation of NH_4^+ in Equation **(8,2)** has the equilibrium constant (K):

$$\frac{[NH_{3(aq)}]\,[H^+]}{[NH_4]} = K = 10^{-9.5} \qquad\qquad (8,5)$$

hence

$$\log \frac{[NH_{3(aq)}]}{[NH_4]} = -9.5 + pH \qquad \text{(8,6)}$$

where $[NH_{3(aq)}]$ is the concentration of NH_3 in solution (Nelson, 1982). From Equation (8,6), approximately 0.0036, 0.36 and 36 per cent of the total soil ammoniacal N is present as $NH_{3(aq)}$ at pH 5, 7 and 9 respectively (Nelson, 1982). NH_3 will diffuse from the soil into the atmosphere according to the difference in NH_3 partial pressure in the soil solution and in the air. As ammoniacal-N is depleted at the soil surface by volatilisation, the pH at the soil surface decreases and for NH_3 volatilisation to continue, more ammoniacal-N in the soil must diffuse to the surface (Rachhpal-Singh and Nye, 1986). These mechanisms are illustrated in Fig. 8.12. The main controls on NH_3 loss from soil are thus the concentration of NH_3 in solution and the soil pH.

Fenn and Kissel (1973) suggested that the reaction of NH_4^+ fertilisers in calcareous soil caused the formation of ammonium carbonate. They formulated the general equation:

$$_A(NH_4)_\gamma X + {}_BCaCO_{3_{(s)}} \rightleftharpoons {}_B(NH_4)_2CO_3 + Ca_\beta X_\alpha \qquad \text{(8,7)}$$

in which: X = the anion of the ammonium salt
α, β, γ depend on the valency of the anions and cations
A and B are stoichiometric variables to account for valencies of cations and anions

For NH_4NO_3 and $(NH_4)_2SO_4$ applications to calcareous soils, this becomes:

$$2NH_4NO_3 + CaCO_3 \rightarrow (NH_4)_2CO_3 + Ca(NO_3)_2 \qquad \text{(8,8)}$$
$$\text{(soluble)}$$

and

$$(NH_4)_2SO_4 + CaCO_3 \rightarrow (NH_4)_2CO_3 + \quad CaSO_4 \qquad \text{(8,9)}$$
$$\text{gypsum (insoluble)}$$

In the case of $(NH_4)_2SO_4$, ammonium carbonate then dissociates in solution according to Equation (8.4). Since NH_4NO_3 forms stable $Ca(NO_3)_2$, $(NH_4)_2CO_3$ production is limited and losses of NH_3 from NH_4NO_3 are fairly small (Terman, 1979). Fenn and Kissel (1973) found that when soluble Ca products were formed, such as $Ca(NO_3)_2$ or $Ca(Cl)_2$, NH_3 volatile losses were small. When insoluble Ca products were formed, such as $CaSO_4$ or $CaHPO_4$, NH_3 volatile losses were high.

Many studies have compared NH_3 losses from different NH_4^+ salts and from urea. Fenn and Kissel (1973) found decreasing NH_3 volatilisation

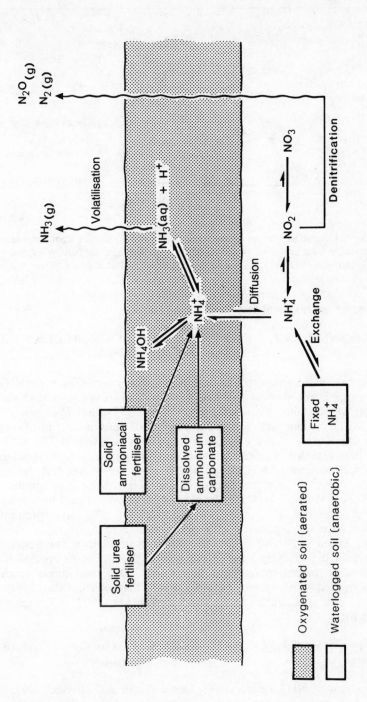

Figure 8.12 Fate of urea fertiliser nitrogen in waterlogged soil

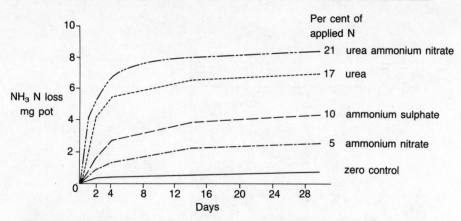

Figure 8.13   NH₃-N losses from an application of 36 mg of N in 30 days from ammonium nitrate, ammonium sulphate, urea, or urea ammonium nitrate solution surface applied to Crofton silty loam (pH 7.8) to which straw residue had been applied. (Source: Meyer, Olson and Rhoades, 1961)

for a range of ammoniacal fertilisers in the order:

$$NH_4F > (NH_4)_2SO_4 > (NH_4)_2HPO_4 > NH_4NO_3 = NH_4Cl > NH_4I$$
$$\text{decreasing } NH_3 \text{ loss} \rightarrow$$

There are two main soil/fertiliser conditions which result in high NH_3 losses: the application of neutral/acid salts to calcareous soils and the application of alkaline fertilisers to neutral/acid soils. The general sequence of increasing NH_3 loss from a range of ammoniacal fertilisers and urea applied to a silty clay loam at pH 7.8 is given in Fig. 8.13. Terman, Parr and Allen (1968) showed that the application of alkaline fertilisers such as urea or NH_4OH to acidic soils gave higher NH_3 losses than when neutral to acid fertilisers such as $(NH_4)_2SO_4$ were applied. They also showed that NH_3 losses from urea could be reduced by mixing it with acid or neutral NH_4^+ salts prior to application. The most common urea acidification treatment is with phosphoric acid to produce urea phosphate. Perhaps a more commonly reported problem is the general observation of higher NH_3 losses from NH_4^+ salts and urea applied to calcareous soils (for example, Fenn and Kissel, 1975) and those which have been treated with lime (for example, Fenn, Matocha and Wu, 1981).

In the presence of exchangeable Ca^{2+} ions in soil, $(NH_4)_2CO_3$ reacts according to:

$$\boxed{}\!\!-\ Ca + (NH_4)_2CO_3 \rightarrow \boxed{}\!\!-\ 2NH_4 +\ CaCO_3 \qquad \textbf{(8,10)}$$
$$\text{(precipitates)}$$

The adsorption of NH_4^+ reduces NH_3 losses (Fenn and Hossner, 1985). These authors further point out that nitrification of the NH_4^+ produces H^+ ions which dissolves the precipitated Ca and further reduces the potential for NH_3 loss. The influence of a high soil CEC on NH_3 loss is also apparent. Where NH_4 is able to exchange onto cation exchange complexes, the most effective reduction in NH_3 losses occurs when the exchanged cation, Ca for example, forms an insoluble precipitate and cannot exchange back onto the complex.

Fenn and Hossner (1985) have reviewed a range of management practices for the reduction of NH_3 losses from ammoniacal and urea fertilisers. NH_3 volatilisation depends on the soil attaining a pH $\geqslant 7$. This explains why either calcareous soils or alkaline conditions caused by fertilisation may enhance NH_3 loss. In theory, any treatment capable of lowering soil pH has potential for reducing NH_3 loss. Fenn and Hossner (1985) list three possible additives to urea-N fertilisers to improve efficiency: (a) inorganic acids, (b) inorganic salts, and (c) urease inhibitors. The addition of acid to urea is potentially most beneficial since urea granules create alkaline microsites. Most commonly, phosphoric acid is used to produce a urea phosphate. Stumpe, Vlek and Lindsay (1984) found that urea phosphate was little superior to urea in reducing NH_3 loss in calcareous soils but was more effective on non-calcareous soils. Fenn and Hossner (1985) suggest that the reason for differences in effectiveness may be due to the ratio of urea to phosphoric acid used in the fertiliser mixture. They recommend a urea:H_2PO_4 molar equivalent ratio $\geqslant 1:1$ for minimum NH_3 loss.

A range of inorganic salts, such as $CaCO_3$, $CaCl_2$ or KCl, have been added to urea to reduce NH_3 loss. Once urea has hydrolysed to ammonium carbonate, the reaction for $CaCl_2$ addition is given by:

$$(NH_4)_2CO_3 + CaCl_2 \quad CaCO_3 + 2NH_4Cl \qquad \textbf{(8,11)}$$
$$\text{precipitated}$$

The rate of NH_3 evolution for NH_4Cl and NH_4NO_3 applied to calcareous soils was shown earlier to be fairly low, so this type of treatment shows potential for reducing ammonia losses. Fenn and Hossner (1985) suggest that a potential saving could be made if, instead of using Ca in the additive salt, an alternative cation which is deficient in the soil could be used instead. In this respect, KCl and KNO_3 have shown some success.

Bremner and Douglas (1971) found that a large number of heavy metals and organic materials were capable of inhibiting urease activity and reducing urea hydrolysis. The efficiency of heavy metals in inhibiting urease activity followed the order:

Ag $\geqslant$ Hg > Cu > Pb > Cd > Zn > Sn > Mn

(Bremner and Douglas, 1971; Tabatabai, 1977). The best inhibitor was found to be $AgNO_3$ which reduced urease activity by 65 per cent in the first five hours. Of the organic materials assessed as urease inhibitors, some success has been achieved with quinones (Gould, Cooke and Bulat, 1978) and with halogen substituted (Cl, Br, F) benzoquinones (Bundy and Bremner, 1973). Despite these successes, the agricultural use of either heavy metals or organic additives is impractical. The large quantities of both types of compounds required for effective urease retardation would be uneconomical and would also lead to soil contamination and biologically toxic conditions.

8.5 FATE OF PHOSPHORUS FERTILISERS APPLIED TO SOIL

Longterm fertiliser experiments at Rothamsted Experimental Station, which have been running for over 100 years, indicate that annual P applications have penetrated no deeper than 40–50 cm downprofile (Cooke, 1980). Despite the apparent utility of longterm P retention in soil, there is much evidence to suggest that its effectiveness in terms of crop uptake decreases through time. This has led to examination of the value of *residual phosphorus* in soil and of its availability to different crops. Efficient retention and longevity of applied phosphorus is due in part to adsorption mechanisms (see Section 5.3.2) and in part to precipitation of insoluble P compounds under certain soil conditions. The dynamics of fertiliser P in soil thus depend on differential dissolution, adsorption and precipitation processes, with the production of various fertiliser reaction products. Three major controls on the fate of applied P fertilisers are thus: (a) the form in which P is applied; (b) the P adsorption capacity; and (c) soil conditions such as pH and Eh which control solubility and precipitation reactions.

Although the budgeting of P fertilisers has received less attention than N fertilisers, it is clear from crop recovery data (see Section 8.2) and longterm fertilisation experiments that there are two major sinks for applied P in the agricultural system: (a) crop uptake, and (b) retention within soil. There are no volatilisation losses and leaching losses occur only when a high rate of soluble P fertiliser applied to freely draining soil is immediately followed by heavy rainfall. Despite this, phosphorus fertiliser loss is heavily implicated in watercourse pollution and freshwater eutrophication. A more important route than leaching for the transport of fertiliser P to drainage waters is the surface runoff of animal slurries and organic manures. Overland flow and erosion of particulate matter, both soil and insoluble surface applied fertiliser, provide an additional P input to drainage waters.

8.5.1 Dissolution and reactions of phosphate fertilisers in soil

The solubility of phosphorus fertilisers applied to soil depends on their mineral form, grain or granule size, and on soil conditions, particularly moisture content and pH. Rock phosphates have been classified by McClellan and Gremillion (1980) into (a) Ca-phosphates (apatites), (b) Ca-Fe-Al phosphates, and (c) Fe and Al phosphates such as strengite ($FePO_4.2H_2O$) and variscite ($AlPO_4.2H_2O$). By far the most important of these commercially are the apatites. Crushed or ground forms of rock phosphates can be used very successfully on acid soils, particularly heathland and peaty forest soils, where the low pH allows adequate solubility. Above pH 6–6.5, rock phosphates are practically insoluble or, at best, sparingly soluble. To improve solubility for agricultural soils, rock phosphates are treated with sulphuric or phosphoric acid to produce superphosphate and triple superphosphate fertilisers respectively (see Table 8.1).

Superphosphate P fertilisers, whose main active constituent is mono-calcium phosphate (MCP), are highly water soluble. Sample, Soper and Racz (1980) suggest that water-soluble P fertiliser granules dissolve fairly rapidly once applied even to soil below field capacity, as they absorb water either by capillarity or by vapour transport. A solution is formed in the immediate vicinity of the granule which is supersaturated with respect to the phosphate salt. This sets up an osmotic potential gradient between the fertiliser solution and the soil water. The result is an outward movement of P in solution and an inward movement of water until the original fertiliser source is depleted and the concentration gradient is dissipated. As it moves away from the granule source, the solution is highly acidic (pH 0.6–1.4 for MCP) and concentrated in both P (3.4–5 M) and Ca (1.0 M) (Lindsay and Stephenson, 1959). In non-calcareous soils, P is removed from solution by reacting with the hydrous oxides of iron and aluminium to form insoluble compounds such as potassium and ammonium taranakites ($H_6K_3Al_5(PO_4)_8.18H_2O$ and $H_6(NH_4)_3Al_5(PO_4)_8.18H_2O$) and Al-Fe phosphate ($H_8K(Al,Fe)_3(PO_4)_6.6H_2O$). In calcareous soils with high pH, reactions between water soluble orthophosphates and soil calcium produce various sparingly soluble or insoluble calcium phosphates ranging from DCPD (dicalcium phosphate dihydrate, $CaHPO_4.2H_2O$) to OCP (octahedral calcium phosphate, $Ca_8H_2(PO_4)_6.5H_2O$) (Bell and Black, 1970).

On the basis of dissolution studies such as those mentioned above, Barrow (1980) simplifies the picture into three reaction zones around a fertiliser granule in soil. A central zone of residual fertiliser phosphate merges into a second zone of phosphate precipitation which merges into an outer zone of phosphate adsorption. In the case of MCP, the central zone of residual fertiliser would be mainly composed of DCPD. Concentrated phosphate solution diffuses out from this central zone into a zone where it may dissolve Ca, Al and Fe minerals which then react to form precipitated mineral phosphates. In the outer zone where concentrations are lower, P reactions are primarily adsorption reactions on the surfaces of soil particles.

DCPD, precipitated at the granule site and in the immediately adjacent soil, is the most commonly reported initial P fertiliser reaction product in agricultural soils. Although DCPD is only sparingly soluble, studies show that it is more soluble and available to plants than is OCP. Increasing soil pH reduces the solubility of Ca phosphates (Fig. 8.14) and Bell and Black (1970) reported that DCPD was even converted to OCP with increased pH and temperature. The ultimate end product precipitated is hydroxyapatite and, if there is sufficient fluorine in soil,

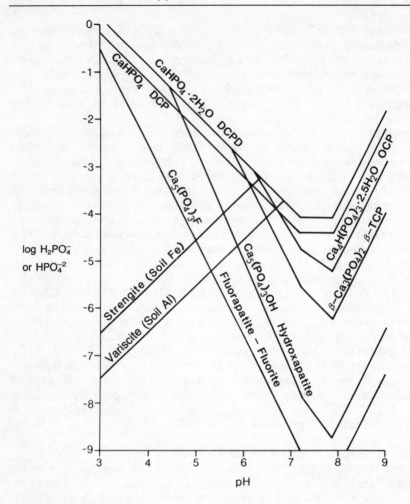

Figure 8.14 The solubility of calcium phosphates compared to strengite and varisicite when Ca^{2+} is $10^{-2.5}$ M or is fixed by calcite and $CO_{2(g)}$ at 0.0003 atm. (Source: Lindsay, 1979)

fluoroapatite, or fluorite (Khasawneh and Doll, 1978). White and Taylor (1977) suggest that at high phosphate concentrations (1000 μM), calcium phosphates will precipitate at values of soil pH > 5.5. Clearly, the formation and subsequent reactions of Ca phosphates in agricultural soils are important controls on the longterm value to crops of residual fertiliser P. In this context, the influence of liming on native and fertiliser P availability has received much attention. In a review of the subject, Haynes (1982) reports a wide range of results, from no effect of liming on phosphate availability, to increased P availability or decreased P availability. This range of experience on different soils is probably due to the complex interaction of several influencing factors relating firstly to

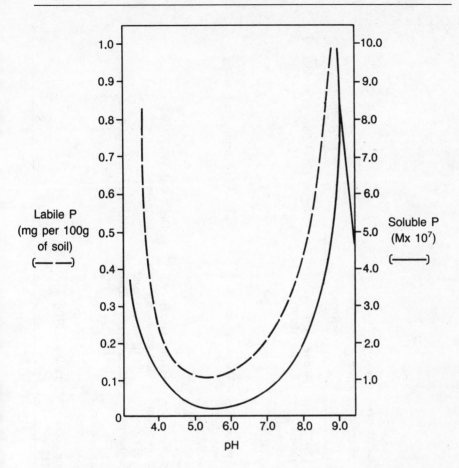

Figure 8.15 Effect of pH on amount of labile P and concentration of P in solution in a Flanagan silt loam in Illinois. (Source: Murmann and Peech, 1969)

increased pH and secondly to increased Ca concentration (Table 8.10).

It is difficult to make generalisations about phosphate availability in calcareous soils and the effect of liming on phosphate availability in acid soils. Haynes (1982) suggests that two main factors lead to increased phosphate availability when acid soils are limed: (a) increased mineralisation of soil organic phosphorus, and (b) air drying after liming causes decreased retention of added phosphate. Barrow (1984) has recently confirmed the results of Murrmann and Peech (1969) which indicate that the effect any increase in soil pH due to liming has on phosphate sorption is dependent on the initial soil pH prior to liming. In both studies, a rise in pH up to 5.5–6.0 decreased P sorption while a rise in pH > 6.0 tended to increase P sorption. Working on soils of the central United States, Murrmann and Peech (1969) found that an increase in soil pH caused both phosphate in solution and labile phosphate (see Section 8.5.2) to

Table 8.10 Some effects of liming on soil phosphate availability

Liming effect	Result	Reference	Phosphate availability
Increased pH	(1) In the short term, Al and Fe phosphates such as strengite and variscite are hydrolysed as the pH increases. This releases phosphate ions into solution	White and Taylor (1977); Lindsay (1979)	Increased (P solubility increased)
	(2) Increasing electrostatic repulsion of ions with increasing pH-dependent surface negative charge as the soil pH is raised by liming	Barrow (1980)	Increased (P sorption reduced)
	(3) If acid soils are limed, the initial rise in pH results in maximum phosphate sorption by hydroxy Al and Fe species at pH 5.5–6.5. Further pH increase above pH 6.5 causes hydroxy Al and Fe species to dissolve with the release of phosphate ions	Haynes (1982); Murrmann and Peech (1969); Sanchez and Uehara (1980); Barrow (1984)	Decreased (maximum P sorption) then Increased (reduced P sorption)
	(4) Calcium phosphates become less soluble with increased pH	Lindsay (1979)	

(Table 8.10 cont...)

Increased calcium concentration	(1) High concentrations of electrolyte cations, including calcium near soil colloid surfaces may reduce their negative charge and hence promote phosphate sorption	Chan, Davey and Geering (1979); Barrow (1984)	Decreased (increased P sorption)
	(2) At high soil phosphorus concentrations, addition of lime causes precipitation of insoluble calcium phosphates	Sample, Soper and Racz (1980)	Decreased (precipitation)
Overall effects	(1) Increased mineralisation of organic phosphorus fraction	Haynes (1982)	Increased (mineralisation)
	(2) (a) Formation of amorphous hydroxy Al-polymers which are highly phosphate-sorbing (stability of these polymers in soil is improved in the presence of fulvic acid, low molecular weight organic acids and the presence of inorganic anions such as phosphate, sulphate and silicate)	Haynes (1982)	Decreased (high P sorption)
	(c) Air drying after liming causes amorphous Al polymers to crystallise. Crystallised Al minerals have approximately four to five times lower phosphate sorbing capacity than amorphous hydroxy Al polymers	Haynes (1982)	Increased (lower P sorption)

initially decrease, pass through a minimum at about pH 5–6 and then increase at pH > 7. Figure 8.15 illustrates this relationship for Flanagan silt loam in Illinois. They concluded that the phosphate concentration in the soil solution was determined by the amount of labile phosphate rather than the solubility of crystalline mineral phosphates with different soil pH.

8.5.2 Labile and residual soil phosphorus

Although initial crop P recoveries of only 10–20 per cent are widely reported, many early texts and research reports describe 'residual' phosphorus and the 'residual value' of phosphatic fertilisers. This terminology has undoubtedly been used to indicate that crops may benefit in the longterm from previous P applications. This is particularly true where insoluble or sparingly soluble P compounds, such as rock phosphates, are applied to perennial crops such as plantation forestry.

Barrow (1978) suggested that P sorption is a two rate process. These rates are divided by White (1980) into (a) *fast reactions* which account for P exchange with other anions and with metal ion ligands, and (b) *slow reactions* which may be explained by a change from physical sorption to chemisorption, the precipitation of insoluble P compounds, the incorporation of P into hydroxy-Al, or Fe polymers or the diffusion of P into the crystal lattice of soil minerals. Evidence from a large number of laboratory studies indicates that a range of mechanisms, including adsorption, chemisorption and precipitation are responsible for fertiliser P retention in the soil (Sample, Soper and Racz, 1980). In most cases, these processes operate concurrently, which tends to mask individual contributions to P retention. Sample, Soper and Racz (1980) point out that it is impossible to quantify adsorption when the potential adsorbing surface is being constantly altered either by dissolution or by surface precipitation. Chemisorption and precipitation have occasionally been used to explain non-conformities in adsorption isotherms (see Section 5.3.3). Larsen (1967) and other workers have suggested that these fast and slow reactions reflect subsequent availability of P for exchange into the soil solution and for plant uptake:

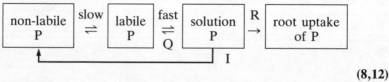

$$(8,12)$$

In this scheme, labile P refers to the pool of surface adsorbed, easily available P. Non-labile P refers to fixed forms which are not immediately available for plant uptake. Using the adsorption isotherm terminology of

Table 8.11 Indices used to characterise labile soil phosphorus

Index	Definition	Objective	Reference
L-value	The amount of soil phosphorus which is exchangeable, when isotopic equilibrium is attained, with labelled $^{32}PO_4^{3-}$ ions added to the soil. The L-value is measured by crop uptake over the growing season	To measure the total quantity of plant available phosphorus	Larsen (1952)
E-value	Since complete isotopic equilibrium cannot be achieved under laboratory conditions, the E-value is the amount of phosphorus exchangeable after an arbitrary shaking time. The E-value refers to a soil suspension from which no phosphorus is removed	'Speedy' laboratory equivalent of L-value	Russell, Erickson and Adams (1954)
A-value	The amount of available nutrient in a particular source measured in terms of a fertiliser standard, assuming that the plant will take up nutrient from two sources (for example, soil phosphorus and added labelled ^{32}P source) in direct proportion to the amounts available	To measure the availability of soil phosphorus relative to a standard source	Fried (1964)

Section 5.3.3, the labile P pool represents Q (quantity) while solution P represents I (intensity). R is the rate factor which is the replenishment of phosphate in solution around the roots.

The true definition of labile soil phosphorus is that fraction of soil phosphorus which can enter the soil solution by isoionic exchange in a specified time period (Larsen, 1967). The measurement of this fraction is only possible using isotope dilution techniques with ^{32}P, such as those outlined by Fried (1964). In a pure system with uniform mixing, isotopic dilution occurs when an exchange is set up between added $^{32}PO_4$ groups and exchangeable soil PO_4 groups, causing a dilution in the added $^{32}PO_4$ concentration. The major assumptions of the technique which cannot be attained in soil are that it is a pure system with uniform mixing of the applied ^{32}P. Three main indices of labile soil P have been developed (Table 8.11). L and E values depend on rates of reaction between soil and solution fractions of soil P while the A value is a plant availability index. It is clear that labile soil P does not represent a precisely defined and clearly distinct fraction of soil P but one that, in particular, has arbitrary boundaries of time and whose magnitude depends on rates of reaction (Olsen and Khasawneh, 1980). All three values appear to be fairly reliable for their specified objectives (see Table 8.11) within one growing season. Reddy, Saxena and Srinivasulu (1982) suggest that all three indices are good predictors of plant available P for maize and three types of pulse. They found very good correlations between E, L and A values and both dry matter production and P uptake.

In an attempt to understand the mechanisms of exchange involved in labile P exchange, Beckett and White (1964) identified two main classes of labile P: one held on 'net exchange sites' and one held on 'surface exchange sites'. Phosphates in the first group can be exchanged with OH^- or other $H_2PO_4^-$ ions, depending on solution $H_2PO_4^-$ concentration, while those in the second group are part of the crystal surface, exchanging isotopically and dissociating upon removal of an equivalent amount of complementary cation. If we accept this categorisation, Olsen and Khasawneh (1980) then query which of these two fractions control the intensity (I) of P in the soil solution. If P exchanged on net exchange sites controls solution P, then the relationship between intensity and labile P can be explained by an adsorption isotherm. If, on the other hand, solution P is controlled by surface exchange mechanisms, then solubility products will govern the relationship between intensity and labile P. While Olsen and Khasawneh (1980) speculate that it is probable that both sites operate simultaneously, several authors have shown that the solubility of the fertiliser reaction products, OCP and DCPD, are the main controls on solution P (see, for example, Sadler and Stewart, 1977; Griffin and Jurinak, 1973).

In the rooting zone, the amendment of soil with a soluble P fertiliser immediately increases the concentration of P in solution and hence the labile P pool. Since the rate of P transformation from labile to non-labile forms is very slow, high enough P intensities are usually maintained long enough after fertilisation to support the vital early stages of crop growth. The transformation of water-soluble phosphate fertiliser to insoluble forms does not necessarily mean that it has lost its fertiliser value. Russell (1973) suggests, as evidence for this statement, the fact that it is becoming increasingly difficult to find previously fertilised soils on which responses to further P fertiliser can be obtained. Quantification of this longterm or residual P may be made using indices of labile P such as those outlined in Table 8.11. Several other estimations of the residual value of phosphate fertilisers have been used. Simplest of these is the time period over which an application remains adequate. More complex are techniques which measure either the residual availability relative to that of a fresh fertiliser application or relative to the original availability of the fertiliser (Barrow, 1980).

The form of fertiliser and the granule or grain size are important controls on both immediate and residual effectiveness. Although a water-soluble phosphate fertiliser can enter the labile pool immediately while an insoluble one cannot, once in the labile pool, both will lose their effectiveness at the same rate. For this reason, soluble fertilisers, such as superphosphates, are most effective one to two years after application while insoluble fertilisers, such as DCP, have a more important residual value after two to three years (Russell, 1973). Similarly, powder form fertilisers may be more effective in the year of application, while granulated fertiliser P will have a longer residual value. This effect is illustrated in the data of Bouldin, DeMent and Sample (1960) for the uptake of P from two phosphate sources by oat crops in two successive years (see Table 8.12). Uptake from the soluble MAP source is much higher in the first year after application. Although the amounts of less soluble DCP taken up overall are lower than for MAP, P uptake from DCP is consistently higher in the second crop and also consistently higher from more finely granulated fertiliser.

Several authors have attempted to quantify the effectiveness of residual soil phosphorus with time after application. The simplest suggestion is that the half-life remains constant, or that the amount remaining effective decreases by a constant fraction each year. For Australian data, Barrow (1980) indicates that a linear decline in effectiveness may hold only for the first few years after application. The accumulation in soil of large reserves of fertiliser phosphate such as reported in longterm fertilisation studies at Rothamsted (Cooke, 1980), also suggests that the decline in effectiveness of fertiliser phosphate slows down with time. Laboratory measurements of change in exchangeable phosphorus confirm that the decrease in effectiveness is exponential, with the half-life varying from about one to six years (Larsen, 1967).

Wagar, Stewart and Moir (1986) used a sequential extraction in six stages to fractionate the phosphorus present in P fertilised soil. After five to eight years of cropping P fertilised soil, Wagar, Stewart and Moir (1986) found that at least half of the P fertiliser residues remained in a form available to plants (Fig. 8.16). It is unfortunate that these authors did not monitor changes in the six fractions during the one to five year period, so the rate of change in these fractions in the critical two to three year period after application could not be studied. There are two clear results from this work. First, only the more available forms of P declined over the one to five year period. Second, significant amounts of applied P were converted into soil organic P. Other authors have indicated the importance of phosphatase enzyme activity in the production of plant available phosphorus from the organic P pool (see, for example, Harrison, 1982; Hedly, Stewart and Chaunan, 1982). Sharpley (1985) clearly demonstrates the build up of organic P after the application of phosphate fertiliser (Fig. 8.17a) and the seasonality of availability of different organic P fractions (Fig. 8.17b). Several authors have estimated plant available P using a single chemical extraction. Commonest of these is the $HCl-NH_4F$ extraction of Bray and Kurtz (1952); often termed Bray-1 phosphorus or Bray-1P. Barber (1979) used the Bray-1P extraction to estimate soil P availability over 25 years of cropping and five P fertiliser application rates. Using the linear regression of Bray-1P versus

Table 8.12 Uptake of phosphate by two successive oat crops from soil treated with five different granule sizes of MAP (monoammonium phosphate, soluble) and DCP (dicalcium phosphate, relatively insoluble) (relative uptake in arbitrary units) (from Bouldin, DeMent and Sample, 1960)

Granule size (mm)	2.0–1.2		1.14–0.84		0.59–0.42		0.42–0.30		0.30–0.25	
Fertiliser	MAP	DCP	MAP	DCP	MAP	DCP	MAP	DCP	MAP	DCP
First oat crop	608	30	601	50	423	90	387	119	331	136
Second oat crop	211	57	168	133	246	253	220	320	217	376

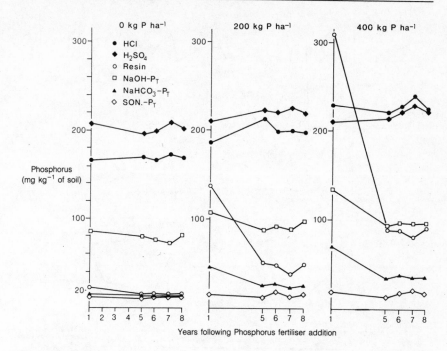

Figure 8.16 The phosphorus content of six soil phosphorus fractions extracted from the surface horizons (0–15 cm) of Waskada soil receiving 0.200 and 400 kg P ha^{-1}, sampled 1, 5, 6, 7 and 8 years following fertiliser application. (Source: Wagar, Stewart and Moir, 1986)

net change in soil P/ha over 25 years, he found that 17 kg P ha^{-1} was required to cause a change in Bray-1P of $1 \mu g^{-1}$. Sharpley (1985) compared the Bray-1P index with inorganic and organic P contents of both fertilised and unfertilised agricultural soils. Fig. 8.18a illustrated the relationship, in unfertilised soil, between the slope of the Bray-1P versus organic P regression and phosphatase enzyme activity. As phosphatase activity increases, the potential for organic P mineralisation and the release of available P increases. Similarly, Fig. 8.18b illustrates the relationship in fertilised soil between the slope of the Bray-1P versus inorganic P regression and the Langmuir P adsorption maximum. As the P adsorption maximum increases, a smaller proportion of applied fertiliser P remains in an available form (Sharpley, 1985).

It is clear from these results that the contribution to plant available P from organic P in non-fertilised soils and inorganic P in fertilised soils requires further study, particularly in soils of widely differing organic matter content and for different P fertilisers. The relative longevity of their contributions may also be important in assessing the longerterm residual value of P fertilisers.

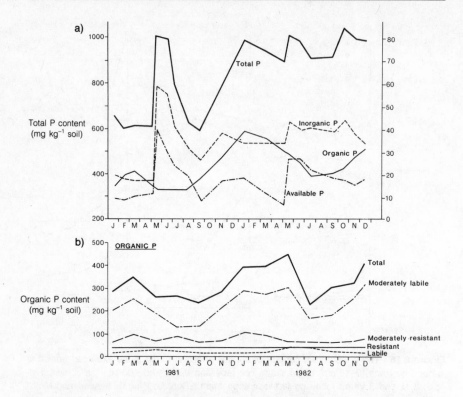

Figure 8.17(a) Seasonal variation in total, inorganic, organic and available P content of surface soil (0–50 mm depth) from fertilized soil. **(b)** Seasonal variations are labile, moderately labile, moderately resistant and resistant organic P content of surface soil (0–50 mm depth) from fertilised soil. (Source: Sharpley, 1980)

8.5.3 Losses of phosphate fertiliser by leaching, runoff and erosion

Although agricultural fertilisers have been heavily implicated in the eutrophication of ground and surface waters, there is scant evidence for phosphate leaching from the majority of mineral agricultural soils. In a fertilised catchment in Pennsylvania, Gburek and Heald (1974) found that less than 2 per cent of the applied phosphate fertiliser was carried into adjacent streams. Phosphorus concentrations in streamwater was markedly seasonal, with highest concentrations in summer months. Substantial P leaching losses have been reported on sandy soils (for example, Ozanne and Shaw, 1961; Mattingly, 1970) and on organic soils (see, for example, Miller, 1979). In the majority of agricultural soils, P sorption retains applied phosphates within the soil profile and acts as a sink rather than a source for P in surface waters.

Since phosphorus is relatively immobile in soil, P is primarily lost from

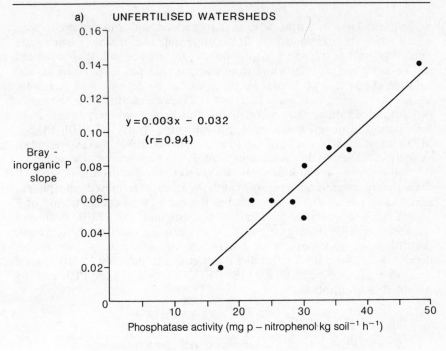

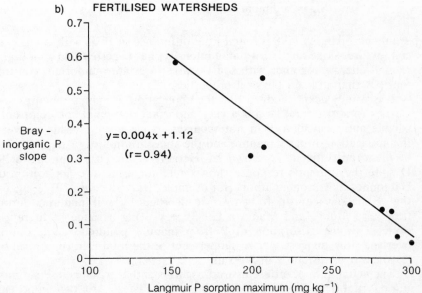

Figure 8.18(a) Relationship between the slope of the regression Bray–organic P content for 1981 and 1982 and phosphatase activity for the unfertilised soils. **(b)** Relationship between the slope of the regression Bray–inorganic P content for 1981 and 1982 and Langmuir P sorption maximum for the fertilised soils. (Source: Sharpley, 1980)

agricultural land in forms adsorbed to eroded soil and transported to watercourses by overland flow. Both inorganic and organic particles are implicated in the transport of phosphorus by surface runoff: phosphates can be adsorbed and desorbed from mineral particles, depending on soil solution chemistry, while organic particles can be decomposed to release soluble inorganic and organic P. There is a clear indication that the major proportion of P in surface runoff to streamwater is sediment bound (see, for example, Burwell et al., 1974; Sharpley and Syers, 1979; Bhatnagar, Miller and Ketcheson, 1985). Gburek and Heald (1974) also suggest that, in their catchment, the main source of P in streamwater was sediment derived from streambank erosion. Since eroded soil is usually richer in P than the remaining surface soil, authors have determined phosphorus enrichment ratios (PER), calculated as the ratio of the concentration of P in the eroded sediment to that in the original soil. PER decreases markedly with increasing intensity of soil erosion, and the observation that PER increases as the sediment concentration in surface runoff decreases has been used to predict P enrichment. Sharpley (1980) reports a negative linear expression for the relationship between log PER and log sediment concentration:

$$\ln_{PER} = 2.4 - 0.27 \ln_{SD}$$

where: PER = total phosphorus enrichment ratio
SD = sediment discharge.

Although Sharpley (1980) found little difference in PER with changes in soil texture, slope angle or rainfall intensity, he suggested that runoff and rainfall energy, together with soil P status, are more important controls on PER than are soil physical properties.

It is widely acknowledged that the application of animal manures and slurries to agricultural land is a very important non-point source of both soluble and particulate P in watercourses. A close relationship between the application of dairy manure and the concentration of orthophosphate in tile drain effluent is shown by Hergert et al. (1981) in Fig. 8.19. Despite the dramatic response shown here for a manure application of 200 tonnes ha^{-1}, some authors (for example, Burwell et al., 1974) suggest that P in surface runoff can be readsorbed by soil material during transport. Moore and Madison (1985), reviewing a range of literature sources on the adsorption of P from manure polluted runoff during overland flow, suggest that 46–88 per cent of the total P in runoff can be removed over distances of 4–33 m.

The actual role of different sized particles and aggregates in P sorption during runoff is debated in the literature. There are two important aspects of P adsorption during surface wash. First, if particles in the surface soil adsorb P from solution during overland transport but they themselves are not detached and eroded, but remain intact, then the concentration of P in surface runoff will be reduced. Thus, Bhatnagar, Miller and Ketcheson (1985) found that larger soil aggregates preferen-

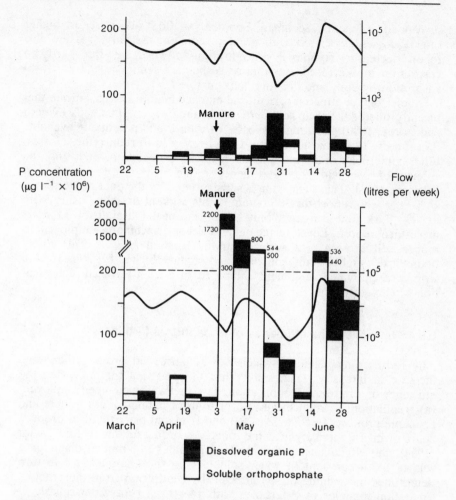

Figure 8.19 Soluble orthophosphate and dissolved organic P in tile drain effluent from soil fertilised with dairy manure. (Source: Hergert *et al.*, 1981)

tially adsorbed the P from liquid manure applications but not from inorganic fertilisers. They suggest that this may explain lower extractable phosphate PER values in sediments form manured plots compared to fertilised plots. Second, if particles carried along in suspension adsorb soluble P, then the P concentration of surface runoff has the capacity to become more concentrated with distance downslope. Sharpley *et al.* (1981) have suggested that the selective erosion of finer soil particles which have a greater P sorption capacity enhances P sorption during overland flow.

There is a clear indication from numerous studies that greater P losses are associated with higher P application rates. The potential for soluble P loss is dependent on the adsorbing capacity of both surface soils and suspended sediments. Schuman, Spomer and Piest (1973) report higher total P losses associated with greater sediment discharges. They also report decreasing soluble P concentrations downstream, due partly to adsorption by increased amounts of suspended sediment and partly to adsorption by gully and channel bank material.

Studies on the effect of agricultural management practices indicate that recently tilled soils (Barisas *et al.*, 1978) and row cropped soils (Wendt and Corey, 1980) show highest losses of total P and potentially available P in runoff. Two main techniques have been used to reduce these losses. First, underfield drainage systems increase percolation, reducing the volume of overland flow and hence reducing P runoff (Baker and Johnson, 1976). This suggestion is substantiated by the data of Hergert *et al.* (1981) who report that P content of tile effluent draining a silty loam in New York State contained only 19 per cent of the total dissolved losses in surface runoff. Level terracing is a second technique employed to reduce soil erosion and surface runoff. Burwell *et al.* (1974) found dramatically higher sediment yield and slightly higher P losses from a strip corn cultivated catchment compared to a level terraced, conservation catchment (Table 8.13).

8.5.4 Improving the efficiency of phosphorus fertilisers

Any fertiliser management which can enhance and prolong the initial increase in intensity (I) in soil P, has the potential for improving the efficiency of a fertiliser P application. Two methods have been employed: (a) granulation, to reduce the rate of transformation to less soluble and insoluble products (see Table 8.12); and (b) band placement, to produce zones of high P intensity near the roots. According to Sample and Taylor (1964), the effectiveness of P fertiliser granules $\leqslant 6$ mm in diameter is related to the amount of water soluble P in the granule. This in turn determines the volume of soil affected by P diffusion from the granule. For 6 mm granules of P fertiliser with 14–70 per cent water-soluble P, they found that P diffusion reached soil volumes of 4.2 and 20.6 cm respectively. The choice of a large granule size is thus based on the probability of roots coming in contact with larger but widely spaced P diffusion zones. A major aim in band placement is to reduce soil–fertiliser contact and hence reduce soil 'fixation' of the applied P. Sleight, Sander and Peterson (1984) studied band placement in a calcareous silt loam in which fertiliser P is relatively immobile due to the high fixing capacity. They found that, for the early, effective use of fertiliser P by oats, improved fertiliser–root contact by band placement was more important than reduced soil–fertiliser contact.

Table 8.13 Sediment and phosphorus content of runoff from a level terraced catchment in South-west Iowa (data from Burwell et al. 1974)

Catchment	Year	Precipitation (cm)	Total runoff (cm)	Sediment yield (kg ha^{-1})	Phosphorus in surface runoff (kg ha^{-1}) solution P	sediment P	Total phosphorus discharged (kg ha^{-1})
Level terraced (157.5 ha)	1970	70.41	6.94	360	0.112	0.093	0.243
	1971	67.79	10.74	1770	0.272	0.329	0.648
Strip cultivated corn (33.6 ha)	1970	78.28	10.49	16 640	0.058	0.464	0.603
	1971	73.71	16.40	29 720	0.186	1.123	1.330

8.6 FATE OF FERTILISER POTASSIUM IN SOIL

Most agricultural soils show fairly high crop uptake efficiencies for K fertilisers (Fink, 1982) and only rarely, on light soils, are high leaching losses associated with K fertilisation (see, for example, Jones and Hinesly, 1986). Cooke (1966) suggested that one of the main inefficiencies in K fertilising is due to 'luxury consumption' when crops take up more K than they need to give a maximum yield. These observations suggest that exchange and fixation of K on soil colloids acts not only to retain potassium in the rooting zone, where plant roots can use it efficiently, but also to prevent potassium leaching losses in drainage waters. This efficient retention can lead to the build up, particularly in clay soils, of substantial residual K fertiliser.

The leachability of K fertilisers depends on the anion with which it is associated. Generally, the order of increased leaching for the more widely applied K salts is: $KH_2PO_4<K_2SO_4<KNO_3=KCl$. While potassium leaching losses on clay soils are frequently reported as insignificant (<2 per cent) (see, for example, Williams, 1971; Jones and Hinesley, 1986), up to 70 per cent leaching losses have been reported on a sandy forest podzol in Dorset (Bolton and Coulter, 1966). A second process for K losses from surface applied fertilisers is by erosion and runoff. Jones and Hinesly (1986) found that large K losses from KCl applied to corn occurred in runoff during heavy summer rains and in winter when rapid snowmelt and heavy rain took place.

We saw in Section 5.2.3 that K^+ and Na^+ monovalent cations, with small hydrated atomic radii, are strongly attracted to cation exchange sites, but may be replaced by divalent cations such as Ca^{2+} and Mg^{2+} if these are present in sufficient quantities in the soil solution. This form of exchangeable K is undoubtedly the main store of plant available K in soil. Additionally, soil minerals such as micaceous clays and orthoclase feldspars, may weather slowly to supply soluble K in the rooting zone. The analagous sorption and fixation of NH_4^+ and K_+ ions in soil was discussed in Section 8.3.1; K^+ ions, like NH_4^+ ions, are held on soil colloids with two different degrees of strength, depending on whether they are fixed in internal clay lattices, primarily vermiculites and degraded illites, or whether they are exchanged onto colloid surfaces. It is thus possible to identify exchangeable (plant available) and non-exchangeable forms of K in soil (Fig. 8.20). Nye and Tinker (1977) suggest that it is convenient to divide K exchange reactions into three groups:

(1) *Rapid cation exchange reactions* with half-times of the order of $\leqslant$ one hour.
(2) *Intermediate K fixation reactions*, occurring in the interlayers of vermiculite and illite clay minerals, with half-times of 1–24 hours.
(3) *Slow mineral weathering reactions*, with half times of the order of $\geqslant$ one week.

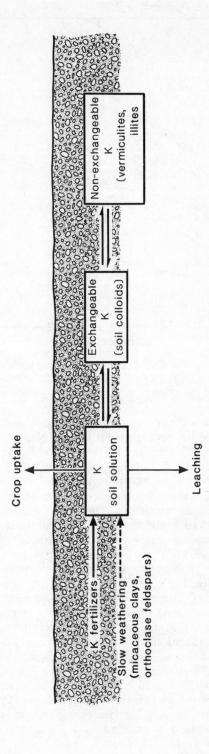

Figure 8.20 Fate of fertiliser K in soil

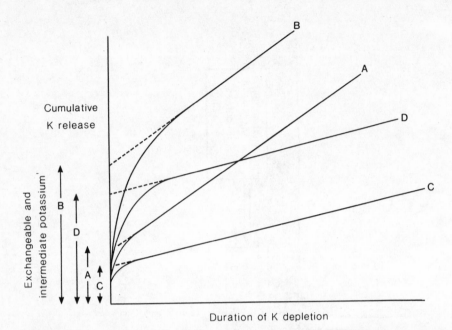

Figure 8.21 Representation of release of potassium from soils under continuous cropping unfertilised A and fertilised B plots on a virgin soil. Unfertilised C and newly fertilised D plots on the same soil, following prolonged exhaustion. (Source: Nye and Tinker, 1977)

The release of K from a range of K fertilised and K unfertilised soils in England was examined by Becket (1969) (Fig. 8.21).

The capacity of clay minerals to fix K was found by Arfin, Perkins and Tan (1973) to range from 0.3 to 0.6 me g^{-1}, increasing in the order: micas > illite > vermiculite > montmorillonite. There is some doubt about how quickly K$^+$ applied to the soil can become fixed in a non-exchangeable form in clay interlayers. The slow diffusion coefficients for K$^+$ in clay interlayers (10^{-17}–10^{-23} cm^2 s^{-1}) accounts for the very slow K adsorption (several months) in clay interlayers, reported by Amberger, Gutser and Teicher (1974). There is a range of evidence to suggest that interlayer fixed K is available for uptake by plant roots (see, for example, Tabatabai and Hanway, 1969), particularly from the weathering edges of clay minerals. It is also clear that interlayer K can become replaced by other cations and H$_2$O. Where replacing cations are larger in size, such as Ca^{2+} and Mg^{2+}, clay layers are wedged open to allow replaced K$^+$ to diffuse out. The concentration of K$^+$ in equilibrium with interlayer K$^+$ decreases as the interlayer K$^+$ is depleted (Newman, 1969).

Greatest root uptake efficiency is obtained when optimum solution K concentrations (5×10^{-5}–5×10^{-3} M) can be maintained in solution adjacent to active root surfaces (Haworth and Cleaver, 1963). The ability of the soil to maintain adequate K$^+$ concentrations around root surfaces depends on (a) mass flow to the root (K$^+$ carried in solution, as determined by H_2O absorption, or transpiration rate); (b) diffusion of K^+ ions in solution; and (c) the ability of the cation exchange complex to supply K^+ to the soil solution. Drew and Nye (1969) have shown that direct contact between root and soil can only account for about 6 per cent of the plant's K requirements. This result illustrates the importance of root elongation and solution transport processes in providing K from beyond the limit of the root hair cylinder.

Since applied K^+ is exchanged rapidly on cation exchange sites and fixed in interlayers of clay minerals, residual K can build up in longterm fertilised soils. Unlike residual phosphorus, not all the effects of residual fertiliser K are advantageous. In 70–100 year long fertilising experiments in England, Goulding and Talibudeen (1984a, 1984b) have shown that residual, inorganic fertiliser K has an important effect in decreasing CEC in some soils and in decreasing K^+ adsorption on K selective sites. They suggest that residual K may gradually modify colloidal exchange sites so that the binding strength of K^+ relative to Ca^{2+} is decreased.

8.7 SUMMARY

Nitrogen, phosphorus and potassium (N, P, K) are the main three nutrients applied extensively to agricultural soils. In examining the fate of these materials in the soil and agricultural system, two main issues attract particular attention: (a) fertiliser efficiency, and (b) environmental pollution. There are five possible 'fates' of fertilisers applied to the soil: (a) uptake by plants and animals (immobilisation); (b) fixation (adsorption and exchange) in the soil; (c) leaching and soluble losses in drainage; (d) volatilisation and gaseous loss; and (e) surface loss as runoff and erosion. The annual percentage recovery of N, P and K fertilisers are about 30–70 per cent, 30 per cent and 75–90 per cent respectively, with higher P recoveries over longer time periods.

The main losses of nitrogen from applied N fertilisers are through nitrate leaching, particularly in coarse textured soils, and by denitrification in anaerobic and waterlogged soils. NO_3^- leaching losses tend to predominate if heavy rainstorms immediately postdate nitrate fertiliser applications, while denitrification losses tend to predominate on heavy clay soils, usually with high summer temperatures and frequent irrigation. N losses of up to 30–35 per cent of that applied have been recorded for both leaching and denitrification. Current nitrate field and modelling studies aim to provide fertiliser management advice to minimise the amounts of nitrates leaching to drainage and watercourses.

Fertiliser phosphates are strongly adsorbed by Fe and Al hydroxyoxides in soil and are retained in the profile for long periods of time: over 100 years in fertiliser experiments at Rothamsted Experimental Station. The effectiveness, in terms of crop uptake, of this retained or 'residual' phosphate, decreases through time. Phosphorus availability to crops has been described in terms of labile (surface adsorbed, easily available P) and non-labile (fixed P, not immediately available for plant uptake).

Since calcium phosphates precipitate at pH > 5.5, soil pH and liming are important controls on both early P availability and longterm residual P value to crops. The recovery by crops of applied potassium fertilisers is usually high, with leaching losses only significant on light soils. Efficient exchange of fertiliser K onto cation exchange sites in soil retains K locally available for plant uptake. Only occasionally can the buildup of residual K cause a reduction in effective soil CEC.

CHAPTER 9

Fate of Synthetic Pesticides Applied to Soil

9.1 INTRODUCTION

The wide range of organism-killing compounds commonly called pesticides includes three main groups of compounds: insecticides, herbicides, and fungicides. Also in common use are molluscicides and rodenticides. Although virtually all of the modern formulations are synthetic organic chemicals, many of the earlier compounds were:

(1) inorganic, such as 'Paris Green' (copper acetoarsenate) or 'Bordeaux Mixture' (copper sulphate, quicklime and water);
(2) derived from coal tar, such as creosote; or
(3) derived from plant materials, such as pyrethrum, derris root or nicotine.

Inorganic compounds are now rarely used, due to the lifespan of their heavy metal or toxic component in the soil. Pesticides made from plant materials are considered to be safer than synthetic compounds and are still in wide use, in response to the environmental conservation lobby. The vast majority of pesticides are organic chemicals, numbering over 450 compounds, in over 10,000 formulations. Two main reasons can be suggested for the continued development of further compounds:

(1) threats to the environment and man through pesticide persistence and toxicity, particularly to non-target organisms; and
(2) the development of pesticide resistance in target organisms.

When applied to soil, these organic compounds behave in a number of ways. They can be:

(1) degraded, either by soil organisms, or by physicochemical processes;
(2) adsorbed by soil organic matter, clay minerals or iron and aluminium sesquioxides;
(3) washed into watercourses through leaching and runoff; or
(4) volatilised, resulting in atmospheric pollution.

In addition, pesticides are known to cause adverse effects on non-target organisms in the soil, resulting in disrupted soil biological processes. Each of these problems is explored in subsequent sections of this chapter. The discussion is confined to the effects on soil processes of compounds widely used as agricultural and domestic insecticides and herbicides, including those landing at the soil surface from foliar wash. Many more compounds are added to soil for weed control than for other kinds of pest control. In many cases the mode of action of herbicides, whether leaf contact or root absorption, for example, determines when and how the chemicals are applied to the soil. This has an important influence on subsequent pesticide behaviour and persistence in soil. A comprehensive classification of herbicides and their mode of operation is given in Ashton and Crafts (1973).

9.2 TYPES OF PESTICIDES USED IN SOIL

Three main groups of insecticides have been used in agriculture and in soil (Fig. 9.1). *Organochlorine insecticides*, including DDT, lindane, dieldrin, aldrin and heptachlor, have the longest persistence in soil of all organic pesticides. Their persistence follows the order: DDT > dieldrin > lindane > heptachlor > aldrin, with half lives ranging from about eleven years for DDT to four years for aldrin. DDT was originally highly successful during the Second World War in controlling the lice that carried typhus and the mosquitoes that carried malaria. Only more recently was its persistence and accumulation realised, when large amounts of DDT residues were discovered in the soil, freshwater sediments, ecological foodchains and in foodstuffs such as fish and cows' milk destined for human consumption. Mellanby (1967) and Carson (1963) give many examples where whole microbial and animal communities were destroyed by the use of DDT, which is now banned in many countries, but not the use of some of its related organochlorine compounds.

The second major group of insecticides, *organophosphates*, are related to nerve gases and are highly toxic to mammalian pests and dangerous to humans. They are used in preference to organochlorines primarily because their half-life in soil is much shorter: around six months for parathion, diazinon and demeton. *Carbamate* pesticides can be used to kill a wider range of organisms than organochlorine and organophosphate insecticides. They are used as molluscicides, fungicides and insecticides. Carbamates are about as persistent in soil as organophosphates, with half-lives of around six months.

Five main groups of organic chemicals are used as herbicides (Fig. 9.2). The *phenoxyacetic acids* are used as selective herbicides, the most important of these being 2,4-D and 2,4,5-T, which are used to control selectively broadleaf herbs and shrubs respectively in grassland swards or cereal crops. Both compounds appear to be degraded fairly quickly in soil but both have recently been banned in many countries because they have caused growth abnormalities in experimental animals. In addition, highly toxic dioxin impurities were discovered in 2,4,5-T manufactured in the United States. This came to light in the aftermath of the war in Vietnam where highly concentrated defoliants, such as 'Agent

Orange' (a 50 : 50 mixture of 2,4-D and 2,4,5-T), have not only destroyed the forest and mangrove vegetation, but are also thought to have caused animal and human growth and reproductive abnormalities.

The *toluidines* and *triazines* are not easily leached from soil. Toluidines are strongly adsorbed, while some triazines are relatively insoluble and thus act over a longer period of time as they dissolve. The *phenylureas* are less persistent than toluidines and triazines since they are very soluble and easily leached from soil. Perhaps the best known of all domestic herbicides are the *bipyridyls*, of which paraquat is the most commonly used. Bipyridyls are total herbicides, strongly adsorbed in soil and very persistent.

9.34   PERSISTENCE OF PESTICIDES IN SOIL

Pesticide persistence in soil is determined by the balance between adsorption on soil colloids, uptake by plants, transformation or degradation processes and losses in liquid or gaseous form (Fig. 9.3). This balance is controlled partly by the type of pesticide used, its amount and method of application, and partly by soil characteristics such as texture and organic matter content. Environmental factors, such as temperature and moisture, through their influence on rates of reactions, are also important controls on pesticide persistence. Hurle and Walker (1980) suggest that the optimal period for persistence of herbicides in soil would be 'long enough to give an acceptable period of weed control but not so long that herbicide residues, after crop harvest, limit the nature of the subsequent crops which can be grown'. This requirement has led to the development and increasing use of pesticides whose active life in the soil can be measured in terms of weeks and months rather than years. The persistence of some commonly used insecticides and herbicides is given in Table 9.1 and Fig. 9.4. Data such as these are typically generated in laboratory incubations or, at best, glasshouse plots, often using simplified soils such as silts and sands. Their application to field conditions requires some care.

Some field data on the persistence, particularly of longterm compounds such as metallic and organochlorine pesticides, are available to substantiate laboratory studies. It is evident in the orchard soils of Massachusetts that 50 years of lead arsenate pesticide application has

Table 9.1   Persistence of pesticides in soil

Group	Persistence	Example
Non-persistent	1–12 weeks	Organophosphates, carbamates
Moderately persistent	1–18 months	Phenoxyacetic acids, triazines
Persistent	2–5 years	Organochlorines
Permanent	not degraded	Heavy metals (Hg, Pb, Cu), As

NAME	EXAMPLES	INSECTICIDE TYPE AND MAIN USE	SOLUBILITY IN WATER AT STP	STABILITY
1. Organochlorines	DDT	persistent contact insecticide for soil pests and other soil insects.	1.2×10^{-3} mg l^{-1}	Fe, light and UV promote decomposition.
	1,1,1-Trichloro-2,2-bis(p-chlorophenyl)ethane (also known as dichlorodiphenyltrichloroethane)			
	ALDRIN	contact insecticide in seed dressing, to prevent wireworm and beetle damage	<0.05 mg l^{-1}	stable in alkali, neutral and weak acidic solutions.
	1,2,3,4,10,10-hexachloro-1,4,4a,5,8,8a-hexahydro-1,4endo-exo-5,8-dimethanonaphthalene			
	HEPTACHLOR	persistent contact insecticide in seed dressings to prevent wireworm and beetle damage.	insoluble	very stable
	1,4,5,6,7,8,8-heptachloro-3a,4,7,7a-tetrahydro-4,7-endomethanoindene			
2. Organophosphates	PARATHION	contact insecticide and acaricide, biting and sucking pests in fruit and vegetable crops	24 mg l^{-1}	hydrolyses rapidly in alkaline solution, to slowly at pH<6

MALATHION

0,0-diethyl O-p-nitrophenyl phosphorothionate

contact insecticide, sucking pests — aphids, mosquitoies, spidermites

145 mg l^{-1}

decomposed by acids and bases; stable in neutral solutions

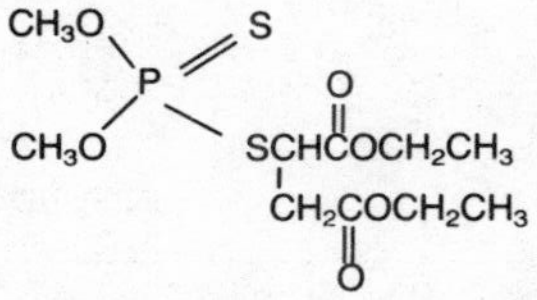

0,0-dimethyl S-(1,2dicarbethoxyethyl) phosphorodithioate

3. Carbamates

CARBARYL

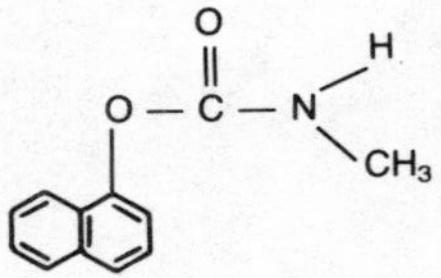

1-naphthyl N-methylcarbamate

contact insecticide, insect control in fruit crops, earthworm killer in turf

<1 mg l^{-1}

stable in neutral and weakly acid solutions hydrolyses in alkaline solutions

CARBOFURAN

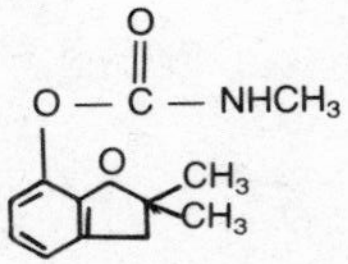

2,3-dihydro-2,2-dimethylbenzofuran-7-yl, methyl carbonate

contact insecticide, wide range of vegetable pests, including Colorado beetle, root flies, nematodes

250–700 mg l^{-1}

unstable in alkaline solution

Figure 9.1 Main types of insecticides used in soils. (Source: Royal Society of Chemistry, 1983)

NAME	EXAMPLES	HERBICIDE TYPE AND MAIN USE	SOLUBILITY IN WATER AT STP	STABILITY
1. Phenoxyacetic Acids	2,4-D 2,4-dichlorophenoxyacetic acid	Selective contact herbicide; Post emergent control of broadleaved weeds in cereal and grass crops	600 mgl^{-1}	stable up to 50 °C for at least 2 years.
	2,4,5-T 2,3,5-trichlorophenoxyacetic acid	Selective contact herbicide for woody weeds in grassland, heather and woody weeds in forestry.	<278 mgl^{-1}	stable up to 50 °C for at least 2 years
2. Toluidines	TRIFLURALIN 2,6-dinitro-N,N-dipropyl-4-trifluoromethylaniline	Selective, post emergent soil herbicide for incorporation into seedbed soil and pre emergent weeds in cereal crops.	<1 mgl^{-1}	very stable, but decomposed by UV light

3. Triazines			
ATRAZINE	Soil and leaf contact herbicide pre emergent and post emergence control of weeds under annual crops. High concentration acts nonselectively on paths.	70 mgl^{-1}	stable in neutral, weakly and weakly alkali solution.

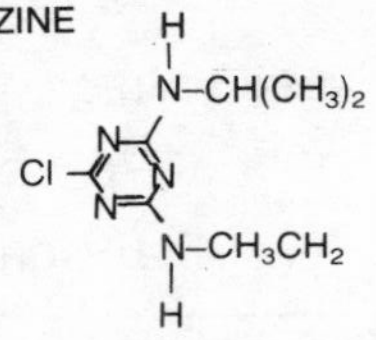

2-chloro-4-ethylamino-6-isopropylamino-1,3,5-triazine

| SIMAZINE | Root absorption herbicide total weed control on non-cropped land, grass and weeds under fruit. | 5 mgl^{-1} | stable in neutral, weakly acid and weakly alkali solution |

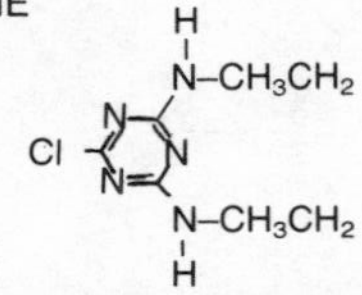

2-chloro-4,6-bis(ethylamino)-1,3,5-triazine

4. Phenylureas			
FENURON	Soil-leaf herbicide with low selectivity, absorbed via roots; woody and deep-rooted perennial weeds, aquatic weeds nonselectively on paths.	3.85 mgl^{-1}	stable; decomposed by strong acids and alkalis

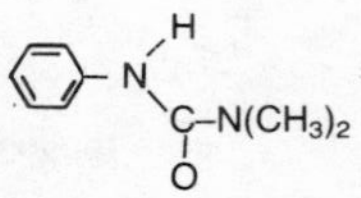

1,1-dimethyl-3-phenurea

Figure 9.2 Main type of herbicides used in soils. (Source: Royal Society of Chemistry, 1983)

(Figure 9.2 continued)

5. **Bipyridyls**			
DIQUAT (dibromide)	non-selective contact herbicide and desiccant, broadleaves in cereal crops; aquatic weeds and algae	700 mgl⁻¹	stable in neutral and acid solution; decomposed by UV light

1,1'-ethylene-2,2'-dipyridylium dibromide

PARAQUAT (dichloride)	short duration contact herbicide pre emergent control in seedbeds and for perennial weeds in non-cropped areas.	readily soluble	photochemically decomposed by UV light in aqueous solution

1,1'-dimethyl-4,4'-bipyridylium dichloride

6. **Glycines**			
GLYPHOSATE	systemic, non-selective post emergent leaf herbicide for grasses, broadleaved weeds and woody shrubs	10 mgl⁻¹	5% loss after storage at 50 °C for 2 years

N-(phosphonomethyl) glycine

Royal Society of Chemistry (1983)

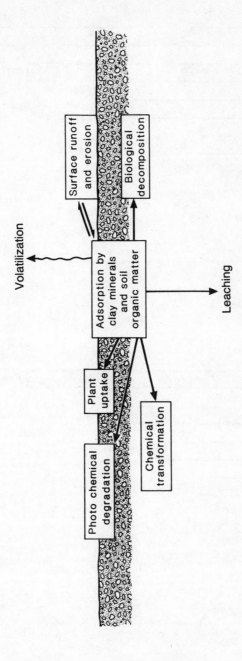

Figure 9.3 Fate of pesticides in soil

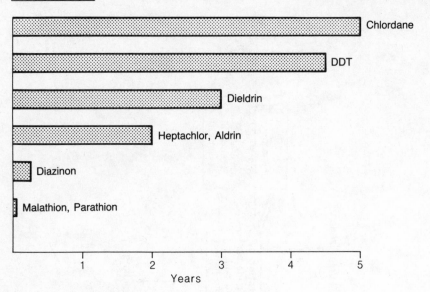

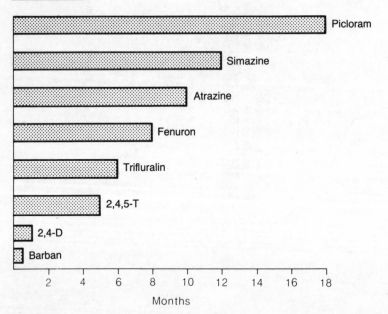

Figure 9.4 Persistence of some commonly used insecticides and herbicides in soil (compiled from a variety of sources)

caused a build up of Pb and As to toxically high concentrations in surface soils: 870 and 120 µg g^{-1} respectively (Veneman, Murray and Baker, 1983). The longterm accumulation and persistence of organochlorine residues has been reported by Ricci, Hubert and Richard (1983) in the sediments of Red Rock Reservoir in Iowa. They relate the amounts of dieldrin and DDT transformation products in lake sediments to the use of a wide range of organochlorine pesticides in agriculture in Iowa during the 1960s and early 1970s. Although in this study there appeared to be no relationship between clay and organic content of sediments and the amounts of pesticide residues, Pionnke and Chesters (1973) related pesticide accumulation to their adsorption onto eroded soil particles, carried to watercourses by surface runoff.

In a review of herbicide persistence in Canadian soils, Smith (1982) reports a large number of results for currently applied compounds, which indicate a range of persistence rates (expressed as per cent applied): 90 per cent after 17 weeks to 10 per cent after 104 weeks for *picloram*; 30 per cent after 22 weeks to 15 per cent after 50 weeks for *trifluralin*; < 5 per cent after 10 weeks for *2,4,-D*; < 10 per cent after 10 weeks for *2,4,5-T*; 15 per cent after 22 weeks to < 10 per cent after 50 weeks for *atrazine*; 35 per cent after 14 weeks to 10 per cent after 50 weeks for *simazine*; 83–86 per cent after 17 weeks to 50 per cent after 60 weeks for *paraquat*; 60–70 per cent after 17 weeks to 15–20 per cent after 72 weeks for *linuron*. These data indicate that picloram is the most persistent of the compounds reviewed, with paraquat and the urea herbicide, linuron, also fairly persistent, and the triazine and phenoxyacetic acid compounds the least persistent.

Several models have been developed to predict the persistence of pesticides in soil. Their varied success is probably due to the need to account for soil and climatic factors as well as an understanding of the relative importance of degradation, transformation, adsorption and losses through leaching, runoff and volatilisation.

9.4 DEGRADATION AND TRANSFORMATION OF PESTICIDES IN SOIL

Both physicochemical and biological processes are important during the degradation of pesticides in soil. The kinetics and optimal conditions for such processes are of particular importance to environmentalists and soil biologists, since many of the degradation end products are less toxic than their original chemical counterparts and some are even biologically inactive. Adsorption onto soil colloids and the presence of water are often important catalysts for degradation. Much attention has been paid to experimental techniques for isolating the relative contributions of biological and chemical degradation, usually through comparisons of pesticide effects on 'natural' and 'sterile' soil. It is, however, difficult to assess the influence of sterilisation effects themselves in altering the chemical and physical properties of the soil, or to account for the fact that sterilants rarely kill the entire soil population. For these reasons it is sometimes difficult to distinguish between chemical and biological transformations. A useful review of sterilisation effects on pesticide transformation in soil is given by Helling, Kearney and Alexander (1971).

9.4.1 Non-biological degradation of pesticides

Non-biological degradation processes include hydrolyis, oxidation/ reduction and photodecomposition. A range of soil factors act to catalyse these chemical processes. Most important of these are the presence of clay mineral surfaces, soil organic matter and metal oxides.

With the dissociation of water into $H^+ + OH^-$ and accompanying pH change, both acid and alkaline hydrolysis of pesticides occurs. The most studied pesticide hydrolysis reactions are for organophosphate insecticides and for triazine herbicides. Of the organophosphorus insecticides, malathion, for example, undergoes alkaline hydrolysis, while diazinon undergoes acid hydrolysis (Konrad, Armstrong and Chesters, 1967; Konrad, Chesters and Armstrong, 1969):

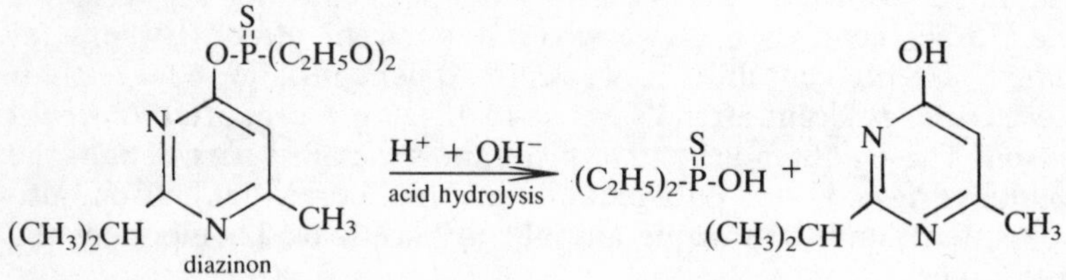

$$\text{diazinon} \xrightarrow[\text{acid hydrolysis}]{H^+ + OH^-} (C_2H_5)_2\text{-}\overset{S}{P}\text{-OH} \; + \; \text{(pyrimidinol)} \tag{9,1}$$

$$\text{malathion} \xrightarrow[\substack{\text{alkaline} \\ \text{hydrolysis}}]{H^+ + OH^-} (CH_3O)_2\text{-}\overset{S}{P}\text{-OH} + HS\text{-CH-}C\overset{O}{\underset{OC_2H_5}{\diagup}} \tag{9,2}$$

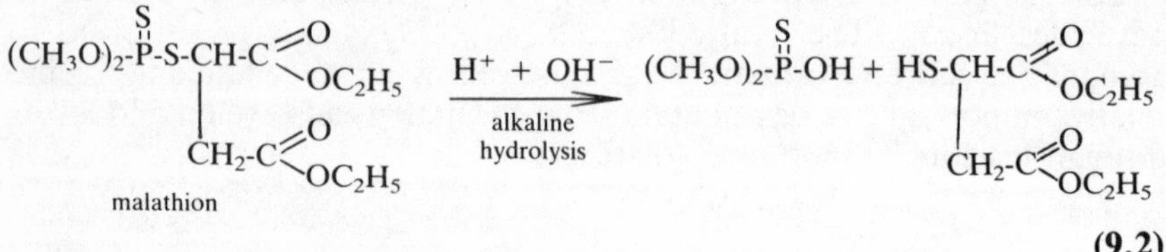

then:

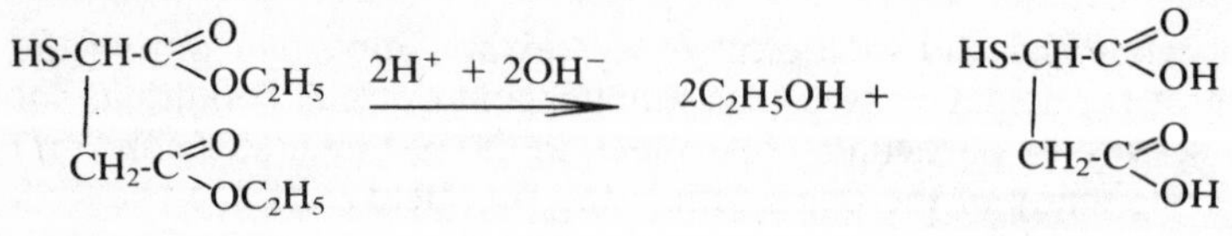

$$HS\text{-CH-}C\overset{O}{\underset{OC_2H_5}{\diagup}} \xrightarrow{2H^+ + 2OH^-} 2C_2H_5OH + HS\text{-CH-}C\overset{O}{\underset{OH}{\diagup}} \tag{9,3}$$

As for many pesticide hydrolysis reactions, both of the above reactions proceed faster in wet soil than in soil-free aqueous solution. Crosby and Tang (1969) found that monuron degradation in dilute aqueous solution in sunlight amounted to < 6 per cent in fourteen days. Experimental evidence suggests that it is the presence of organic matter and clay mineral surfaces which catalyse pesticide hydrolysis reactions. Basic amino acids and amino acid functional groups have been shown to catalyse the hydrolysis of organophosphorus compounds (Gatterdam, Casida and Stoutamire, 1959). The presence of moist clay mineral surfaces has been shown by Armstrong and Chesters (1968) to catalyse the hydrolysis of atrazine herbicide. The low surface pH of clay minerals may be an important aid in atrazine hydrolysis. Russell *et al.* (1968) suggested montmorillonite as a likely adsorption surface for the catalysis of atrazine hydrolysis, following the sequence (Crosby, 1976):

chlorotriazine	protonated (either on ring or side chain)	Cl-bearing ring becomes e⁻ deficient	OH⁻ dissociated from water displaces Cl⁻ ion	hydroxytriazine stabilised by transformation into keto-(amide) form

$$(9,4)$$

The detailed examination of the hydrolysis of atrazine adsorbed to montmorillonite using infrared absorption spectra has confirmed Equation (**9,4**) as a likely sequence of events (Skipper *et al.*, 1978). Several other adsorption catalysed reactions for pesticides have been suggested: for organophosphates on Cu^{2+} ions (Mortland and Raman, 1967), for organophosphates on clay minerals (Mingelgrin, Saltzman and Yaron, 1977), for parathion on Ca-kaolinitic clays (Saltzman, Mingelgrin and Yaron, 1976; Mingelgrin, Yariv and Saltzman, 1978), and for atrazine on humic acids (Li and Felbeck, 1972) and fulvic acids (Khan, 1978). Gamble and Khan (1985) suggest that it is the hydrogen ions and undissociated carboxyl groups of soil fulvic acids which catalyse atrazine degradation.

Many organic compounds can potentially act as reducing agents, particularly in waterlogged and oxygen-deficient soil. While many pesticide molecules are known to be decomposed by oxidation – parathion to paraoxon, or aldrin to dieldrin – it is not clear whether these reactions are entirely chemical or whether they are biologically mediated.

Photodecomposition is an important mechanism whereby pesticides at the soil surface are chemically transformed by absorbing ultraviolet (UV) radiation of wavelengths 290–450 nm. Ultraviolet wavelengths <290 nm are adsorbed to varying degrees by ozone in the atmosphere and thus tend to be less important for pesticide degradation at the soil surface. Some important herbicide absorption maxima are given in Table 9.2. There are many reports of pesticide degradation in aqueous solution, adsorbed to clay minerals and in the presence of organic matter, such as humic acids.

Most common and rapid of pesticide photolysis reactions are photo-oxidations. Crosby and Tang (1969) found that paraquat and diquat herbicides were photolysed within a matter of hours under laboratory

Table 9.2 Ultraviolet wavelengths of maximum absorption for some important herbicides (from Bailey and White, 1965)

Herbicide		Wavelength absorption maximum (nm)
Phenoxyacetic acids	2,4-D	283
	2,4,5-T	289
Substituted ureas	Monuron	244
Toluidines	Trifluralin	376
Amides	Dicryl	258
Phenols	DNBP	375
Bipyridyl	Paraquat (in solution)	257
	Paraquat (adsorbed)	275
	Diquat	310

ultraviolet irradiation. Photo-oxidations are particularly important in the breakdown of phenoxyacetic acids and substituted urea herbicides. Crosby (1976) suggests two important pathways for the decomposition of monuron:

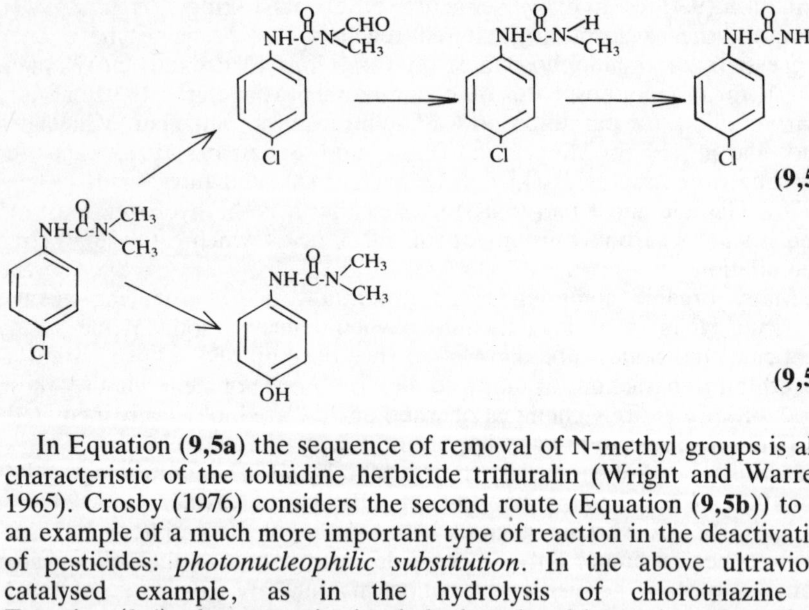

(9,5a)

(9,5b)

In Equation (9,5a) the sequence of removal of N-methyl groups is also characteristic of the toluidine herbicide trifluralin (Wright and Warren, 1965). Crosby (1976) considers the second route (Equation (9,5b)) to be an example of a much more important type of reaction in the deactivation of pesticides: *photonucleophilic substitution*. In the above ultraviolet catalysed example, as in the hydrolysis of chlorotriazine in Equation (9,4), the aromatic ring is hydroxylated by replacement of the ring Cl with OH. Crosby (1976) suggests that photonucleophilic substitution reactions, particularly hydroxylation of chlorinated aromatic rings, may prove significant for the disappearance of pesticides from the environment. These reactions are particularly important for the phenoxyacetic acids, such as 2,4-D and 2,4,5-T and the substituted ureas, such as fenuron and monuron.

Although the organochlorines are the most persistent pesticides in soil, all are broken down to some degree. Endrin is decomposed very rapidly in dry soil at normal soil temperatures. Asai, Westlake and Gunther (1969) applied endrin to a sandy loam at 25°C and 3°C. After only sixteen hours, 90 per cent and 65 per cent of the endrin had been degraded at these temperatures. Temperatures > 100°C and the presence of iron oxides such as geothite and haematite, catalyse the conversion of DDT to DDE (Birrell, 1963). A high clay content and high allophane content also increase DDT degradation, while a high soil organic matter content seems to retard DDT and heptachlor degradation (Helling, Kearney and Alexander, 1971).

The presence of other substances in solution appears to aid the photolysis of some pesticides. Crosby and Wang (1973) found that by adding an organic solvent, such as acetone, the photolysis of 2,4,5-T in soil could be increased more than ten-fold. In such reactions, the solvent is thought to act as a photosensitiser. Naturally occurring photosensitisers may be present in surface waters. Harvey and Han (1978) found that oxamyl (an acetamide insecticide) was photolysed more rapidly in river water than in distilled water. Khan (1980) suggests that humic acids in surface waters may act as photosensitisers. Fulvic acids and their degradation products also appear to aid the photolysis of atrazine in soil (Khan and Schnitzer, 1978).

Although Yaron, Gerstl and Spencer (1985) found little photo-decomposition of pesticides adsorbed onto soil particles, Slade (1966) reported quite different results for paraquat behaviour. Paraquat in aqueous solution has a sharp radiation absorption maximum at 257 nm, which lies outside the range of sunlight received at the soil surface. When paraquat becomes adsorbed onto soil colloids, its radiation absorption maximum is broadened and shifted to about 275 nm. This just includes the lower end of sunlight wavelengths and photodecomposition occurs. Diquat is rapidly photolysed in both aqueous media and on soil colloids since its radiation absorption maximum lies at 310 nm.

The photolysis of a large number of commonly applied pesticides, mainly in solution, but also adsorbed on soil organic matter and clay particles, suggests that optimal conditions for photochemical degradation may occur in aqueous suspensions such as irrigation waters and overland flow. Crosby (1976) notes several examples where dieldrin, trifluralin and monuron have been shown to be degraded under such conditions. These findings give some optimism that herbicides applied to irrigated systems or 'herbigation', the application of herbicides in irrigation water (Yaron, Gerstl and Spencer, 1985), may increase herbicide degradation and deactivation before drainage waters reach adjacent watercourses.

9.4.2 Microbial decomposition of pesticides

The difficulty in distinguishing between physical, chemical and biological breakdown of pesticide molecules has already been indicated.

Alexander's (1964) 'principle of microbial infallibility', or the concept that some micro-organism exists in nature which is capable of metabolising and destroying *any* organic compound, led soil scientists to assume automatically that the disappearance of pesticide molecules from soil was microbially mediated. While it is now clear that many other soil factors, such as temperature, ultraviolet light, pH or cation exchange, can catalyse pesticide breakdown, the majority of processes appear to have some biological contribution. Extracellular enzymes in soil also catalyse pesticide lysis.

The loss of most pesticide compounds from soils follows a first order reaction such as the exponential decay illustrated in Fig. 2.7b, in which the rate of loss is proportional to the original concentration of pesticide in the soil. As Kaufman and Kearney (1976) point out, this simple decay curve is frequently modified by the influence of other losses such as photodecomposition, volatilisation or surface runoff. Pesticide loss curves are typified by an initial lag phase in which decomposer organisms can become adapted to the pesticide substrate. This lag is followed by a period of rapid loss as the substrate-adapted microbial population builds up (Fig. 9.5a and b). At this stage, the soil is said to be 'enriched' with organisms capable of decomposing the substrate. After enrichment occurs, subsequent additions of the same substrate will be broken down more quickly, as found for the insecticide, parathion (Nelson, Yaron and Nye, 1982) and for the phenoxyacetic acid herbicides 2,4-D (Kunc, Tichy and Vancura, 1985) and MCPA (Torstensson and Rosswall, 1977). An interesting process in the microbial decay of pesticides is that once organisms become adapted to degrade one particular pesticide compound, they also have an ability to break down compounds of similar molecular structure. In pesticide decomposition this process is known as *cometabolism*. Golovleva and Skryabin (1978) believe that a number of degradation reactions may be carried out cometabolically, for example: oxidation of the methyl groups on aromatic of heterocyclic compounds; hydroxylation of aromatic rings; and acetylation of amino groups. A schematic illustration of pesticide cometabolism is given in Fig. 9.5c. Pretreatment of soil with 2,4-D results in enhanced MCPA degradation and vice versa (Soulas, Codaccioni and Fournier, 1983) and pretreatment of soil with carbaryl enhanced the degradation of carbofuran and vice versa (Rajagopal *et al.*, 1983; Harris *et al.*, 1983).

Biological degradation of pesticides can occur throughout the soil, but this activity is concentrated in surface soil and in the rhizosphere where microbial numbers are higher and where there may also be more extracellular enzymes. Other soil factors which influence biodegradation include (a) temperature, (b) moisture, (c) aerobic/anaerobic conditions, (d) pH, (e) soil organic matter, and (f) clay mineral surfaces.

Two different effects of temperature on herbicide degradation have been reported. Parker and Doxtader (1983) found that the lag, or the slow build-up phase prior to enrichment, for the decomposition of 2,4-D was reduced at high temperature and was lost all together at 37°C. As with soil organic matter, the rate of microbial decomposition of organic

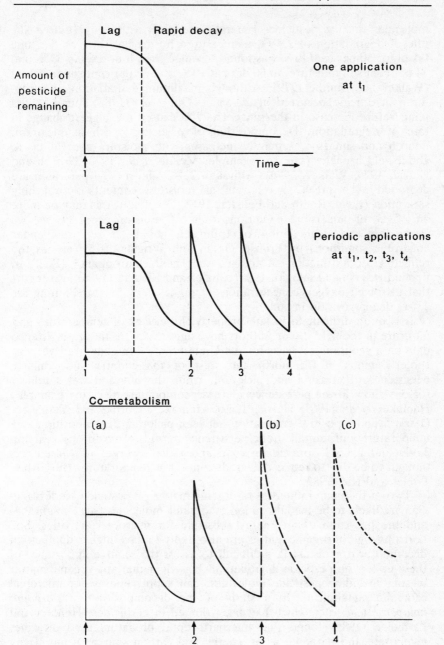

Figure 9.5 Similar decay response to pesticides — (b) and (c) — once the microbial population has adapted to pesticide decay (a). (Source: Kearney *et al.*, 1964)

molecules such as pesticides generally increases with temperature. The rate of degradation of 2,4-D, as measured by the half life, or the time taken for 50 per cent loss, was found to range from four days at 35°C and 34 per cent soil moisture, to 60 days at 10°C and 20 per cent soil moisture (Walker and Smith, 1979). Although maximum degradation of carbofuran insecticide occurred at 27–35°C, Ou *et al.* (1982) found that a temperature increase in the range 15–27°C caused the biggest change in rate of degradation. Decomposition of pesticides, such as picloram, linuron and simazine, generally increases with moisture content up to about field capacity (see, for example, Meikle *et al.*, 1973; Usoroh and Hance, 1974; Walker, 1976). Although 2,4-D and methyl parathion are degraded at a much slower rate at moisture contents approaching saturation (Lavy, Roeth and Fenster, 1973; Ou, 1985), this may be more an effect of anaerobic conditions than of moisture content *per se.* Reduction and dealkylation of trifluralin has been reported under anaerobic conditions by Probst, Golab and Wright (1975), as has the reductive dechlorination of lindane and pentachlorophenol (PCP) in flooded rice soils (Tsukano, 1986). Shaler and Kléckla (1986) also report that dissolved oxygen concentrations $< \text{mg l}^{-1}$ may be rate-limiting for 2,4-D decomposition in soil.

It is quite difficult to isolate properly the effects of temperature and moisture in the field. Some authors have aggregated pesticide persistence data on a geographical basis for the United States. In general, dryer and cooler regions in the north and west of the country show higher persistence of atrazine and picloram, while the hot and wet southern regions show lower persistence of these compounds (see, for example, Hamaker *et al.*, 1967; Harris, 1969). Hamaker, Goring and Youngson (1966) found a positive correlation between picloram decomposition and temperature and rainfall. Long persistence periods of up to one year for 2,4-D and 2,4,5-T sprayed on road and railside verges in Alaska are thought to be due to reduced degradation at low temperature (Burgoyne, Triblet and Ric, 1983).

Two of the main influences of organic matter on pesticide biodegradation are likely to be the adsorption on organic molecules (see Section 9.4) and the provision of an energy substrate for microbial activities. Soil horizons high in organic matter are also likely to have higher numbers of decomposer organisms. It is difficult to assess the relative importance of these effects on pesticide degradation. It is clear that any organic matter in soil, including pesticide molecules, can supply energy for microbial catabolic metabolism. In a study of the decomposition of *Sorghum halopense* leaf litter which had been dipped in herbicides, Hendrix and Parmelee (1985) found that microarthropod predators used atrazine, glyphosate and paraquat as an energy and carbon source. Kunc, Tichy and Vancura (1985) found that when glucose and 2,4-D were added to soil glucose was mineralised before the herbicide, but then once the glucose was used up 2,4-D was degraded at a faster rate than when it was applied without glucose. Similar results in flooded rice soils indicated that a range of organic amendments in the order farm yard manure > rice straw > glucose inhibited the hydrolysis of parathion (Rajaram and Sethunathan, 1975). The reverse order of organic materials promoted the nitrogroup reduction of parathion.

The effect of clay minerals on the biodegradation of pesticides is due to the inaccessibility of adsorbed pesticide molecules to potential decomposer organisms. This effect appears to be particularly important for paraquat. According to Summers (1980), paraquat adsorbed on kaolinitic and vermiculite clays remains exchangeable and available to plants and micro-organisms, but is bound very strongly to montmorillonite and smectite clays so that it is unavailable to micro-organisms, microarthropods or earthworms (Riley, Wilkinson and Tucker, 1976).

The influence of pH on pesticide degradation is complex. Chapman and Cole (1982) divided a range of insecticides into two groups: (a) those whose degradation was unaffected by changes in pH over the range pH 4.5–8.0; and (b) those whose degradation rate was affected over this range. The first group included heptachlor, aldicarb and the phosphorodithioate insecticides phorate, terbufos and fensulfothion. Insecticides whose degradation was affected by pH included the carbamates carbamyl, carbofuran and oxamyl, and the organophosphate insecticide trichlorofon. Faster rates of hydrolysis at low pH have been suggested as a reason for faster degradation in acid soils of some pesticides, such as the triazine herbicides, atrazine (Hiltbold and Buchanan, 1977) and simazine (Nearpass, 1965). Higher degradation of 2,4-D and MCPA at higher pH has been reported by Tortstensson (1975) and for linuron by Hance (1979). Complex and often conflicting results from pH studies have sometimes been attributed to unrealistically simple pH adjustment techniques in laboratory studies (Hamaker, 1972).

Biodegradation processes

Biodegradation processes can be divided into those in which the pesticide is used as the carbon and energy source in microbial catabolism, and a range of other processes in which microbes may take part, such as interaction with extracellular enzymes, pH changes and chemical or photochemical reactions. Only the first group of catabolic processes will be considered here. A wide range of catabolic processes are reviewed in detail by Hill (1978) and Kaufman and Kearney (1976). Some of the more common biodegradation processes are given in Fig. 9.6. In many cases, a pathway of several different degradation mechanisms appear to operate in sequence, carried out by a range of different soil organisms, primarily bacteria and fungi. In the case of 2,4-D degradation, oxidation, hydroxylation, hydrolytic dechlorination, ring cleavage and reductive dechlorination have all been identified (Fig. 9.7).

As far as degradation of pesticide compounds is concerned, there are two main topics for further concern. Firstly, Kaufman and Kearney (1976) have drawn attention to the large number of pesticide metabolites which can be produced in soil as a result of degradation. It is clearly vital to examine the toxicity, adsorption, bioaccumulation and persistence in

Mechanism	Example	Reaction	Organisms involved
1. Oxidation	(a) *β-oxidation* phenoxyacetic acids	$R-C-C-C-C\overset{O}{\underset{OH}{\parallel}} \xrightarrow{\text{acetyl group}} -C-C-C\overset{O}{\underset{OH}{\parallel}} + H-C-C\overset{O}{\underset{CoA}{\parallel}}$ sequential removal of two carbon groups	many soil organisms
	(b) *Epoxidation* cyclodiene organochlorides	aldrin $\rightarrow$ dieldrin	*Aspergillus, Penicillium* *Trichoderma*
	(c) *N-Dealkylation* s-triazines toluidines substituted ureas	$R-N(CH_3)_2 \xrightarrow{O_2} R-N(CH_2OH)(CH_3) \rightarrow R-N(H)(CH_3) + HCHO$ sequential removal of alkyl groups	soil fungi of the genera: *Aspergillus* *Fusarium* *Paecilomyces* *Trichoderma*
2. Reduction	(a) *Reductive dehydrogenation* organochlorines	$R-C(Cl)_2 \xrightarrow{H} -C(H)(Cl)- + Cl$	soil bacterial of the genus: *Enterobacter, Arthrobacter* soil fungi of the genus: *Trichoderma*
	(b) *Nitro-reduction* toluidines (trifluralin) organophosphates (parathion)	$R-N\overset{O}{\underset{O}{}} \rightarrow [R-N\overset{O}{} \rightarrow R-N(OH)_2] \rightarrow R-NH_2$	nitroreductase soil microorganisms

3. **Hydrolysis**	(a) Carbonates	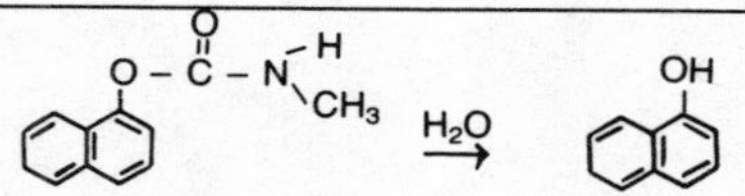	many soil organisms with: phosphatases, esterases, nitrilases, amidases.
	(b) Organophosphates	$R - P - S - CH_2 - S - R \xrightarrow{H_2O} R - P + HS - CH_2 - S - R$	
4. **Hydroxylation**	*Hydrolytic dehalogenation* (a) *aliphatic:* dichloropropionate	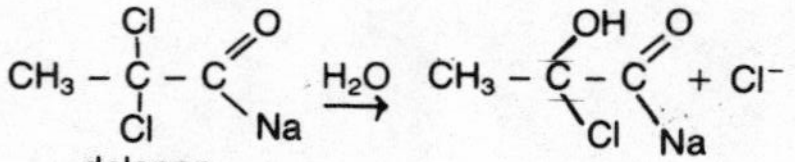 dalapon	*Arthrobacter*
	(b) *aromatic:* phenoxyacetic acids organochlorines	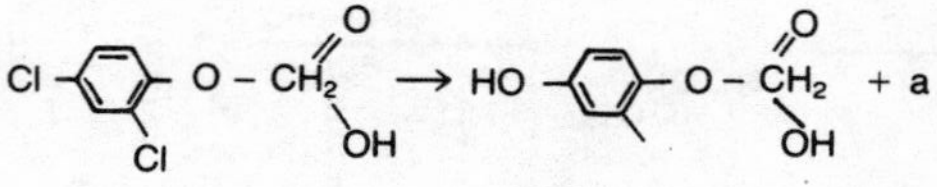 2,4 – D	*Aspergillus* *Clostridium, Escherichia coli*
5. **Aromatic Ring Cleavage**	(a) *Non-heterocyclic ring cleavage* (e.g. phenoxyacetic acids) (b) *Heterocyclic ring cleavage* (e.g. cyclodiene organochlorines)		*Arthrobacter* *Bacillus* *Nocordia*

Figure 9.6 Major processes of pesticide biodegradation in soils

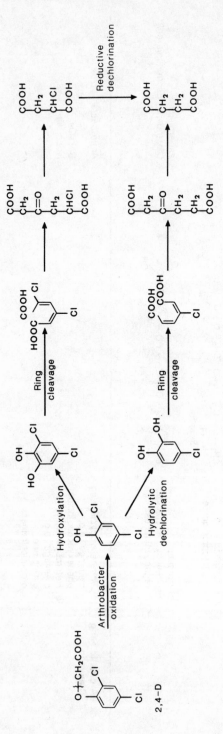

Figure 9.7 Biological degradation pathways for 2, 4-D in soil. (Source: modified from Kaufman, 1974)

soil of these compounds as well as their pesticide precursors. The scale of this problem is potentially massive, considering the very large number of compounds currently applied to soil as pesticides (round 450 pure chemical compounds listed by the Royal Society of Chemistry, 1983). It is secondly important to note that the vast majority of degradation mechanisms and pathways have only been studied under controlled, model conditions, usually with isolated organisms. Their applicability to complex field conditions must be properly assessed.

9.5   ADSORPTION OF PESTICIDES ON SOIL CONSTITUENTS

Two main characteristics of pesticides influence their adsorption capacity in soil: their surface charge (if any) and their degree of aqueous solubility. Both of these may be affected by the pH of the soil solution. Bailey and White (1970) list four structural factors which characterise the chemistry of a pesticide molecule and hence its ability to be absorbed on soil constituents:

(1) Nature of *functional groups* (for example, carboxyl-, carbonyl-, amino and alcoholic hydroxyl- groups). Amino groups can protonate and be adsorbed as cations. Both amino and carbonyl groups can be involved in hydrogen bonding. Adsorption is generally increased with functional groups such as $-OH$, $-NH_4$, $-CONH_4$.
(2) Nature of *substituting groups* that can alter the behaviour of functional groups.
(3) *Position of substituting groups* with respect to functional groups. Positioning may enhance or hinder intramolecular bonding.
(4) Presence and magnitude of *unsaturation in the molecule* which affects the lyophilic–lyophobic balance.

If there is a charge on the molecule surface, it may be permanent, as in the case of bipyridylium compounds, or it may be due to uneven distribution of electrons, the protonation of weak bases (such as triazines) or deprotonation of weak acids (such as phenoxyacetic acids), such as described for soil organic matter in Section 5.2 and Table 5.1. For non-ionic pesticides, the aqueous solubility is thought to be the most important molecular parameter affecting adsorption. Attempts to find a general relationship between pesticide water solubility and adsorption have, however, provided contradictory results. Bailey, White and Rothbert (1968) found a positive relationship between solubility and absorbability for some s-triazine herbicides, but a negative relationship for some substituted urea herbicides. Other authors have found no relationship at all (see, for example, Briggs, 1969).

Pesticide adsorption mechanisms, adsorption isotherms, and factors controlling rates and magnitudes of adsorption for different pesticide types will be examined in the following sections.

9.5.1 Pesticide adsorption mechanisms

Calvet (1980) groups bonding or adsorption mechanisms of pesticides on soil constituents into two main groups, depending on whether the bonds are high (> 80 kJ mol^{-1}) or low (< 80 kJ mol^{-1}) energy types. Ionic bonds and ligand exchange are examples of high energy bonding, while hydrogen bonding, van der Waals attractions and charge transfers are examples of low energy bonding (Table 9.3). Pesticides can be simply divided into ionic and non-ionic types. There are three main groups of ionic bonding pesticides: (1) *permanent charge*, such as the cationic herbicides paraquat and diquat; (b) *weak bases*, such as the s-triazines, which become cationic by protonation; and (c) *weak acids*, such as 2,4-D and picloram, which become anionic by dissociation and loss of a proton.

The overriding control on the sorption of cationic pesticides and weak bases is the magnitude of the cation exchange capacity, usually dependent on the amount and type of clay in the soil. As with inorganic cations, the negative surface charge of clay minerals can be balanced by cationic pesticide molecules. This was confirmed when Philen, Weed and Weber (1970) found a linear relationship between the surface charge density of vermiculite and mica and the amount of paraquat and diquat adsorbed onto the mineral surfaces. Knight and Tomlinson (1967) suggested that paraquat was bound onto soil clay minerals with two different strengths of adsorption. Their H-shaped adsorption isotherm showed an initial, strong adsorption capacity (SAC) in which paraquat interacts with lattice layer clay minerals, to make montmorillonite, for example, non-expanding (Knight and Denny, 1970). In this condition, sorbed paraquat self-exchanged with ^{14}C labelled paraquat but not with inorganic cations. Riley, Wilkinson and Tucker (1976) suggests that tightly bound paraquat is not exchangeable and is not available for plant and organism uptake, while less tightly adsorbed paraquat at the higher concentration plateau region of the adsorption isotherm (Fig. 9.8) is available for uptake.

Cation exchange mechanisms are also responsible for the adsorption of protonated weak bases, such as the s-triazine herbicides, but these reactions are pH dependent. Weber (1970) reported that the adsorption of a large number of triazines increased with decreasing solution pH, reaching a maximum at the pH value close to the pKa (see Section 1.3.1) for each compound. Related to pH is the ionic balance of the system. Nearpass (1967) found that atrazine and simazine adsorption declined as the base saturation of the soil increased, but that this did not depend on the type of cation (Na, K, Ca or Mg) on the exchange complex. Triazines are adsorbed to a greater extent on clays with high cation exchange capacities. Chopra, Bala and Bhardwaj (1984) found that the adsorption of simazine depended on clay type and increased in the order: bentonite $>$ pyrophyllite $>$ illite $>$ kaolinite. Sorption increased with increasing simazine concentration and decreased with rising temperature up to 35 °C. Triazines become adsorbed on interlamellar surfaces of expanding clays. As a result of this process, Weber, Perry and Upchurch

Table 9.3 Mechanisms of pesticide adsorption on soil constituents

Energy status	Bond type	Examples	Reference
High (Calvet, 1980) (> 80 kJ mol^{-1})	Cationic	paraquat, diquat on clay minerals	Dixon et al. (1970), Knight and Tomlinson (1967), Weed and Weber (1969)
		paraquat, diquat on organic matter	Schnitzer and Khan (1972)
		s-triazine on clay and organic matter	Gaillardon (1975)
	Anionic	2,4-D	Weber, Perry and Upchurch (1965)
		picloram on H-montmorillonite	Bailey, White and Rothbert (1968)
		picloram on a range of oxides and clays	Hamaker, Goring and Thompson (1966)
			Hance (1972), Torstensson (1985)
		glyphosate on unoccupied phosphate adsorption sites	
	Ligand exchange	linuron, atrazine on clays	Hance (1976)
		s-triazine on heavy metals of humic acids	Hamaker and Thomson (1972)
Low (Calvet, 1980) (< 80 kJ mol^{-1})	Hydrogen bonding	carbamate	Bailey, White and Rothbert (1968)
		atrazine on montmorillonite	Calvert and Terce (1975)
		s-triazine on organic matter and humic acids	Senesi and Testini (1980)
		malathion on Na-saturated clay	Bowman, Adams and Feton (1970)
		2,4-D on montmorillonite	Dieguez-Carbonell and Pascual (1975)
	Charge transfer	monuron, fenuron, s-triazine on humic acids	Senesi and Testini (1980)
		bipyridylium herbicides on clays and humic acids	Khan (1973a)
	van der Waals attractions	carbaryl, parathion, picloram on organic matter	Nearpass (1976)
		s-triazine, substituted ureas on humic acids	Senesi and Testini (1980)
		MCPA on goethite	Kavanagh, Posner and Quirk (1980)

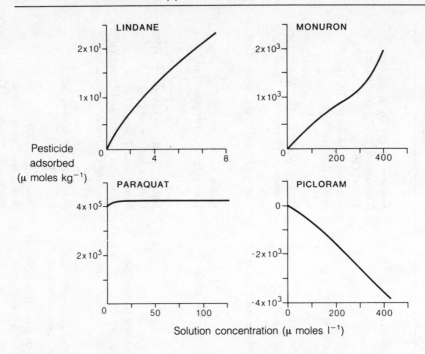

Figure 9.8 Typical absorption isotherms for four commonly used pesticides. Paraquat: ion exchange type reaction; lindane and monuron: constant partition of absorbate between solvent and absorbent; picloram: negative sorption. (Source: Green, 1974)

(1965) found that the basal spacing in montmorillonite was expanded by the adsorption of prometone. By sequentially removing soil organic matter and Fe/Al sesquioxides, Huang, Grover and McKercher (1984) showed that both of these fractions played an important role in atrazine adsorption. Triazines appear to be adsorbed on cation exchange sites of soil organic matter, particularly humic acids (McGlamery and Slife, 1966; Gaillardon, 1975). Infrared spectra indicate that protonated s-triazine herbicides formed ionic bonds with the carboxylate anions of soil humic acids (Senesi and Testini, 1980).

Although crystallised and amorphous Fe and Al hydroxides are generally poor adsorbents, they are important in the sorption of weak acid compounds such as phenoxyacetic acids. The stronger adsorption of picloram and 2,4,5-T by Fe and Al oxides than by clays was reported by Hamaker, Goring and Youngson (1966). O'Connor and Anderson (1974) also found a positive correlation between soil Fe content and the adsorption of 2,4,5-T. Examination of the adsorption of MCPA, 2,4-D and 2,4,5-T on goethite suggests that van der Waals attractions may be the main mechanism operating, particularly for MCPA (Kavanagh, Posner and Quirk, 1980). Weak acids become dissociated at high pH. We might therefore expect greatest adsorption of anionic herbicides to occur at high pH. The opposite is true for picloram adsorption, which increased as soil pH decreased (Arnold and Farmer, 1979). This implies that it is adsorbed in molecular rather than ionic form. Nearpass (1967) showed that picloram was adsorbed onto humic acids and humin in the form of uncharged molecules. A rather different mechanism operates for glyphosate adsorption. Glyphosate is adsorbed on 'unoccupied' phosphate sorption sites (Hance, 1976; Torstensson, 1985). The main control on glyphosate adsorption is the availability of unoccupied P adsorption sites and low soil phosphate concentrations, since inorganic phosphate excludes glyphosate from sorption sites (Hance, 1976).

For most other types of non-ionic pesticide molecules, it is the amount and composition of soil organic matter which determines sorption. Considering that the vast majority of pesticides are non-ionic, a disproportionately small amount of information is available for the actual mechanisms, quantities and rates of reaction on soil organic matter. Since soil organic matter contains carboxyl, hydroxyl and amino groups, hydrogen bonding should operate with pesticide compounds containing like groups. Some compounds in soil organic matter, such as fats, waxes and resins, are water-repellent. The term 'hydrophobic bonding' is sometimes used to describe the accumulation of non-polar molecules at these surfaces where polar water molecules are excluded. Walker and Crawford (1968) described the partitioning of s-triazines out of solution onto hydrophobic surfaces.

It is difficult to generalize about the adsorption of non-ionic pesticides, so the three main types belonging to this group, organochlorine and organophosphate insecticides and the phenylurea herbicides will be discussed in a little detail. Several authors have reported a positive relationship between organochlorine adsorption, particularly DDT, and soil organic matter content (see, for example, Peterson, Adams and Cutkomp, 1971). DDT is adsorbed more strongly on highly humified soil organic matter (Shin, Chodan and Wolcott, 1970), and its movement in forest soils has been associated with humic and fulvic acids (Ballard, 1971). Other organochlorines, such as lindane, are also adsorbed on soil organic matter (see, for example, Adams and Li, 1971). Organochlorines are non-polar molecules with very low aqueous solubility and may well be adsorbed by 'hydrophobic bonding' (Weed and Weber, 1974) or adsorbed onto lipids of soil organic matter (Pierce, Olney and Felbeck, 1971).

Although organophosphate pesticides are adsorbed by both clay and organic matter fractions, in aqueous solution, they show a greater affinity for soil organic matter (Saltzman, Kliger and Yaron, 1972). Felsot and Dahm (1979) found that aldicarb and a range of organophosphates, including phorate, parathion and chlorpyrifos, were increasingly adsorbed with increasing soil organic matter content. As much as 2 µg chlorpyrifos per g air dried soil could be completely adsorbed in peat. Several organophosphates including malathion and parathion are adsorbed in the interlayers of montmorillonite (see, for example, Bowman, Adams and Feton, 1970; Biggar, Mingelgrin and Cheung, 1978). Sánchez Camazano and Sánchez Martin (1983) show that this intercalation causes different degrees of lattice expansion. They found that organophosphates with

P=O and P=S groups and those with long functional groups tended to cause greatest basal expansion. Thus, dichlorvos forms a three molecular layer complex within the lattice space and causes high expansion, demeton-s-methyl (metasystox) forms a two molecular layer complex and causes intermediate expansion, while malathion forms a single molecular layer complex and causes the lowest amount of expansion.

The adsorption of phenylurea herbicides is also related to the organic matter content of the soil (see, for example, Carringer, Weber and Monaco, 1975). Madhun, Young and Freed (1986) showed that diuron and chlorotoluron were bound to humic acids and water-soluble organic matter of molecular weights resembling fulvic acids. They suggest that this binding may be an important factor in the mobility of phenylurea herbicides in the soil. Although Hance (1969) found that the adsorption of phenylureas was independent of pH, Weber (1972) found conflicting results for fluometuron under laboratory conditions. Fluometuron was adsorbed strongly on a weakly basic anion exchange resin and only very weakly on a strongly acid cationic exchange resin. Weber (1972) thus suggests that phenylureas are weakly acidic, with no basic properties.

9.5.2 Pesticide adsorption isotherms

As for inorganic solutes in the soil solution, many pesticide adsorption reactions have been described by Langmuir and Freundlich adsorption isotherms (see Section 5.3.3). Giles *et al.* (1960) classified pesticide adsorption isotherms into four types (Fig. 9.9):

(1) *L-type*, or Langmuir isotherm – the most common shape of adsorption isotherm, indicating high affinity between the solid and the solute pesticide at low concentration, with decreasing affinity as vacant sites become filled at high concentration.
(2) *S-type* – indicates what Giles *et al.* (1960) called 'cooperative adsorption'. Adsorption becomes easier as concentration increases. It may indicate competition for sorption sites between solvent molecules and other like molecules.
(3) *C-type* – represents the 'constant partition' of solutes between the solution and the sorbing surface which is constant over the concentration range. At maximum possible adsorption, the plot would abruptly change to a horizontal plateau.
(4) *H-type* – an uncommon adsorption isotherm, representing very high affinity between the pesticide solute and the sorbing surface.

Some examples of these adsorption isotherm classes are illustrated by Green (1974) in Fig. 9.8. By far the most common shape for pesticide adsorption isotherms is the L-type curve. Although Giles *et al.* (1960) called this the Langmuir type plot, these types of isotherms are usually

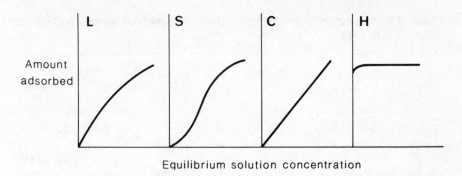

Equilibrium solution concentration

Figure 9.9 Classification of adsorption isotherms according to Giles *et al.* (1960) (L = normal or 'Langmuir' isotherms; S = cooperative adsorption; C = constant partition; H = high affinity.) (Source: Giles *et al.*, 1960)

described by Freundlich adsorption equations. Adsorption isotherms with L-type plots have been reported for several pesticides in soil, for example: the s-triazines, prometryn (Kozak, Weber and Sheets, 1983), prometone (Weber, 1972), and simazine (Chopra, Bala and Bhadwaj, 1984); diuron (Peck, Corwin and Farmer, 1980); 2,4-D and picloram (Khan, 1973c) and for paraquat and diquat (Gamar and Mustafa, 1975), despite some reports of H-type plots for the bipyridilium herbicides (see, for example, Knight and Tomlinson, 1967; Weber, Perry and Upchurch, 1965). Khan (1980) describes the H-type isotherm as a special case of the L-type curve, in which the solute has such a high affinity that in dilute solutions it is completely adsorbed, giving a vertical limb to the initial part of the isotherm.

It is possible to compare the adsorption of different pesticides, without showing individual isotherms, by using their distribution coefficients (K_d) where:

$$K_d = \frac{\text{pesticide adsorbed } (\mu\text{mol kg}^{-1})}{\text{pesticide in solution } (\mu\text{mol kg}^{-1})} \tag{9,6}$$

K_d values for a range of pesticides are given by Green (1974) in Table 9.4. Although K_d values allow the comparison of relative amounts of different pesticide adsorbed at known solution concentrations, nothing can be implied about the different shapes of the isotherm plots which probably indicate different adsorption mechanisms and rates. For this reason, it is sometimes useful to compare pesticide adsorption using the empirical Freundlich adsorption isotherm coefficients K_f and l/n from the log of the Freundlich equation:

Table 9.4 Adsorption (distribution coefficients) of pesticides on soil constituents (from Green, 1974)

Class	Chemical group	Example	Adsorption (K_d) Montmorillonite low pH	high pH	Kaolinite
Cationic	Bipyridium	Paraquat	–	4.2×10^4	1.7×10^3
Basic	Triazine	Atrazine	1.5×10^3	1.5×10^3	3.0 (low pH)
Acidic	Phenoxyacetic acid	2,4-D	0	0	0
		Picloram	50	0	–
Non-ionic	Substituted urea	Monuron	27	5	–
Non-ionic	Organochlorines	Aldrin	–	2.5	2.5
		Lindane	–	2.5	0.6
Non-ionic	Organophosphate	Parathion	–	106	3.4

$$\log \frac{x}{m} = \log K_f + \frac{1}{n} \log C \qquad (9,7)$$

where: K_f = intercept
$1/n$ = slope
x/m = amount of pesticide adsorbed ($\mu g\ g^{-1}$)
C = concentration of pesticide in the equilibrium solution ($\mu g\ ml^{-1}$)

A linear isotherm is obtained with values of n (or $1/n$) close to unity, allowing useful comparison of K_f values for different systems (Yaron, Gerstl and Spencer, 1985). These authors advocate extreme care in comparing systems whose $1/n$ values differ from unity.

The general importance of soil organic matter in the adsorption of non-ionic pesticides has led to the use of an adsorption coefficient which is based on the soil organic matter content (K_{oc}):

$$K_{oc} = \frac{K \times 100}{\text{per cent organic carbon}} \qquad (9,8)$$

K_{oc} coefficients have generally been found to vary less than K_f values. This may be because K_{oc} values are theoretically independent of the properties of the mineral soil or sediment. Nkedi-Kizza, Rao and Johnson (1983) measured the K_{oc} for diuron and 2,4,5-T on different soil particle sizes. They suggest that soil particles can be separated into two fractions: fine ($< 50\ \mu m$) and coarse ($> 50\ \mu m$) and a particular K_{oc} assigned to each fraction. They found that the fine fraction adsorbed three times more 2,4,5-T than the coarse fraction. An understanding of the organic carbon enrichment of runoff sediments has allowed the use of K_{oc}

coefficients to predict pesticide adsorption and transport associated with sediments in soils for which no adsorption data is available.

9.5.3 Factors affecting pesticide adsorption in soil

Many of the factors affecting pesticide adsorption in soil have already been mentioned in Section 9.5.1 in relation to individual pesticides. Some general points concerning pH and temperature effects on pesticide adsorption are worth noting. The influence of soil and solution pH on pesticide adsorption is primarily in controlling the protonation or dissociation of clay and soil organic colloids as well as the dissociation of functional groups of pesticide molecules; pH may also alter pesticide aqueous solubility. Ward and Weber (1968) reported a substantial increase in solubility of a range of s-triazine herbicides with decreasing pH. The interaction of H^+ ions and pesticide ions is also pH dependent. Narine and Guy (1982) found that the adsorption of diquat on humic acid was severely depressed as solution pH decreased from pH 9.0 to pH 6.0–2.0, and suggested that at low pH, H^+ ions competed with the cationic herbicide for cation exchange sites. To simplify this range of possible pH effects, Calvet (1980) provides a very useful summary of the three main types of relationship found between amount of pesticide adsorbed and pH (Fig. 9.10).

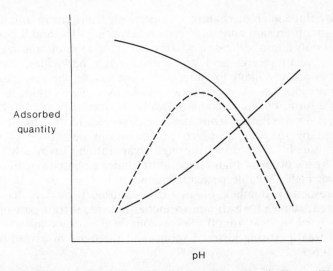

Adsorbed quantity

pH

Figure 9.10 Relationships between pesticide adsorption and soil pH. _____ Weak bases absorbed on electrically neutral surfaces (such as s-triazines on charcoal) ```} These two shapes of curve are observed under three circumstances: (a) weak bases absorbed on negatively charged surfaces (such as s-triazines on clays or soil organic matter); (b) weak acids absorbed on positively charged surfaces (such as phenoxyactic acids on oxides and hydroxides); (c) neutral molecules absorbed on clays (such as phenylureas on montmorillonite). (Source: Calvet, 1980)

Although there are many exceptions, the general influence of temperature on pesticide adsorption appears to be reduced adsorption with increased temperature. Reduced adsorption of diuron on freshwater sediments (Peck, Corwin and Farmer, 1980), simazine on bentonite, illite and kaolinite (Chapra, Bala and Bhardwaj, 1984), diquat on humic acid (Narine and Guy, 1982), 2,4,5-T in silt loam soil (Koskinen and Cheng, 1983) and lindane on montmorillonite (Mills and Biggar, 1969) with increasing temperatures are all examples. Such results indicate that these pesticide adsorption reactions are exothermic. Mills and Biggar (1969) point out that solution concentration changes with increased temperature may be responsible for altered adsorption at higher temperature. They compare normal adsorption isotherms, plotting x/m against C, with corrected isotherms (x/m against C/Co) in which the concentration of pesticide in solution at a given temperature (C) is calibrated against the solubility of the pesticide at the same temperature (Co). For pesticides whose solubility increases with increasing temperature, such as atrazine and parathion, normal isotherms show reduced adsorption with increased temperature (exothermic adsorption reaction) while corrected isotherms show increased adsorption with increased temperature (endothermic adsorption reaction).

9.6   TRANSPORT OF PESTICIDES IN SOIL

There are three main mechanisms for pesticide transport in soil: leaching, surface runoff/erosion and volatilisation. From Figs 9.1 and 9.2 it can be seen that only a few pesticides, notably the organophosphorus insecticides and the phenoxyacetic acid and bipyridylium herbicides, have high solubilities and are likely to be leached from soil. In the vast majority of cases, pesticides are washed to watercourses by surface runoff, sometimes in soluble form, but more commonly adsorbed to suspended soil organic and mineral particles. Gaseous losses of pesticides to the atmosphere are controlled by the rate at which pesticides can be moved to the soil surface, usually in solution, through evaporation. Invariably, highest volatile losses occur at high temperature under irrigated conditions with the use of highly volatile pesticide compounds.

A substantial number of pesticide transport models have been formulated, some of which aim to model individual transport processes such as leaching and runoff. Some combined process models aim to estimate total pesticide losses. Examples of these are discussed below in Section 9.6.1.

9.6.1   Transport of pesticides in the vapour phase

Volatilisation is an important pesticide dissipation process, particularly for surface applied, low solubility and low vapour pressure compounds such as the organochlorine insecticides. The potential volatility of any pesticide is related to its vapour pressure at the soil–air interface. Two of the most important controls on the vapour pressure is the degree to which the pesticide becomes (a) dissolved in the soil solution, and (b) adsorbed on soil particles. Generally, volatilisation losses are minimised in dry soils with low clay and organic matter contents. The water : air partition ratio, calculated from the pesticide aqueous solubility/its vapour pressure, gives a useful idea of the likely equilibration between these two phases (Hance, 1980). This is particularly useful in wet soil and in watercourses. Pesticide adsorption on soil constituents becomes a more important controlling factor in dry soils. When considering pesticide volatilisation hazard, it is common to estimate adsorption for a 'standard soil' containing 2 per cent organic matter (Anon., 1978). This consideration is important since several authors have reported an inverse relationship between soil organic matter content and pesticide volatilisation. Such a relationship has been reported by Guenzi and Beard (1970) for lindane and DDT. The volatilisation rate depends on environmental factors such as temperature and windspeed or air turbulence at the soil surface, which control the rate of pesticide diffusion from the evaporating soil surface. Farmer et al. (1972) found that either raising the temperature from 20 to 30°C, or increasing the wind speed from 2 to 8 ml s^{-1} caused almost a doubling of the volatile loss of dieldrin from a silt loam soil. Temperature exerts three effects on volatilisation. First, vapour pressure is affected, according to:

$$\log p = A - B/T$$

where: p = vapour pressure, T = temperature and A and B are constants (Hance, 1980). Second, enhanced evaporation and evapotranspiration leads to increased soil moisture potential gradients which increase the mass movement of water from depth to the soil surface. Third, heating causes the soil to dry out, thus enhancing pesticide adsorption on soil particles. Pesticide movement by diffusion and mass flow to evaporating surfaces occurs in the same manner as inorganic solutes (see Section 5.4). Hance (1980) provides a useful review of these processes.

Many studies have measured rates of pesticide volatilisation and estimated percentage losses both in the field and under laboratory conditions. Some examples of these are given in Table 9.5. Perhaps one of the most important findings is that substantially larger losses of pesticides occur directly from the leaves of sprayed plants than from the soil, particularly under warm summer conditions when the soil surface is dry (Willis et al., 1983).

9.6.2 Transport of pesticides in solution

As suggested above, movement of pesticides in the soil solution occurs through the processes of mass flow and diffusion, as described in Section 5.4 for inorganic solutes. Convection, or mass flow, describes the flow of water and thus of solute pesticides carried in soil water. Diffusion

Table 9.5 Pesticide volatilisation losses from soils

Pesticide type	Vapour pressure	Conditions	Rates of loss $(g\ ha^{-1}\ d^{-1})$	Per cent loss	Authority
Insecticides					
Organochlorines:					
DDT	7.3×10^{-7}	Sprayed onto moist soil in enclosed chamber	0.5-1.5		Nash (1983)
Dieldrin	9.9×10^{-6}		0.5-6.0		
Endrin	2.0×10^{-7}		1.0-4.5		
Lindane	1.3×10^{-4}		1.0-36		
Heptachlor	3.0×10^{-4}		5.0-180		
DDT		Gila silt loam, 30°C, 10 per cent soil moisture, 100 per cent relative humidity, 8 ml s^{-1} airflow; pesticide concentrations of 1-50 µg g^{-1} soil; losses recorded over 7 days	0.77-12.88	0.9-9.2	Farmer et al. (1972)
Dieldrin			3.84-60.00	4-40	
Lindane			9.04-550.68	25-60	
Herbicides					
Trifluralin	2.0×10^{-4}	Sprayed onto moist soil in enclosed chamber	5.0-70		Nash (1983)
2,4-D		Wheat field soil over 3 days		30	Grover et al. (1973)
Diazinon		Turf over 3 weeks		2	Branham et al. (1986)

Table 9.6 Relative mobility of pesticides in soils (after Helling, Kearney and Alexander, 1971)

| | Increasing mobility classes $\longrightarrow$ | | | |
| Least mobile | | | | Most mobile |
1	2	3	4	5
Lindane[i]	Terbutryn[h]	Fenuron[h]	Picloram[h]	Dalapon[h]
Parathion[i]	Diuron[h]	2,4,5-T[h]	MCPA[h]	Tricamba[h]
Diquat[h]	Linuron[h]	Terbacil[h]	2,4-D[h]	Dicamba[h]
Dieldrin[i]	Dichlorobenil[h]	Thionazin[i]	Bromacil[h]	
Paraquat[h]	Diazinon[i]	Monuron[h]		
Trifluralin[h]		Atrazine[h]		
Heptachlor[i]		Simazine[h]		
Endrin[i]				
Aldrin[i]				
Chlordane[i]				
Toxaphene[i]				
DDT[i]				

Decreased mobility (vertical label at left, with downward arrow)

[i] insecticide
[h] herbicide

of pesticides in solution follows Fick's First Law of Diffusion (see Section 5.4.2). As for inorganic solutes, convective-dispersion transport models with resultant breakthrough curves are used to study pesticide leaching. The two main routes for pesticide transport to watercourses are leaching and surface runoff. Each of these will be examined, together with models used to predict pesticide losses in solution.

Pesticide leaching

Two main techniques, laboratory columns and field tile drain studies, have been used in assessing pesticide leaching. Although rates of pesticide leaching have been examined in column breakthrough curves under controlled conditions, it is much more difficult to assess leaching rates under field conditions. More common are assessments of the relative leachabilities of different pesticides and pesticide groups. Table 9.6 lists relative pesticide mobilities for some more commonly used compounds (Helling, Kearney and Alexander, 1971). It appears from this review that most insecticides, particularly the organochlorines, are relatively immobile in soil, while herbicides are more likely to be mobile.

As a general rule, Rao, Hornsby and Jessup (1985) suggest that pesticides with solubilities > 10 mg l^{-1} and half lives > 50 days show the highest potential for leaching. They included in this group the herbicides simazine, cyanazine and bromacil and the insecticides aldicarb and carbofuran. It is clear that solubility alone is not the only criterion for pesticide mobility. Ivey and Andrews (1965) found that simazine was more readily leached than diuron, despite the fact that diuron is about eight times more soluble. They attributed this response to the fact that diuron was more readily adsorbed by soil colloids. For strongly adsorbed pesticides such as chlorosulfuran (Mersie and Foy, 1986) and atrazine (Wehtje et al., 1984), leaching is only likely in soil lacking clay and organic colloids. Helling, Kearney and Alexander (1971) suggested that reduced pH may restrict leaching of triazine herbicides by increasing their adsorption. The influence of adsorption and soil bulk density on pesticide movement from a point source was demonstrated by Gerstl and Yaron (1983). They showed that napropamide, which is moderately adsorbed, showed predominantly lateral movement at high application rates. Bromacil, a more weakly adsorbed herbicide, showed a more uniform distribution in soil (Fig. 9.11). After several wetting and drying cycles, bromacil was completely leached from around the emitter and became concentrated at the outer edges of the wetted zone.

Using leaching experiments to study the behaviour of bromacil and napropamide in laboratory cores of Evesham clay, White et al. (1986) found that continuous leaching resulted in greater herbicide breakthrough compared to intermittent leaching. They suggest that during discontinuous leaching, herbicides diffuse from conducting channels into adjacent soil aggregates, thus becoming less susceptible to leaching. To explain the fast breakthrough of highly adsorbent herbicides, they invoke preferential percolation down cracks and macropores. The resultant short travel times allow less adsorption on soil particle surfaces. Field examination of the leaching of napropamide indicates that macropore flow may be an important process in deep soil contamination (Jury, Spencer and Farmer, 1983). Perhaps the most disturbing aspect of this result is that any contaminant percolating below the rooting zone may be degraded only very slowly due to the paucity of micro-organisms and the lack of biological activity at depth.

Pesticide runoff

In a comprehensive review of pesticides in surface runoff and drainage from agricultural fields, Wauchope (1978) divided pesticides into three categories, depending on their susceptibility to runoff loss:

(1) *Wettable powders* applied to the soil surface (for example, s-triazine herbicides). These show the highest losses of any group of herbicides. On moderate slopes (10–15 per cent), losses of up to 5 per cent can occur; on shallow slopes (< 3 per cent), losses of 2 per cent can occur.

(2) *Water insoluble* pesticides which are usually applied as emulsions to leaves (for example, organochlorine insecticides). These show losses of ≤ 1 per cent. Losses of DDT can be higher (2–3 per cent) because it remains in soil and available for runoff for longer.

(3) *Water soluble* pesticides applied as aqueous solutions and

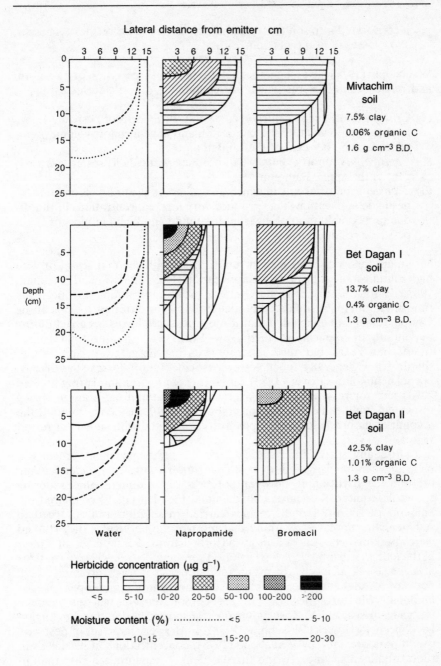

Figure 9.11 Effect of soil type on the distribution of water, napropamide and bromacil, 24 hours after application of herbicide solution at 4 l h⁻¹. B.D. = bulk density (Source: Gerstl and Yaron, 1983)

incorporated into soil (for example, bipyridilium herbicides). These show losses of ≤ 0.5 per cent.

Wauchope (1978) also suggests that there are three types of rainstorm and related runoff events which are conducive to pesticide losses:

(1) *Critical* runoff events which occur within two weeks of pesticide application, have at least 1 cm of rain and have a runoff volume which is ≥ 50 per cent of the rainfall input.
(2) *Catastrophic* runoff events which produce pesticide losses of ≥ 2 per cent of the applied amount.
(3) *Transient* runoff events in which a small amount of rain, soon after pesticide application, may produce very high concentrations in runoff due to a combination of low runoff volume and high amounts of pesticide residues in soil.

Maximum losses of pesticides in runoff occur with the first storm or with high rainfall soon after pesticide application (Hall, Hartwig and Hoffman, 1983; Haith, 1986). This has geographical implications for pesticide runoff. Haith (1986) simulated pesticide runoff for different regions of the United States. He concluded that if most pesticide losses occurred during the month of application, then New York, with greater total annual runoff than Texas, but much less runoff during the month of application, should show generally much lower pesticide runoff losses than Texas. Nicholaichuk and Grover (1983) found that for winter and spring applied 2,4-D and for residues carried over from an autumn application, spring snowmelt could remove a similar amount of 2,4-D as heavy rainfall; the amount of herbicide loss being positively correlated with snowmelt runoff volume.

Before any successful remedial practice or management prevention measures can be instigated for pesticide runoff control, an understanding of pesticide partitioning between solution and sediment phases during runoff is required. Leonard, Langdale and Fleming (1979) compared the amounts of six different herbicides carried in solution and adsorbed to sediments in runoff. They found that most of the herbicides they studied were predominantly carried in an adsorbed form. Percentages of runoff herbicides that were carried in solution amounted to 83–99 per cent for atrazine, 87–88 per cent for cyanazine, 99 per cent for propazine, 92–98 per cent for diphenamide, 89–95 per cent for trifluralin and 0 per cent for paraquat. Measures designed to reduce soil erosion will also reduce pesticide losses due to adsorption on transported sediments, or 'piggy-back' pesticide losses (Wauchope, 1978). In the very large set of pesticide runoff data provided by Wauchope (1978), concentrations of pesticides in runoff sediments can be two to three orders of magnitude higher than in water. However, because of the much larger volumes of water transported during runoff, more pesticides are lost in the water phase. The concentration of pesticide in runoff usually decreases exponentially through time after application as the amount of pesticide in the

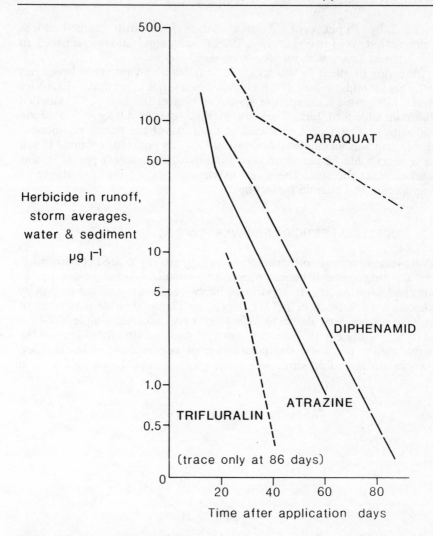

Figure 9.12 Herbicide concentration in runoff with time after application. (Source: Leonard, Langdale and Fleming, 1979)

runoff–active zone at the soil surface decreases (Fig. 9.12). As Wauchope points out, soil erosion prevention will do little to prevent these aqueous losses. This is borne out in a study in Iowa in which soil erosion was decreased by soil conservation measures, but runoff losses of alachlor and cyanazine were virtually unaffected (Baker, Johnson and Laflen, 1976). Other management practices designed to reduce runoff pesticide losses include incorporation with topsoil and strip planting of a vegetation buffer zone at the foot of hillslopes to reduce water and sediment loss. Hall, Hartwig and Hoffman (1983) showed that by strip planting an oat crop at the base of a herbicided slope, water and soil losses could be reduced by 76 per cent. Atrazine losses from strip planted slopes amounted to 0.33 per cent of a 2.2 kg ha^{-1} application, compared to 3.5 per cent losses without strip cropping.

A major problem in assessing the importance of pesticide losses has been the inability to budget for all losses in a single experiment. Branham *et al.* (1986) used a microcosm system to budget the fate of ^{14}C labelled diazinon added to turf. They found that, of the 4.9 kg h^{-1} diazinon application, 47 per cent remained in the form of the parent compound, 22 per cent was metabolised and lost as CO_2, 28 per cent remained in soil as a metabolite, 2 per cent was volatilised and only 1 per cent was leached from the soil. These results clearly indicate the importance of degradation in pesticide budgeting.

9.7   MODELLING PESTICIDE BEHAVIOUR IN SOIL

Two prime objectives of pesticide modelling are (a) to predict adsorption and degradation, and hence to predict accumulation and persistence in soil; and (b) to predict transport and likely watercourse contamination by toxic organic substances (TOS). Since the degradation of a number of pesticides have been shown to follow first order kinetics, simple half-lives have frequently been used as an index of their persistence. Many of the early studies predicted the persistence of organochlorine insecticides, such as aldrin and dieldrin, with quite good success. Using field sites in

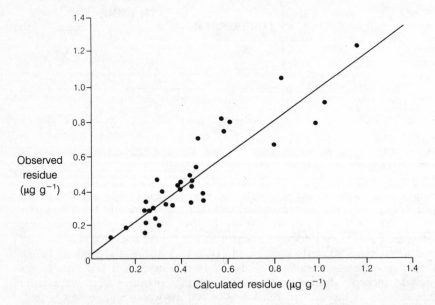

Figure 9.13 Observed and calculated (aldrin and dieldrin) residues in soil. (Source: Dekker, Bruce and Biggar, 1965)

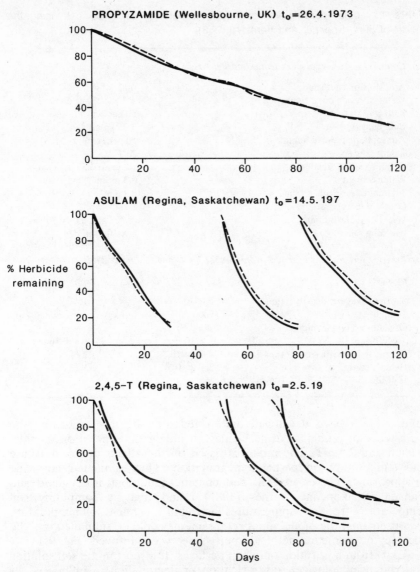

Figure 9.14 Actual (——) and simulated (– –) herbicide residues in soil; simulations carried out using the temperature modification of Walker and Barnes (1981); t_o = date of start.

Illinois, Dekker, Bruce and Biggar (1965) measured (aldrin + dieldrin) levels in soils which had received pesticide applications over the previous thirteen years. Their ability to predict pesticide persistence (Fig. 9.13) was particularly good, considering that no account had been made of different soil and climatic conditions. Walker and coworkers (for example, Walker, 1974, 1976; Smith and Walker, 1977; Walker and

Table 9.7 Input parameters required in simulating pesticide loss using the model of Jury, Spencer and Farmer (1983)

(a) *Common soil physical and chemical parameters*

Air diffusion coefficient	$0.43 \text{ m}^2 \text{ d}^{-1}$
Water diffusion coefficient	$4.3 \times 10^{-5} \text{m}^2 \text{ d}^{-1}$
Porosity	50 per cent
Bulk density	1.35 kg m^{-3}
Atmsopheric relative humidity	50 per cent
Temperature	$25\,^{\circ}\text{C}$
Organic carbon fraction	1.25 and 2.50 per cent
Water content	$0.3 \text{ m}^3 \text{ m}^{-3}$
Amount of pesticide applied	0.1 g m^{-2}
Depth of incorporation	1 and 10 cm
Water evaporation rate	0, 2.5 and $5.0 \times 10^{-3} \text{m}^3 \text{d}^{-1}$
Leaching rate	$5 \times 10^{-3} \text{ m}^3 \text{ d}^{-1}$

(b) *Pesticide physical and chemical properties for 2,4-D and lindane at 25°C*

Property		2,4-D	Lindane
Saturated vapour density	(gm^{-3})	5×10^{-6}	10^{-3}
Solubility	(gm^{-3})	900	7.5
Organic carbon partition coefficient (K_{oc})	(m kg^{-1})	$2 \times 10^{-2} \pm 7$ per cent	1.3 ± 16 per cent
Degradation coefficient	(per day)	4.62×10^{-2}	2.67×10^{-3}
Henry's constant		5.5×10^{-9}	1.33×10^{-4}
Half-life	(days)	15	260

Smith, 1979) have developed and tested a model for predicting the persistence of a number of herbicides, including simazine, propyzamide, asulam and 2,4,5-T. The model is based on the effects of soil moisture and temperature on pesticide degradation. They simulated herbicide disappearance curves for soil and climatic conditions in England and Canada. Walker and Barnes (1981) found that by computing soil temperatures from air temperatures instead of using field soil temperature measurements, the model input requirements could be simplified and the accuracy of predicting herbicide persistence was improved (Fig. 9.14).

Apart from adsorption onto soil particles, transport in the soil solution and chemical/biological transformation or degradation as influenced by temperature changes, the mathematical model presented by Soulas (1982) for pesticide degradation in soil accounts for the additional problem of reduced biological degradation due to pesticide effects on sensitive soil decomposer organisms.

Two main groups of pesticide transport models are available. Simple models for pesticide runoff are based on soil loss or erosion models (for example, Haith, 1986). Their use has been justified under conditions where surface wash is thought to be the most significant pathway for pesticide entry into aquatic systems, such as in irrigated semiarid areas.

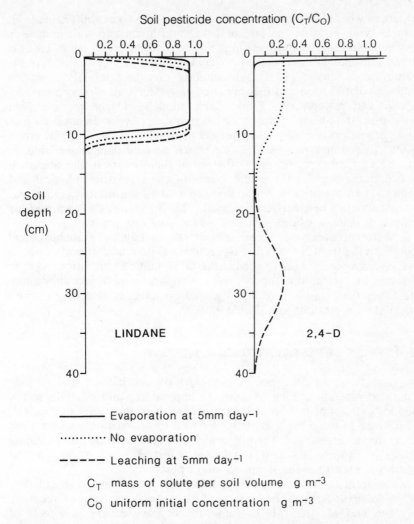

Soil pesticide concentration (C_T/C_O)

Figure 9.15 Calculated lindane and 2,4-D concentrations after 30 days of evaporation or leaching at 5 mm^{-1}. ___: Evaporation at 5 mm day^{-1}; ......: no evaporation; C_T: mass of solute per soil volume (gm^{-3}); C_O: uniform initial concentration (gm^{-3}). (Source: Jury, Spencer and Farmer, 1983)

Other models of pesticide transport are usually based on soil water/solute transport simulation (see Section 5.5) with routines for adsorption and degradation. A pesticide submodel of this type is available in CREAMS, the USDA's field scale model for *c*hemicals, *r*unoff and *e*rosion, from *a*gricultural *m*anagement *s*ystems (Knisel, 1980). Many models of this type are designed to predict, or 'screen' for, leachability of pesticides for which no field data is available. To this end, Jury, Spencer and Farmer (1983) compiled a list of pesticide 'benchmark' physical and chemical properties which are required to predict their loss from soil (Table 9.7). In their simulations of the fate of 2,4-D and lindane in soil, lindane is adsorbed and remains localised in the zone of application, while 2,4-D is mobile and can be transported to depth in the soil (Fig. 9.15). Jury, Spencer and Farmer (1983) subsequently use their model to predict pesticide volatilisation and the influence of surface soil incorporation and organic carbon content. While such models clearly require field verification, the ability to use available pesticide physicochemical data as input parameters provides a simple way of estimating the potential effect on the environment of pesticides for which no field data is available. It must be remembered, however, that since there exists in the literature some disagreement on the actual values of some pesticide physical and chemical characteristics, it is vital that the model's sensitivity to individual input parameters be clearly understood. This is particularly important for pesticide half-lives, whose published values can vary greatly.

A different objective is the subject of a study by Wauchope and Leonard (1980). They used past empirical pesticide runoff data to develop a simple formula to predict the maximum concentration likely to be found in agricultural runoff, given a particular pesticide application rate. They then classified commonly used compounds according to their 'availability' for runoff loss (Table 9.8).

9.8 EFFECTS OF PESTICIDES ON SOIL BIOLOGY

Two aspects of the influence of pesticides on soil biology and ecology require consideration. First, compounds applied to plants or to the soil in one location do not remain at the site where their effects are intentional, but migrate to other regions of the soil and to watercourses where their effects are unintentional. Second, many pesticide compounds, including herbicides, insecticides and fungicides, exert a toxic effect on soil organisms other than the intended target pests.

Some pesticide impacts are reversible while others are persistent. For reversible impacts, a useful ecological criterion is the rate of recovery after the end of stress. It is useful to compare ecological effects of extreme environmental, or natural soil conditions with effects caused by pesticide toxicity. According to Greaves and Malkomes (1980), 50 per cent depression in microbial numbers and process rates can be expected after drought, waterlogging, freezing or heating. Only chemical toxicities resulting in depressions greater than this should be considered as unusual. For assessment of recovery rates, Greaves and Malkomes (1980) suggests that if a realistic microbial cell doubling rate of ten days is used, a biomass reduction of 90 per cent would take three doublings (three × ten days) to bring the population back to the original level. Periods of about 30 days have been observed in soil populations recovering from freezing/thawing or drying/wetting (Greaves and Malkomes, 1980). Accordingly, Domsch, Jagnow and Anderson (1983) classified reversible ecological impacts of pesticide applications into three groups, according

Table 9.8 Pesticide availability for runoff loss (from Wauchope and Leonard, 1980)

Availability Index (A)	Assigned value for A (ppb ha^{-1} kg^{-1})	Properties of pesticide or application conditions	Examples
I	10,000	Wettable powders applied to soil surface	Cyanazine, simazine, atrazine, propazine, diphenamide, linuron
		Soluble salts which bind to clay	Paraquat
		Soluble salts applied to foliage	2,4-D, 2,4,5-T, picloram
II	3000	Soluble salts applied to soil	2,4-D, 2,4,5-T, picloram
		Granular and pelleted pesticides, regardless of solubility – even if incorporated	Picloram, endrin, dieldrin, carbonyl, carbofuran, diazinon
III	1000	Insoluble, persistent pesticides applied to foliage	Endrin, DDT, diuron
		Incorporated but persistent	Dieldrin
IV	300	Insoluble pesticides applied to the soil surface	2,4-D, alachlor
		All incorporated but not persistent pesticides except granular/pelleted soluble salts	Atrazine, trifluralin
		Insoluble and non-persistent pesticides applied to foliage	Parathion

to the time period required after application for biological process rates to fully recover: *negligible* impact ($\leqslant$ 30 days), *tolerable* impact (31–60 days) and *critical* impact (> 60 days). They found that only about 1 per cent of pesticide responses fell into the critical category. These reactions were predominantly the effects of fumigants such as chloropicrin and ethylene dibromide on nitrification processes.

Many studies of the effects of pesticides on soil biology have been carried out under controlled laboratory conditions, using pure substrates. A review of such techniques is given by Greaves and Malkomes (1980), while the problems of comparing laboratory versus field assessment of pesticide impacts is discussed by Eijsackers and van der Drift (1976). A related problem is the fact that soil ecological impacts depend on soil type and on key soil factors such as moisture content, aeration and temperature. This is primarily because these factors influence pesticide degradation and adsorption. The aim in this section is to discuss some of the effects of pesticides on (a) soil microbial numbers, and (b) microbial processes which effect soil fertility.

9.8.1   Effects of pesticides on soil micro-organisms

The method of pesticide application is often an important influence on subsequent microbial effects. Methods which maximise microbial contact have the most devastating effects. Generally, surface applied compounds and aerosols landing on the soil surface have a more harmful effect on the soil population than do incorporated compounds. This is probably because the majority of soil organisms spend a substantial proportion of their time on the soil surface or burrowing into the soil from the surface. The effects of surface applications are probably less important for soil microflora. Generally, the more persistent rather than the more toxic pesticides have the greatest effect on microbial numbers. This is partly because the more transient compounds have time to disappear before organisms can accumulate a lethal dose.

Soil microflora

Few if any herbicides are capable of any great or prolonged effect on total bacterial numbers in soil (Anderson, 1978) and some herbicides, such as paraquat, can even stimulate bacterial activity since they are easily biodegraded (Tu and Bollen, 1968). The effects of many pure compounds on particular groups of the soil microflora have been examined in detail under controlled conditions, often in the laboratory. It is rather difficult to relate these figures to field soils with heterogeneous microbial populations. Authors have also differentiated pesticide effects on the rhizosphere microflora from those on the soil population generally. Gunner *et al.* (1966), for example, found that, after the application of diazinon to the leaves of bean plants, an initial depression in numbers of soil microflora was followed by a rise in microbial numbers to levels higher than before pesticide application. This was due to selective enrichment by one particular type of bacterium. A major reason for selective enrichment by particular organisms is the selective elimination or reduction of competing organisms. There is evidence to suggest that herbicide effects on microbial numbers may be dependent on soil nutrition. Balicka and Bilodub-Panterra (1964) correlated the ability of bacteria to overcome the effects of atrazine with the amounts of organic nutrients present in the soil. Available nutrition may be a key factor in determining competitive abilities. Differences in nutrient availability, organic energy resources, moisture or temperature in different soil types may account for the variety of microbial responses found for individual pesticide types. Simon-Sylvestre and Fournier (1979) show that s-triazine herbicides can have no effect on soil bacteria (see, for example, Freney, 1965; Houseworth and Tweedy, 1973); they can stimulate the soil bacteria (see, for example, Micev, 1979); or they can depress the soil bacteria (see, for example, Simon-Sylvestre, 1974). Explanations for these findings probably relate to the influence of different soil–crop combinations on the relative competitive abilities of soil organisms to survive toxic pesticide effects.

The influence of a pesticide application on microbial numbers depends to some extent on the concentration used. While many compounds are harmless when applied at low rates, they can become stimulatory or depressive at high concentrations. Simon-Sylvestre and Fournier (1979) cite phenoxyacetic acid herbicides as compounds which are harmless towards bacteria when applied at normal field rates, but can become stimulating in controlled laboratory conditions or depressive in field soils when applied at rates 100 and 1000 times higher than normal. Roslycky (1982) used respiration studies to show the severe and longterm depressive effect of high glyphosate concentrations on actinomycetes and soil bacteria (Fig. 9.16). These data show that respiration recovery to near normal rates can occur within two weeks of application if glyphosate concentrations are kept below 100 μg ml^{-1}.

As long as insecticides are applied at normal field rates, they appear to be generally harmless to the soil microflora. A few authors have noted some depressive effects of DDT and other organochlorines on soil bacteria (see, for example, Stojanovic, Kennedy and Shuman, 1972) but generally any initial depressive effect is usually replaced by enhanced microbial numbers. Lethbridge and Burns (1976) found that although the microflora in a sandy loam and a silt loam were totally inhibited by fensulfothion and malathion at 100 and 200 ppm respectively, resistance was developed after three weeks. A comprehensive review of these effects is given by Prasad Reddy, Dhanaraj and Narayana Rao (1984).

Soil fauna

Two main effects of pesticides on the soil fauna have been fairly extensively examined: (a) the effect on organism numbers, and (b) the ability of organisms to take up compounds and accumulate them in body tissues. Only a few studies have examined the effects on soil faunal behaviour, such as the study by van Rhee (1972) on earthworm locomotion.

Rather less information is available for the influence of herbicides on soil fauna than the effects of insecticides. According to Eijsackers and van der Drift (1976), phenoxyacetic acids exert a mild effect on soil fauna while the triazines, substituted ureas and phenolic herbicides have inhibitory influences. These generalities appear to be confirmed in an experiment on meadow soil in which the population densities of *Collembola* were reduced by about 80 per cent by both atrazine and pentachlorophenol (PCP) (Conrady, 1986). Earthworm mortality was also increased by applications of these chemicals. In a study of the effects

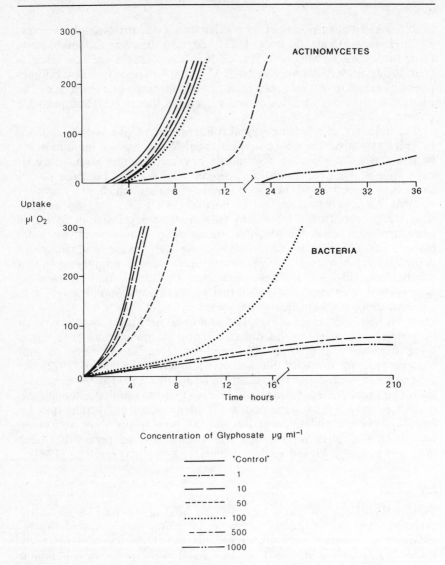

Figure 9.16 Respiration of soil actinomycetes and bacteria at various concentrations of glyphosate. (Reproduced with permission from Roslycky, 1982, *Soil Biology and Biochemistry*, 14, © Pergamon Books Ltd.)

of 2,4-D and MCPA on the soil faunal community, Rapoport and Cangioli (1963) found that while the total community response appeared to be little changed, individual faunal groups responded quite differently (Table 9.9); *Acari* numbers were stimulated, but *Collembola* numbers were almost halved. These results indicate that the generalities given by Eijasackers and van der Drift (1976) may mask more specific responses by individual faunal groups.

Table 9.9 Changes in proportional numbers of soil organisms after treatment with 2,4-D + MCPA (from Rapoport and Cangioli, 1963)

Before application		Per cent	After application	Per cent
Collembola	(springtails)	59.1	Acari	40.8
Acari	(mites)	30.9	Collembola	29.1
Symphyla	(myriapods)	4.5	Homoptera	19.5
Formicidae	(ants)	1.9	Formicidae	6.6
Protura		1.2	Corrodentia	1.7
Other groups		2.4	Other groups	2.3

The effects of insecticides, particularly the organochlorines and organophosphates, on soil fauna have been more widely studied. In an extensive review, Thompson and Edwards (1974) indicate how complicated the soil faunal response to pesticide application can be. They show that since the Acari exhibit a number of trophic preferences and life cycles, effects on preditor/prey interactions, larval stages or on algae, herbage and plant roots, may have important 'knock-on' effects on different groups of soil mites. A series of effects of DDT and aldrin application are illustrated in Fig. 9.17 (Edwards, 1969). The columns representing total arthropod weight show that aldrin kills 70 per cent of the organisms overall, but doubles the number of centipedes. DDT doubles the number of springtails. This upsurge in springtail numbers can be explained from the second set of columns by suggesting that DDT kills more of the mites which prey on springtails, reducing the predation pressure (Edwards, 1969). The total weight of beneficial arthropods is 15 per cent higher in soil treated with DDT. The last set of columns show that both pesticides killed fly and beetle larvae pests effectively but were less successful in controlling symphylids. Thompson and Edwards (1974) suggest that DDT is more toxic to predatory mites than to other mite species. Their review of the literature suggests that other organochlorines, such as heptachlor, chlordane and endrin are toxic to soil mites and springtails, but not to soil nematodes. Organophosphate insecticides are more toxic to nematodes, and again compounds such as diazinon and parathion appear to be more toxic to predaceous soil mites than to other types of mites which tend to flourish after the selective removal of predaceous groups.

The importance of earthworms to soil fertility, through their effects on organic matter decomposition and the maintenance of soil structure, together with their fairly numerous distribution in many soils, has made them particularly important organisms for pesticide studies. In general, insecticides appear to be much more harmful to earthworms than herbicides. According to Thompson and Edwards (1974), the organochlorine aldrin has much less effect on earthworm numbers than other members of the group, such as endrin, heptachlor and chlordane which

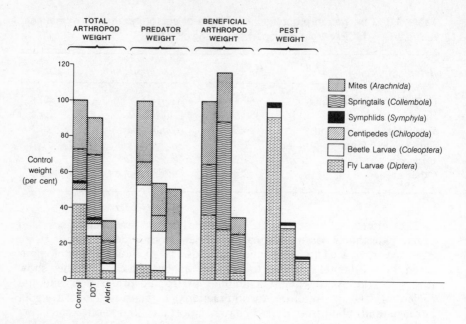

Figure 9.17 Effect of DDT and aldrin on soil arthropods, measured over 1 year. (Source: Edwards, 1969)

are very toxic. Haque and Ebing (1983) examined the effects on ten different pesticides on the common species, *Lumbricus terrestris*. Not surprisingly, the carbamate insecticides, carbofuran and aldicarb had the most deleterious effects on earthworm numbers while atrazine herbicide had a less toxic effect and paraquat the least toxic effect. van Rhee (1972) found that paraquat substantially reduced the activity and locomotion of both *L. terrestris* and *Allolobophora calignosa* but completely killed the surface dwelling species *L. castaneus*. Bharathi and Subba Rao (1986) confirmed that organophosphate insecticides were also toxic to the earthworm *Lampito mauritii*, causing body lesions, body coiling and reduced locomotion.

Uptake and accumulation of pesticides by soil organisms

There is much evidence to suggest that soil invertebrates take up pesticides and concentrate them in body tissues. This is of particular concern when invertebrates, particularly the larger organisms such as earthworms and snails, provide the main food source for birds and some higher vertebrates such as foxes. Organochlorine residues have been found in the tissues of earthworms, slugs and snails. In the data supplied by Gish (1970), it is clear that concentrations in body tissues far exceed those in surrounding soil (Table 9.10). Slugs, which are both saprovores and herbivores, may show highest accumulations since they take up

Table 9.10 Residues of organochlorine insecticides in earthworms, slugs and snails* (from Gish, 1970)

Insecticide	Residues ($\mu g\ g^{-1}$)			
	Soil	Earthworms	Slugs	Snails
DDT	0.08–5.4	1.1 –54.9	10.3– 36.7	0.32–0.38
Aldrin	–	0.02– 0.2	0.2	–
Dieldrin	0.01–0.02	0.04– 0.82	0.2– 11.1	0.02–0.07
Endrin	0.01–3.5	0.4 –11.0	1.1–114.9	2.72

* Insectide treatments of 3.3–18.2 kg ha^{-1}

pesticides and residues from both soil organic matter and live algae and vegetation. Earthworms living near or at the soil surface will concentrate largest amounts of residues when high concentrations exist in surface litter and organic matter soon after spraying. Although less information is available for the uptake of pesticides and residues by soil micro-arthropods, Thompson and Edwards (1974), in their review of the subject, suggest that some *Collembola* and *Acari* do take up organo-chlorines and some may be able to break down DDT to DDE.

9.8.2 Effects of pesticides on soil microbial processes

It is clear that some biological processes are more affected by pesticides than others. In an empirical analysis of the actual responses of 25 microbiological processes to 71 different pesticides in 734 experiments, Domsch (1984) found that the most sensitive indicators of pesticide toxicity were organic matter decomposition, nutrification and acid phosphatase activity, while the least sensitive indicators were denitrification, non-symbiotic nitrogen-fixation and urease activity. Two main areas of interest in pesticide studies are organic matter decomposition, often monitored by soil respiration studies, and nitrogen dynamics, monitored through nitrification, denitrification and nitrogen fixation. An attempt will be made in the following sections to evaluate the effects of different pesticides on each of these processes.

Organic matter decomposition and soil respiration

Oxygen consumption and CO_2 evolution are usually used as indicators of microbial respiration in soil and hence as indicators of microbial activity and organic matter decomposition. There are two main problems associated with this technique: (a) that, in the field, it is difficult to separate microbial respiration from root respiration; and (b) that selective elimination of some species and enhancement of others cannot be assessed. Despite these problems, respiration studies have been used quite extensively to assess pesticide effects on the soil biomass generally and the microflora in particular (see, for example, Alexander, Armstrong and Smith, 1981). These authors found that fungicides tended to suppress the activity of soil fungi with resultant increases in soil bacteria Grossbard (1976) showed that various herbicides have an inhibitory effect on O_2 uptake and CO_2 evolution, even at normal field concentrations. Paraquat, asulam, diuron and simazine inhibit CO_2 production, while paraquat, 2,4-D, diuron, bromacil and trifluralin inhibit O_2 consumption at field concentrations. As suggested earlier, microbial response often depends on the concentration of pesticide applied, with commonly little effect on respiration at low field concentrations. Simazine at a concentration of $2 \mu g g^{-1}$ appears to have no effect on respiration (Gaur and Misra, 1977) while at $10 \mu g g^{-1}$ causes increased CO_2 evolution (Smith and Weeraratna, 1974). Many insecticides appear to depress rates of soil respiration. Bartha, Linsilotta and Pramer (1967) found that carbaryl, malathion and parathion all depresses CO_2 evolution. Several organo-chlorine insecticides, including lindane, endrin and dieldrin have also been shown to inhibit CO_2 production (see Bardiya and Gaur, 1968), the degree of inhibition increasing with increasing insecticide concentra-tion. Respiration-stimulating effects have also been recorded for the application of pesticides such as 2,4-D at high concentration (Bliev, 1973), simazine (Smith and Weeraratna, 1974), glyphosate (Carlisle and Trevors, 1986), dieldrin (Eno and Everett, 1958) or organophosphates (Tu, 1970). These results confirm the earlier suggestion that microbial responses to pesticides are variable and may be related to differences in soil properties.

Apart from respiration studies, the abilities of soil organisms to decompose pure organic compounds such as cellulose or vegetation litter have been used as indices of pesticide toxicity. Several herbicides have been shown to inhibit cellulose decomposition, particularly at high concentration. Szegi (1972) found that paraquat inhibited cellulolytic organisms at concentrations as low as $0.003 \mu g g^{-1}$ but that much higher concentrations of 2,4-D were required to cause similar inhibition. Atrazine has a greater effect on cellulolytic organisms than simazine which is less water-soluble and penetrates less deeply into soil (Klyuchnikov et al., 1964).

The effects of herbicides on the breakdown of vegetation litter was studied by Hendrix and Parmelee (1985) using a litter bag technique. They monitored weight loss from mesh bags containing dried *Sorghum halepense* litter which had been dipped in three herbicides: atrazine, paraquat and glyphosate, at concentrations ten times higher than field application rates. They obtained two main results: (a) overall slower rates of decay for glyphosate and paraquat compared to untreated litter; and (2) faster losses of P, Ca and Mg from herbicide-treated litter. Their interpretation of these results is that decomposer organisms initially use the herbicide as a carbon source, increasing the importance of microarthropod grazing relative to comminution, eliminating or reducing the importance of predaceous microarthropods and increasing the loss of soluble nutrients from the litter via microbial and microarthropod activity. Their general conclusion is that herbicides greatly simplify the breakdown route in decomposing litter by reducing the number of recycling loops, accelerating soluble nutrient loss and slowing down decay of carbon from the leaf tissue.

Soil nitrogen transformations

The effects of pesticides on ammonification, nitrification, denitrification, and nitrogen fixation have received much attention. In general, nitrification is much more sensitive to the toxicity of pesticides than is ammonification. Simon-Sylvestre and Fournier (1979) note that it is really only at high application rates that herbicides such as 2,4-D, atrazine, simazine and linuron can inhibit ammonification. At field concentrations, some herbicides, including those listed above, are more likely to stimulate ammonification, depending on soil conditions. Reviewing the effects of insecticides on ammonification, Prasad Reddy, Dhanaraj and Narayana Rao (1984) suggests that whether stimulation or inhibition occurs may depend on soil moisture content. Generally, low insecticide concentrations stimulate ammonium production while higher levels, up to $100–400$ g ha^{-1}, inhibit the process. Audus (1970) found that demeton was the most toxic insecticide, inhibiting ammonification at concentrations below those normally applied in the field.

Nitrifying bacteria are particularly sensitive to changes in soil properties and are also sensitive to applied pesticides. Both stimulation and inhibition of nitrification processes have been recorded. These results have sometimes masked the effects of pesticides on component nitrification processes. The conversion of NH_4^+ to NO_2^- by *Nitrosomonas* has been shown to be more susceptible to toxic pesticides than the oxidation of NO_2^- to NO_3^- by *Nitrobacter*. This result has been found for substituted urea herbicides generally (Grossbard and Marsh, 1974), for diuron and linuron (Corke and Thompson, 1970), monuron (Casely and Luckwill, 1965) and also for paraquat (Yamanaka, 1983). Inherent differences in the nitrifying capacity of soils play an important part in delaying the effects of pesticides of nitrification. Dubey and Rodriguez (1970) found that herbicides inhibited nitrification less in soils with high nitrifying capacity than in soils with low nitrifying capacity. In forestry soils in Greece, Nakos (1980) found that asulam, applied at a realistic field rate of 10 μg g^{-1}, reduced nitrification by 88 per cent. He considers this side effect of herbiciding as advantageous, since it helps to conserve soil N and to avoid NO_3^- leaching or denitrification.

Many insecticides exert an initially depressive effect on nitrification. The temporary inhibition in mineralization of added $(NH_4)_2SO_4$ fertiliser was caused by $\geqslant 25$ μg g^{-1} levels of organochlorine insecticides in the order lindane > dieldrin > aldrin (Bardiya and Gaur, 1968). Chandra (1967) reported an eight week inhibition of nitrification in a heavy clay, a sandy loam and two loam soils after the application of dieldrin and heptachlor. Nitrification inhibition by organochlorines was least in soils with high organic matter, clay and CEC. Differences have also been noted in the susceptibilities of *Nitrosomonas* and *Nitrobacter* to insecticides; Garretson and San Clemente (1968), for example, reporting that lindane, malathion and baygon were more toxic to *Nitrosomonas* than to *Nitrobacter*. It is important to remember that the magnitude of nitrification inhibition in pure culture studies may not be directly applicable to field soils where adsorption and both chemical and biological degradation may act to (a) reduce inhibitory effects at low concentrations, and (b) speed up recovery rates. This may explain why many authors generally agree that field rates of pesticide application rarely have adverse effects on nitrification.

If altered NO_3^- concentrations are being monitored as a symptom of pesticide application, inhibited or enhanced denitrification, as well as nitrification, may be an equally important control. Evidence suggests that 2,4-D (Bollag and Henninger, 1976), diquat (Atkinson, 1973), atrazine and simazine (McElhannon, Mills and Bush, 1984) are capable of inhibiting denitrification. Yeomans and Bremner (1985a) tested the effects of 20 herbicides, including 2,4-D, diuron, monuron, atrazine, simazine and trifluralin, applied at 10 and 50 $\mu g\ g^{-1}$, on denitrification processes. They found no significant inhibitory effects with any of the herbicides studied. Insecticides do not appear to inhibit denitrification if they are applied at normal field rates. Bollag and Henninger (1976) examined pesticide effects on denitrification in pure culture. They found that carbofuran, DDT, diazinon and endrin had no effect on denitrifying organisms when applied at 100 $\mu g\ g^{-1}$ of culture medium, although at the same concentration, carbaryl slightly inhibited denitrification. Yeomans and Bremner (1985b) reported no significant inhibition of denitrification after the application of seven insecticides, including lindane, malathion and carbofuran, at a concentration of 10 $\mu g\ g^{-1}$. When applied at 50 $\mu g\ g^{-1}$, lindane, fonofos and malathion enhanced denitrification.

Although at field application rates, most herbicides have no effect on the free-living nitrogen-fixing bacteria *Azotobacter* and *Clostridium*, several compounds have caused depressive effects. These include PCP, simazine, atrazine, linuron and paraquat (Simon-Sylvestre and Fournier, 1979). Again, some inhibitory effects can last for an initial period of up to two months, after which normal activity can be achieved. An extensive review of these processes is given by Grossbard (1976). The effects of pesticides on symbiotic nitrogen fixing bacteria are twofold: those affecting the bacterium and those affecting the host plant. In pure culture, the compounds most toxic to *Rhizobium* are probably dinoseb and linuron, although a wide range of responses have been reported (Grossbard, 1976). The nutrient status of the culture or soil medium is an important control on inhibited N_2-fixation by pesticides. According to Grossbard (1976), *Rhizobia* in media high in Mg^{2+} and Ca^{2+}, with neutral to alkaline pH, are most likely to resist the toxic effects of pesticides. Alexander (1985) actually advocates the use of pesticides to enhance both symbiotic and non-symbiotic nitrogen fixation by preferentially killing competitors, preditors and other soil micro-organisms that otherwise would be harmful to N_2-fixing organisms or to the development of an efficient symbiotic relationship.

9.9   SUMMARY

Materials described as pesticides include three main groups of compounds: insecticides, herbicides and fungicides. Also commonly in use are molluscicides and rodenticides. When applied to soil, these compounds can be (a) degraded, biologically or photochemically; (b) adsorbed by soil organic matter, clay minerals or Fe and Al hydroxyoxides; (c) washed into watercourses by leaching and surface runoff; or (d) volatilised, resulting in atmospheric pollution. Pesticides also have adverse effects on non-target organisms in the soil, resulting in disruption of soil biological processes. Pesticides made from heavy metals and organochlorines persist for the longest time in soil, up to tens of years. Adsorption of pesticide, such as the triazines, onto clay mineral surfaces catalyses degradation while exposure to ultraviolet light decomposes the phenoxyacetic acids. Most organic pesticides, irrespective of their complexity, can eventually be decomposed biologically. Two main characteristics of pesticide molecules influence their adsorption capacity in soil: (a) their surface charge (if any); and (b) their degree of aqueous solubility. Adsorbed pesticides are either permanently charged, weak bases or weak acids. These different forms are adsorbed with varying degrees of strength, depending on soil pH and the cation and anion exchange capacities of the soil. Adsorption and desorption are described using Langmuir and Freundlich adsorption isotherms. Apart from soil pH and the clay and organic matter content of soil, moisture content and temperature are important controls on pesticide adsorption. As well as transport in solution, volatilisation is an important pesticide dissipation process, particularly for surface applied, low solubility and vapour pressure compounds such as the organochlorines. Pesticides with solubilities > 10 mg l^{-1}, such as simazine, bromacil and aldicarb, show the highest potential for leaching. Surface applied pesticides are washed off in runoff both in soluble form and adsorbed to fine colloidal particles. If runoff sediments have high adsorption capacities, pesticide concentration in runoff waters increase downslope. Apart from watercourse pollution by pesticides, environmentalists and soil ecologists are concerned about the effect pesticides have on non-target organisms. Not only can soil microbial numbers be adversely affected, but biological processes, including organic matter decomposition and nitrogen fixation can be disrupted.

Postscript: New Directions

A very large number of interacting processes have been outlined in the preceeding chapters. Clearly, an understanding of soil variability at both matrix and field scales requires an understanding of the superimposition of component processes. The systematic approach adopted here aimed to introduce the idea of a series of levels of process interaction. Necessarily it is a selection of examples. Three main areas of interest are, regrettably, omitted.

Fundamental to the study of soil formation and development are the *influences of geomorphology*, particularly hydrological processes and soil erosion. The simplest representation of topographic influences on soil development is given in the soil catena. Although several authors (for example, Furley, 1968) have related soil properties to slope angle or to position on slope, regression models of this type encounter serious autocorrelation problems (Richards, Arnett and Ellis, 1985). Examination of the variation in soil properties on slopes led Furley (1971) to identify slope positions which are predominantly erosional and slope positions which are predominantly depositional. These simple 'explanations' do not invoke processes either of slope or soil profile development. Clearly, an understanding of the links between denudational (weathering, usually solutional processes) and both soil and slope transport processes is required. The influence of soil profile development on slope hydrology requires further study. Anderson and Richards (1987), for example, suggest that progressive B horizon illuviation could result in reduced surface infiltration caused by the development of a perched water table during rainstorms. It is not difficult to envisage a range of soil profile conditions and soil management practices which could impose equally important changes on rates and routes of soil hydrological processes. The influence of plant roots on soil and slope processes also required further study. There is a need to examine links between mechanical effects of roots in soil, the stabilising effect of root exudates and added organic litter from both shoots and roots, and the hydrological effects of interception and evapotranspiration. A further area of interest in the study of vegetational effects on soil processes is their role in the provision of biochannels which act as macropores to enhance 'quickflow' or bypassing flow. Particularly on slopes, these may prove to be an important transport route for both water and solutes during storm events.

A second major omission in this survey of soil processes is consideration of *non-temperate conditions*. Arctic, humid tropical or arid conditions impose a whole series of different profile development processes on the basic soil formative processes of weathering and decomposition. Accompanying these different climatic processes are a series of different soil management practices, including irrigation, burning and the addition of organic slurries and manures.

Of some current concern are the influences of *environmental inputs* on the directions and rates of basic soil processes. Much research is currently directed towards the understanding of acid rain effects on soil chemistry. Ross (1987) used chemical thermodynamics to calculate the weathering reactions of albite and anorthite to kaolinite under acidified conditions. Calculations indicated that neither mineral would weather spontaneously under non-acid conditions. Including the effects of $(H_2O + CO_2)$ on these reactions, yielded highly negative ΔG_r^o values, indication that both reactions would occur spontaneously, with albite→kaolinite occurring preferentially to anorthite→kaolinite. Elemental budgets for small catchments appear to corroborate this idea of enhanced weathering under the influence of acid rain inputs to soil. In a long term study of water quality in the River Elbe catchment in Czechoslovakia, Paces (1985) attributed the increased acidity and mineral concentrations to accelerated weathering and desorption of exchangeable ions from soil. The consumption of H^+ by ion exchange in soil and hydrolysis of bedrock was not enough to neutralise the acidifying inputs. Calculating the H^+ ion (proton) budgets in acid rain-stressed soils and vegetation systems, through an understanding of the proton consumption and production of component soil processes, is an important new step in examining the effect of altering the balance between internal and external proton cycling in natural systems. Using this approach to examine proton fluxes in a range of different ecosystems, van Breemen, Driscoll and Mulder (1984) show that acid rain increases external loading to soils, such that they exceed internal loadings. This condition results in soil acidification by aluminium dissolution, with aluminium export to drainage waters. The effects on soil chemical processes of the leaching of soluble organic matter and nutrients from vegetation canopies by acidified rain has yet to be studied in any detail.

A second type of atmospheric pollution to affect soil processes is that of radioactive fallout. The major problem in estimating effects of such inputs, as with acid rain effects, is the fairly long timescale (years to tens of years) over which processes may be altered, the degree of alteration often changing through time. This problem has led many workers to use models to simplify and speed up rates of processes. Examples of hardware approaches are leaching columns and incubation studies. While approaches such as these may give an insight into the order, direction and route of particular processes, there are problems in relating results to field conditions. Relating rates of processes monitored under controlled conditions to those operating in the field can be particularly problematical. A large number of soil process simulation models have been presented in previous chapters. Modelling objectives range from simulating the development of soil profiles in 'natural' environments, to the predicting adsorption and routing of applied agrochemicals in agricultural soils. All require stochastic routines to handle variability in the relatively few, key soil properties.

A range of spatial and temporal scales of approach to the study of soil processes has been outlined. Many conceptual and simulation models have greatly aided understanding of inherently complex soil systems. While empirical field studies have also been valuable in estimating rates of reactions, as well as seasonal and other environmental effects, they often suffer from the problem that field instrumentation alters the speed and direction of processes. The future use of models for studying the effects of soil management practices on existing soil properties and processes makes environmental as well as economic sense, but the main requirement for further progress is the field verification of applied soil process models.

References

Abbott, I., Parker, C.A. and Sills, I.D. (1979), Changes in the abundance of large soil animals and physical properties of soils following cultivation. *Australian Journal of Soil Research*, 17, 343–353.

Adams, R.S. Jr. and Li, P. (1971), Soil properties influencing sorption and desorption of lindane. *Soil Science Society of America Proceedings*, 35, 78–81.

Addiscott, T.M. (1977), A simple computer model for leaching in structured soils. *Journal of Soil Science*, 28, 554–563.

Addiscott, T.M. and Cox, D. (1976), Winter leaching of nitrate from autumn-applied calcium nitrate, ammonium sulphate urea, and sulphur-coated urea in bare soil. *Journal of Agricultural Science, Cambridge*, 87, 381–389.

Addiscott, T.M. and Wagenet, R.J. (1985), Concepts of solute leaching in soils: a review of modelling approaches. *Journal of Soil Science*, 36, 411–424.

Adu, J.K. and Oades, J.M. (1978), Physical factors influencing decomposition of organic materials in soil aggregates. *Soil Biology and Biochemistry*, 10, 109–115.

Agboola, A.A. and Fayemi, A.A.A. (1972), Fixation and excretion of nitrogen by tropical legumes. *Agronomy Journal*, 64, 409–412.

Akkermans, A.D.L. and van Dijk, C. (1976), The formation and nitrogen fixing activity of the root nodules of *Alnus glutinosa* under field conditions. In Nutman, P.S. (ed.) *Symbiotic Nitrogen Fixation in Plants. International Biology Programme*, 7, 511–520, Cambridge University Press, Cambridge.

Akkermans, A.D.L. and Houwers, A. (1979), Symbiotic nitrogen fixers available for use in temperate forestry. In Gordon, J.C., Wheeler, C.T. and Perry, D.A. (eds) *Symbiotic Nitrogen Fixation in the Management of Temperate Forests*, pp. 23–35. Proceedings of the Workshop, April 1979, Corvallis, Oregon. National Science Foundation.

Alexander, J.P.E., Armstrong, R.A. and Smith, S.M. (1981), Methods to evaluate pesticide damage to the biomass of the soil microflora. *Soil Biology and Biochemistry*, 13, 149–153.

Alexander, M. (1964), Microbiology of pesticides and related hydrocarbons. In *Principles and Applications in Aquatic Microbiology*, John Wiley & Sons, Chichester, pp. 15–38.

Alexander, M. (1977), *Introduction to Soil Microbiology*, 2nd Edition, John Wiley & Sons, Chichester.

Alexander, M. (1985), Enhancing nitrogen fixation by use of pesticides: a review. *Advances in Agronomy*, 38, 267–282.

Alexander, V., Billington, M. and Schell, D.M. (1978), Nitrogen fixation in Arctic and Alpine tundra. In Tieszen, L.L. (ed.) *Vegetation and Production Ecology of an Alaskan Arctic Tundra. Ecological Studies Vol. 29*. pp. 539–558, Springer Verlag, New York.

Allen, M.F., Smith, W.K., Moore, T.S. and Christensen, M. (1981), Comparative water relations and photosynthesis of mycorrhizal and non-mycorrhizal *Boutelona gracilis. New Phytology*, 88, 683–693.

Allen, S.E. and Grimshaw, H.M. (1960), Effect of low temperature storage on the extractable nutrient ions in soils. *Journal of the Science of Food and Agriculture*, 13, 525–529.

Allen, S.E., Grimshaw, H.M., Parkinson, J.A., and Quarmby, C. (eds) (1974), *Chemical Analysis of Ecological Materials*. Blackwell Science Publishers, Oxford.

Allison, F.E. (1955), The enigma of soil nitrogen balance sheets. *Advances in Agronomy*, 7, 213–250.

Allison, F.E. (1966), The fate of nitrogen applied to soils. *Advances in Agronomy*, 18, 219–258.

Allmaras, R.R., Rickman, P.W., Ekin, L.G. and Kimsall, B.A. (1977), Chiseling influences on soil hydraulic properties. *Soil Science Society of America Journal*, 41, 796–803.

Amberger, A., Gutser, R., Teicher, K. (1974), Kaliumernährung der Pflanzen und Kaliumdynamik auf Kaliumfixierendem Boden (Potassium nutrition of plants and potassium dynamics on a potassium-fixing soil). *Plant and Soil*, 40, 269–284.

Anderson, H.A., Berrow, M.L., Farmer, V.C., Hepburn, A., Russell, J.D. and Walker, A.D. (1982), A reassessment of the podzol formation process. *Journal of Soil Science*, 33, 125–136.

Anderson, J.M. (1973), Carbon dioxide evolution from two temperate, deciduous woodland soils. *Journal of Applied Ecology*, 10, 361–378.

Anderson, J.R. (1978), Pesticide effects on non-target soil microorganisms. In Hill, I.R. and Wright, S.J.L. (eds) *Pesticide Microbiology*, pp. 313–533. Academic Press, London.

Anderson, J.W. (1978), *Sulphur in Biology. Studies in Biology No 101*, Edward Arnold, London.

Anderson, M.G. and Richards, K.S. (1987), Modelling slope stability: the complementary nature of geochemical and geomorphological approaches. In Anderson, M.G. and Richards, K.S. (eds) *Slope Stability*, pp. 1–9. John Wiley, Chichester.

Anderson, M.P. (1979), Using models to simulate the movement of contaminants through ground water flow systems. *Chemical Rubber Co. Critical Reviews in Environmental Control*, 9, 97–156.

Anon. (1978), Environmental Protection Agency: guidelines for registering pesticides in the United States. *Federal Register*, 43, No 132.

Ardakani, M.S. Volz, M.G. and McLaren, A.D. (1975), Consecutive steady state reactions of urea, ammonium and nitrite nitrogen in soil. *Canadian Journal of Soil Science*, 55, 83–91.

Arfin, H., Perkins, H. and Tan, K.H. (1973), Potassium fixation and reconstitution of micaceous structures in soils. *Soil Science*, 116, 31–35.

Armstrong, D.G. and Chesters, G. (1968), Adsorption catalyzed chemical hydrolysis of atrazine. *Environmental Science and Technology*, 2, 683–689.

Arnold, P.W. (1978), Surface–electrolyte interactions. In Greenland, D.J. and Hayes, M.H.B. (eds), *The Chemistry of Soil Constituents*, pp. 355–404, John Wiley, Chichester.

Arnold, J.S. and Farmer, W.J. (1979), Exchangeable cations and picloram sorption by soil and model adsorbents. *Weed Science*, 27, 257–262.

Arnott, R.A. and Clements, C.R. (1966), The use of herbicides in alternate husbandry as a substitute for ploughing. *Weed Research*, 6, 142–157.

Asai, R.I., Westlake, W.E. and Gunther, F.A. (1969), Endrin decomposition on air dried soils. *Bulletin of Environmental and Contamination Toxicology*, 4, 278–284.

Ashton, F.M. and Crafts, A.S. (1973), *Mode of Action of Herbicides*. Wiley-Interscience, Chichester.

Atkinson, G. (1973), Effects of diquat on the microbiological organisms in the soil of lakes. *Proceedings of the Pollution Research Conference, New Zealand, DSIR Information Series No 97*, pp. 529–538.

Aubertin, G.M. (1971), Nature and extent of macropores in forest soils and their influence on subsurface water movement. *Forest Service Research Papers NE (U.S.)* 192PS.

Audus, L.J. (1970), The action of herbicides and pesticides on the microflora. *Mededelingen van de Faculteit Landbouw wetenschappen, Rijksuniversiteit Gent*, 35, 465–492.

Aulakh, M.S. and Rennie, D.A. (1984), Transformations of fall-applied nitrogen-15-labelled fertilizers. *Soil Science Society of America Journal*, 48, 1184–1189.

Ausmus, B.S., Edwards, N.T. and Witkamp, M. (1976), Microbial immobilisation of carbon, nitrogen, phosphorus and potassium: implications for forest ecosystem processes. In Anderson, J.M. and MacFadyen, A. (eds) *The Role of Terrestrial and Aquatic Organisms in Decomposition Processes*, pp. 397–416, Blackwell, Oxford.

Avery, B.W. (1980), Soil Classification for England and Wales. [Higher Categories]. *Soil Survey of England and Wales, Technical Monograph No 14*.

Baes, C.F. and Mesmer, R.E. (1976), *Hydrolysis of Cations*, John Wiley, New York.

Baeumer, K. and Bakermans, W.A.P. (1973), Zero-tillage. *Advances in Agronomy*, 25, 77–123.

Bailey, G.W. and White, J.L. (1965), Herbicides: a compilation of their physical, chemical and biological properties. *Residue Reviews*, 10, 97–122.

Bailey, G.W. and White, J.L. (1970), Factors influencing the adsorption, desorption and movement of pesticides in soil. *Residue Review*, 32, 29–92. Single Pesticide Vol: The Triazine Herbicides. Springer Verlag.

Bailey, G.W., White, J.L. and Rothbert, T. (1968), Adsorption of organic herbicides by montmorillonite, role of pH and chemical character of adsorbate. *Soil Science Society of America Proceedings*, 32, 222–234.

Bain, D.C. (1977), The weathering of ferruginous chlorite in a podzol from Argyllshire, Scotland. *Geoderma*, 17, 193–208.

Baker, J.L. and Johnson, H.P. (1976), Impact of subsurface drainage on water quality. *Proceedings of the Third National Drainage Symposium Publication*, 1–77, pp. 91–98. Amer. Society of Agricultural Engineering, St. Joseph, Michigan.

Baker, J.L., Johnson, H.P. and Laflen, J.M. (1976), Completion report – effect of tillage systems on runoff losses of pesticides: a simulated rainfall study. *Iowa State Water Resource Research Institute. Report No ISWRRI-71*, Ames, Iowa.

Balandreau, J.P., Rinaudo, G., Oumarov, M.M. and Dommergues, Y.R. (1976), Asymbiotic N_2 fixation in paddy soils. In Newton, W.E., and Nyman, C.J. (eds). *Proceedings of the First International Symposium on Nitrogen Fixation*, Vol 2, pp. 611–628, Washington State University Press, Pullman.

Balicka, N. and Balodub-Panterra, P. (1964), The influence of atrazine on some soil bacteria. *Acta microbiol. Polon.*, 13, 149–151.

Ball, B.C. and O'Sullivan, M.F. (1982), The characterisation of pores in ploughed and direct-drilled soils in Scotland. *Ninth Conference of ISTRO, Yugoslavia, 1982*, pp. 396–401.

Ballard, T.M. (1971), Role of humic carrier substances in DDT movement through forest soil. *Soil Science Society of America Proceedings*, 35, 145–147.

Barber, D.A. and Gunn, K.B. (1974), The effect of mechanical forces on the exudation of organic substances by the roots of cereal plants grown under sterile conditions. *New Phytology*, 73, 39–45.

Barber, S.A. (1979), Soil phosphorus after 25 years of cropping with five rates of phosphorus application. *Communications in Soil Science and Plant Analysis*, 10, 1459–1468.

Barclay, W.R. and Lewin, R.A. (1985), Microalgal polysaccharide production for the conditioning of agricultural soils. *Plant and Soil*, 88, 159–169.

Bardiya, M.C., and Gaur, A.C. (1968), Effect of some chlorinated hydrocarbon insecticides on nitrification in soil. *Zentralblatt für Bakteriologie, Parasitenkunde, Infectionskrankheiten und Hygiene* (Abt. II), 124, 552–555.

Barisas, S.G., Baker, J.L., Johnson, H.P. and Laflen, J.M. (1978), Effect of tillage systems on runoff losses of nutrients, a rainfall simulation study. *Transactions of the American Society of Agricultural Engineers*, 21, 893–897.

Barraclough, D., Geens, E.L. and Maggs, J.M. (1984), Fate of fertilizer nitrogen applied to grassland: II: Nitrogen-15 leaching results. *Journal of Soil Science*, 35, 191–199.

Barraclough, D., Hyden, M.J., Davies, G.P. (1983), Fate of fertilizer nitrogen applied to grassland 1. Field leaching results. *Journal of Soil Science*, 34, 483–497.

Barrow, N.J. (1974), Effect of previous additions of phosphate on phosphate adsorption by soils. *Soil Science*, 118, 82–89.

Barrow, N.J. (1978), The description of phosphate adsorption curves. *Journal of Soil Science*, 29, 447–462.

Barrow, N.J. (1980), Evaluation and utilisation of residual phosphorus in soils. In Khasawneh, F.E., Sample, E.C. and Kamprath, E.J. (eds), *The Role of Phosphorus in Agriculture*, pp. 333–359.

Barrow, N.J. (1983), A mechanistic model for describing the sorption and desorption of phosphate by soil. *Journal of Soil Science*, 34, 733–750.

Barrow, N.J. (1984), Modelling the effects of pH on phosphate sorption by soils. *Journal of Soil Science*, 35, 283–297.

Barrow, N.J. (1986a), Testing a mechanistic model: I: The effects of time and temperature on the reaction of fluoride and molybdate with a soil. *Journal of Soil Science*, 37, 267–275.

Barrow, N.J. (1986b), Testing a mechanistic model: II: The effects of time and temperature on the reaction of zinc with a soil. *Journal of Soil Science*, 37, 277–286.

Barrow, N.J. and Shaw, T.C. (1979), Effect of solution : soil ratio and vigour of shaking on the rate of phosphate desorption by soil. *Journal of Soil Science*, 30, 67–76.

Barrow, N.J., Malajczuk, N. and Shaw, T.C. (1977), A direct test of the ability of vesicular-arbuscular mycorrhiza to help plants take up fixed soil phosphate. *New Phytology*, 78, 269–276.

Bartha, R., Linsilotta, R.P. and Pramer, D. (1967), Stability and effects of some pesticides in soil. *Applied Microbiology*, 15, 67–75.

Bascomb, C.L. (1968), Distribution of pyrophosphate-extractable iron and organic carbon in soils of various groups. *Journal of Soil Science*, 19, 251–268.

Bauer, A. and Black, A.L. (1981), Soil carbon, nitrogen and bulk density comparisons in two cropland tillage systems after 25 years and in virgin grassland. *Soil Science Society of America Journal*, 45, 1166–1170.

Beckett, P.H.T. (1969), Residual potassium and magnesium: a review. *Technical Bulletin No 20*, MAFF, HMSO (London), pp. 183–196.

Beckett, P.H.T. and White, R.E. (1964), Studies on the phosphate potential of soils: III: pool of labile inorganic phosphate. *Plant and Soil*, 21, 253–282.

Becking, J.H. (1979), Environmental requirements of azolla for use in tropical rice production. In *Nitrogen and Rice*, International Rice Research Institute, Philippines, pp. 345–373.

Bell, F. and Nutman, P.S. (1971), Experiments on nitrogen fixation by nodulated lucerne, In Lie, T.A. and Mulder, E.G. (eds), *Biological Nitrogen Fixation in Natural and Agricultural Habitats. Plant & Soil Special Volume*, 231–264.

Bell, L.C. and Black, C.A. (1970), Transformation of dibasic calcium phosphate dihydrate and octacalcium phosphate in slightly acid and alkaline soils, *Soil Science Society of America Proceedings*, 34, 583–587.

Benecke, U. (1970), Nitrogen fixation by *Alnus viridis* (Chaix) DC. *Plant and Soil*, 33, 30–48.

Berg, B. and Bosatta, E. (1976), A carbon–nitrogen model of a pine forest soil ecosystem – a development paper. *Swedish Coniferous Forest Project, Internal Report 42*, 61pp.

Bevan, K. and Germann, P. (1982), Macropores and water flow in soils. *Water Resource Research*, 18, 1311–1325.

Bharathi, Ch. and Subba Rao, B.V.S.S.R. (1986), Toxic effects of two organophosphate insecticides, monocrotophos and dichlorovos, to common earthworm *Lampito mauritii* (Kinberg). *Water, Air and Soil Pollution*, 28, 127–130.

Bhat, K.K.S. and Nye, P.H. (1973), Diffusion of phosphate to plant roots in soil. I. Quantitative autoradiography of the depletion zone. *Plant and Soil*, 38, 161–175.

Bhat, K.K.S. and Nye, P.H. (1974), Diffusion of phosphate to plant roots in soil: III: Depletion around onion roots without root hairs. *Plant and Soil*, 41, 383–394.

Bhat, K.K.S., Nye, P.H. and Baldwin, J.P. (1976), Diffusion of phosphates to plant roots in soil: IV: The concentration–distance profile in the rhizosphere of roots with root hairs in a low P-soil. *Plant and Soil*. 44, 63–72.

Bhatnagar, V.K., Miller, M.H. and Ketcheson, J.W. (1985), Reaction of fertilizer and liquid manure phosphorus with soil aggregates and sediment phosphorus enrichment. *Journal of Environmental Quality*, 14, 246–251.

Biggar, J.W. and Nielsen, D.R. (1976), Spatial variability of the leaching characteristics of a field soil. *Water Resources Research*, 12, 78–84.

Biggar, J.W., Mingelgrin, U. and Cheung, M. (1978), Equilibrium and kinetics of adsorption of picloram and parathion with soils. *Journal of Agricultural and Food Chemistry*, 26, 1306–1312.

Birch, H.F. (1958), The effect of soil drying on humus decomposition and nitrogen availability. *Plant and Soil*, 10, 9–31.

Birch, H.F. (1959), Further observations on humus decomposition and nitrification. *Plant and Soil*, 11, 262–286.

Birch, H.F. (1960), Nitrificaton in soil after different periods of dryness. *Plant and Soil*. 12, 81–96.

Birkeland, P.W. (1974), *Pedology, weathering and geomorphological research*. Oxford University Press.

Birrell, K.S. (1963), Thermal decomposition of DDT by some soil constituents. *New Zealand Journal of Science*, 6, 169–178.

Blevins, R.L., Doyle Cook, S.H., Phillips, S.H. and Phillips, R.E. (1971), Influence of no-tillage on soil moisture. *Agronomy Journal*, 63, 593–596.

Bliev, Th. K. (1973), Effect of herbicides on biological activity of soils. *Pochvovedenie*, 7, 61–68.

Black, C.A. (1968), *Soil and Plant Relationships*. Wiley, New York.

Bloomfield, C. (1950), Some observations on gleying. *Journal of Soil Science*, 1, 205–211.

Bloomfield, C. (1951), Experiments on the mechanism of gley formation. *Journal of Soil Science*, 2, 196–211.

Bloomfield, C. (1981), The translocation of metals in soils. In Greenland, D.J. and Hayes, M.H.P. (eds), *The Chemistry of Soil Processes*, pp. 463–504. John Wiley & Sons Ltd, Chichester

Bloomfield, C. (1952), Translocation of iron in podzol formation. *Nature (London)*, 170, 540.

Bloomfield, C. (1955), Leaf leachates as a factor in pedogenesis, *Journal of Science and Food Agriculture*, 6, 641–651.

Bloomfield, C. (1953–55), A study of podzolisation. I–V. *Journal of Soil Science*, 4, 5–16; 17–23; 5, 39–45; 46–49; 50–56.

Bloomfield, C. (1957), The possible significance of polyphenols in soil formation. *Journal of Science and Food Agriculture*, 8, 389–392.

Bloomfield, C. and Zahari, A.B. (1982), Acid sulphate soils. *Outlook on Agriculture*, 11(2), 48–54.

Boddey, R.M. and Dobereiner, J. (1984), Nitrogen fixation associated with grasses and cereals. In Subba Rao, N.S. (ed.) *Current Developments in Biological Nitrogen Fixation*, pp. 277–313. Edward Arnold, London.

Bohlool, B.B. and Schmidt, E.L. (1974), Lectins: a possible basis for specificity in the *Rhizobium*–legume root nodule symbiosis. *Science*, 185, 269–271.

Bohn, H., McNeal, B. and O'Connor, G. (1985), *Soil Chemistry*, 2nd edition, Wiley-Interscience, Chichester.

Bollag, J.-M. and Henninger, N.M. (1976), Influence of pesticides on denitrification in soil and with an isolated bacterium. *Journal of Environmental Quality*, 5, 15–18.

Bolt, G.H. (1978), Surface interaction between the soil solid phase and the soil solution. In Bolt, G.H. and Bruggenwert, M.G.M. (eds) *Soil Chemistry: Basic Elements*, pp. 43–53. Elsevier Science, Amsterdam.

Bolton, J. and Coulter, J.K. (1966), Distribution of fertilizer residues in a forest manuring experiment on a sandy podzol at Wareham, Dorset, Rep. Forest Res. Cond., for 1965, 90–92.

Bouldin, D.R., DeMent, J.D. and Sample, E.C. (1960), Interaction between dicalcium and monoammonium phosphates granulated together. *Journal of Agricultural and Food Chemistry*, 8, 470–474.

Boulter, D., Jeremy, J.J. and Wilding, (1966), Amino acids liberated into the culture medium by pea seedling roots. *Plant and Soil*, 24, 121–127.

Bowen, G.D. (1973), Mineral nutrition of ectomycorrhizae, In Marks, G.C. and Kozlowski, T.T. (eds) *Ectomycorrhizae. Their Ecology and Physiology*, 151–205. Academic Press, London, New York.

Bowen, G.D. and Rovira, A.D. (1973), Are modelling approaches useful in rhizosphere biology? In Rosswall, T. (ed.) *Modern Methods in the Study of Microbial Ecology, Bulletin of Ecology and Research Communications (Stockholm)*, 17, 443–450.

Bowen, G.D. and Rovira, A.D. (1976), Microbial colonisation of plant roots. *Annual Review of Phytopathology*, 14, 121–144.

Bowman, B.T., Adams, R.S. Jr and Feton, S.W. (1970), Effect of water upon malathion adsorption onto five montmorillonite systems. *Journal of Agricultural and Food Chemistry*, 18, 723–727.

Brady, N.C. (1974), *The Nature and Properties of Soils*, 8th edition. Macmillan Pub. Co. Inc., New York.

Branham, B.E., Wehner, D.J., Torello, W.A. and Turgeon, A.J. (1986), A microecosytem for fertilizer and pesticide fate research. *Agronomy Journal*, 77, 176–180.

Bray, J.R. and Goreham, E. (1964), Litter production in forests of the world. *Advances in Ecology Research*, 2, 101–157.

Bray, R.H. and Kurtz, L.T. (1952), Determinations of total, organic and available forms of phosphorus in soils. *Soil Science*, 59, 39–45.

Bremner, J.M. and Douglas, L.A. (1971), Inhibition of urease activity in soils. *Soil Biology and Biochemistry*, 3, 297–307.

Bresler, E., Bielorai, H. and Laufer, A. (1979), Field test of solution flow models in a heterogeneous irrigated cropped soil. *Water Resources Research*, 15, 645–652.

Bresler, E., McNeal, B.L. and Carter, D.L. (1982), *Saline and Sodic Soils. Principles and Dynamics in Modelling*. Springer-Verlag, New York.

Briggs, G.G. (1969), Molecular structure of herbicides and their sorption by soils. *Nature (London)*, 223, 1288–1289.

Bristow, A.W., Ryden, J.C. and Whitehead, D.C. (1987), The fate at several time intervals of ^{15}N-labelled ammonium nitrate applied to an established grass sward. *Journal of Soil Science*, 38, 245–254.

Broadbent, F.E. (1962), Biological and chemical aspects of mineralisation. *Transactions of the Joint Meeting of Commissions IV and V, International Society of Soil Science, New Zealand*, 220–229.

Brouzes, R. and Knowles, R. (1973), Kinetics of nitrogen fixation in a glucose-amended, anaerobically incubated soil. *Soil Biology and Biochemistry*, 5, 223–229.

Brownlee, C., Duddridge, J.A., Malibari, A., and Read, D.J. (1983), The structure and function of mycelial systems of ectomycorrhizal roots with special reference to their role in forming interplant connections and providing pathways for assimilate and water transport. *Plant and Soil*, 71, 433–443.

Brunsden, D. (1979), Weathering. In Embleton, C. and Thornes, J. (eds) *Process in Geomorphology*. Edward Arnold, London.

Bundy L.G. and Bremner, J.M. (1973), Effects of substituted *p*-benzoquinones on urease activity in soils. *Soil Biology and Biochemistry*, 5, 847–853.

Bunnell, F.L., Tait, D.E.N., Flanagan, P.W. and Van Cleve, K. (1977a), Microbial Respiration and Substrate Weight Loss: I: A general model of the influences of abiotic variables. *Soil Biology and Biochemistry*, 9, 33–40.

Bunnell, F.L., Tait, D.E.N. and Flanagan, P.W. (1977b), Microbial Respiration and Substrate Weight Loss: II: A model of the influences of chemical composition. *Soil Biology and Biochemistry*, 9, 41–47.

Buresh, R.J., Casselman, M.E. and Patrick, W.H. Jr (1980), Nitrogen fixation in flooded soil systems, a review. *Advances in Agronomy*, 33, 149–192.

Buresh, R.J. and Patrick, W.H. Jr (1978), Nitrate reduction to ammonium in anaerobic soil. *Soil Science Society of America Journal*, 42, 913–918.

Burgoyne, W., Triblet, T. and Ric, H.R. (1983), The persistence of picloram and 2,4-D in soils at latitudes north of 60° north. *Weeds Today*, 14, 10–11.

Burnett, E. and Tackett, J.L. (1968), Effect of soil profile modification on plant root development. *Transactions of the Ninth International Congress in Soil Science, Australia, Vol III*, 329–337.

Burns, I.G. (1974), A model for predicting the redistribution of salts applied to fallow soils after excess rainfall or evaporation. *Journal of Soil Science*, 25, 165–178.

Burns, I.G. (1976), Equations to predict the leaching of nitrate uniformly incorporated to a known depth or uniformly distributed throughout a soil profile. *Journal Agricultural Science Cambridge*, 86, 305–313.

Burns, I.G. (1980), A simple model for predicting the effects of leaching of fertilizer nitrate during the growing season on the nitrogen fertilizer need of crops. *Journal of Soil Science*, 31, 175–185.

Burris, R.H. (1980), The global nitrogen budget – science or seance? In Newton, W.E. and Orme-Johnson, W.H. (eds) *Nitrogen Fixation, Vol I*, pp. 7–16. University Park Press, Baltimore.

Burwell, R.E., Schuman, G.E., Piest, R.F., Spomer, R.G. and McCalla, T.M. (1974), Quality of water discharged from two agricultural watersheds in southwestern Iowa. *Water Resources Research*, 10, 359–365.

Busenberg, E. and Clemency, C.V. (1976), The dissolution kinetics of feldspars at 25°C and 1 atm. CO_2 partial pressure. *Geochemica et Cosmochimica Acta*, 40, 41–49.

Buurman, P. (1985), Carbon-sesquioxide ratios in organic complexes and the transition ablic–spodic horizon. *Journal of Soil Science*. 36, 255–260.

Calvet, R. (1980), Adsorption–desorption phenomena. In Hance, R.J. (ed.) *Interactions Between Herbicides and the Soil*. Academic Press, London.

Calvet, R. and Terce, M. (1975), Adsorption de l'atrazine par des montmorillonites-Al. *Annales Agronomiques*, 26, 693–707.

Cameron, D.A. and Klute, A. (1977), Convective-dispersive solute transport with a combined equilibrium and kinetic adsorption model. *Water Resources Research*, 13, 183–188.

Cameron, D.R., Kowalenko, C.G. and Ivarson, K.C. (1978), Nitrogen and chloride leaching in a sandy field plot. *Soil Science*, 126, 174–180.

Cameron, K.C. and Wild, A. (1982), Prediction of solute leaching under field conditions: an appraisal of three methods. *Journal of Soil Science*, 33, 659–669.

Cameron, R.S., Thornton, B.K., Swift, R.S. and Posner, A.M. (1972), Molecular weight and shape of humic acid from sedimentation and diffusion measurements on fractionated extracts. *Journal of Soil Sciences*, 23, 394–408.

Campbell, C.A., Jame, Y.W. and Winkleman G.E. (1984), Mineralisation rate constants and their use for estimating nitrogen mineralisation in some Canadian prairie soils. *Canadian Journal of Soil Science*, 64, 333–343.

Campbell, D.J., Dickson, J.W. and Ball, B.C. (1982), Controlled traffic on a sandy clay loam under winter barley in Scotland. *9th Conference of ISTRO, Yugoslavia*, 189–194.

Campbell, R. and Rovira, A.D. (1973), The study of the rhizosphere by scanning electron microscopy. *Soil Biology and Biochemistry*, 6, 747–752.

Cannell, R.Q. and Finney, J.R. (1974), Effects of direct drilling and reduced cultivation on soil conditions for root growth. *Outlook on Agriculture*, 7, 184–189.

Cannell, R.Q., Belford, R.K., and Beetlestone, G.R. (1977), Uptake of fertilizer nitrogen by winter wheat and losses of nitrogen by leaching from two soils. *Letcomb Laboratory Annual Report for 1976*, 75–76.

Cannell, R.Q., Davies, D.B. and Pigeon, J.D. (1979), The suitability of soils for sequential drilling of combine-harvested crops in Britain: a provisional classification. In Jarvis, M.G. and Mackney, D. (eds) *Soil Survey Applications. Soil Survey Technical Monograph*, 13, 1–20.

Carlisle, S.M. and Trevors, J.T. (1986), Effect of the herbicide glyphosate on respiration and hydrogen consumption in soil. *Water, Air and Soil Pollution*, 27, 391–401.

Carringer, R.D., Weber, J.B. and Monaco, T.H. (1975), Adsorption–desorption of selected pesticides by organic matter and montmorillonite. *Journal of Agricultural and Food Chemistry*, 23, 568–572.

Carson, R. (1963), *Silent Spring*. Hamish Hamilton, London.

Carter, M.R. and Rennie, D.A. (1982), Changes in soil quality under zero tillage farming systems: distribution of microbial biomass and mineralizable C and N potentials. *Canadian Journal of Soil Science*, 62, 587–597.

Caseley, J.C. and Luckwill, L.C. (1965), The effect of some residual herbicides on soil nitrifying bacteria. *Report of the Agricultural and Horticultural Research Station, University of Bristol*, pp. 78–86.

Cassel, D.K. (1982), Tillage effects on soil bulk density and mechanical impedance. In Unger, P.W. and Van Doren, D.M. Jr. (eds), *Predicting Tillage Effects on Soil Physical Properties and Processes, ASA Special Publication No 44*, 45–67. American Society of Agronomy and Soil Science Society of America.

Catchpool, V.R. (1975), Pathways for losses of fertilizer nitrogen from a Rhodes grass pasture in Southeastern Queensland. *Australian Journal of Agricultural Research*, 26, 259–268.

Chan, K.Y., Davey, B.G. and Geering, H.R. (1979), Adsorption of magnesium and calcium by a soil with variable charge. *Soil Science Society of America Journal*, 43, 301–304.

Chandra, P. (1967), Effect of two chlorinated insecticides on soil microflora and nitrification process as influenced by different soil temperatures and textures. In Graff, O. and Satchell, J.E. (eds), *Progress in Biology*, Vieweg, Braunnschweig.

Chaney, K. and Swift, R.S. (1984), The influence of organic matter on aggregate stability in some British soils. *Journal of Soil Science*, 35, 223–230.

Chao, T.T. and Zhou, L. (1983), Extraction techniques for selective dissolution of amorphous iron oxides from soils and sediments. *Soil Science Society of America Journal*, 47, 225–232.

Chapman, R.A. and Cole, C.M. (1982), Observations on the influence of water and soil pH on the persistence of insecticides. *Journal of Environmental Science and Health*, B17, 487–504.

Cheshire, M.V., Sparling, G.P. and Mundie, C.M. (1983), Effect of periodate treatment of soil carbohydrate constituents and soil aggregation. *Journal of Soil Science*, 34, 105–112.

Chiariello, N. Hickman, J.C. and Mooney, H.A. (1982), Endomycorrhizal role for interspecific transfer of phosphorus in a community of annual plants. *Science*, 217, 941–943.

Chichester, F.W. and Smith, S.J. (1978), Disposition of ^{15}N labelled fertilizer nitrate applied during corn culture in field lysimeters. *Journal of Environmental Quality*, 6, 211–216.

Cho, C.M. (1982), Oxygen consumption and denitrification kinetics in soil. *Soil Science Society of America Journal*, 46, 756–762.

Cho, C.M. (1985), Ionic transport in soil with ion-exchange reaction. *Soil Science Society of America Journal*, 49, 1379–1386.

Cho, C.M. and Sakdinan, L. (1977), Mass spectrometric investigation on denitrification. *Canadian Journal of Soil Science*, 58, 443–457

Cho, C.M. and Sakdinan, L. (1978), Mass spectrometric investigation on denitrification. *Canadian Journal of Soil Science*, 58, 443–457.

Chopra, S.L., Bala, K., and Bhardwaj, S.S. (1984), Adsorption of simazine by some clay minerals. *Agrochemica*, 28, 220–227.

Clark, F.E. and Rosswall, T. (1981) (eds), Terrestrial nitrogen cycles – processes, ecosystems, strategies and management impacts. *Ecology Bulletin* No 3.

Clymo, R.S. (1965), Experiments on breakdown of *Sphagnum* in two bogs. *Journal of Ecology*, 53, 747–758.

Coale, F.J., Meisinger, J.J. and Wiebold, W.J. (1985), Effects of plant breeding and selection on yields and nitrogen fixation in soybeans under two soil nitrogen regimes. *Plant and Soil*, 86, 357–367.

Colbourn, P. (1985), Nitrogen losses from the field: denitrification and leaching in intensive winter cereal production in relation to tillage method of a clay. *Soil Use and Management*, 1, 115–120.

Colbourn, P., Harper, I.W. and Iqbal, M.M. (1984), Denitrification losses from ^{15}N-labelled calcium nitrate fertilizer in a clay soil in the field. *Journal of Soil Science*, 35, 539–547.

Colbourn, P., Iqbal, M.M. and Harper, I.W. (1984), Estimation of the total gaseous nitrogen losses from clay soils under laboratory and field conditions, *Journal of Soil Science*, 35, 11–22.

Cole, D.W., Gessel, S.P. and Turner, J. (1978), Comparative mineral cycling in red alder and Douglas fir. In Briggs, D.G., Debell, D.S. and Atkinson, W.A. (eds) *Utilisation and Management of Alder*, pp. 327–336, USDA, Forest Service, Gen. Tech. Rept. PNW-70, Pacific N.W. Forest and Range Experimental Station, Oregon.

Collins, V.G., D'Sylva, B.T. and Latter, P.M. (1976), Microbial populations in peat. In Heal, O.W. and Perkins, D.F. (eds) *Production Ecology of British Moors and Montane Grasslands. Studies in Ecology*, 27, 94–112.

Conrady, D. (1986), Ecological studies on the effect of pesticides on the animal community of a meadow. [Okologische Untersuchungen Über die Wirkung von Umweltchemikalien auf die Tiergemeinschaft eines Grunlandes], *Pedobiologia*, 29, 273–284.

Cooke, G.W. (1966), Phosphorus and potassium fertilizers: their forms and their places in agriculture. *Proceedings of the Fertilizer Society*, 92, 45pp.

Cooke, G.W. (1980), *Fertilizing for Maximum Yield*. 3rd edition, Granada Publishing Ltd, London.

Cooke, G.W. (1981), The fate of fertilizers. In Greenland, D.J. and Hayes, M.H.B. (eds) *The Chemistry of Soil Processes*, 563–592. John Wiley & Sons Ltd., Chichester.

Cooke, G.W. and Cunningham, R.K. (1958), Soil nitrogen: III: Mineralisable nitrogen determined by an incubation technique. *Journal of Science and Food Agriculture*, 9, 324–330.

Coote, D.R. and Ramsey, J.F. (1983), Quantification of the effects of over 35 years of intensive cultivation on four soils. *Canadian Journal of Soil Science*, 63, 1–14.

Corke, C.T. and Thompson, F.R. (1970), Effect of some phenylamide herbicides and their degradation products on soil nitrification. *Canadian Journal of Microbiology*, 16, 567–571.

Côté, B. and Gamiré, C. (1985), Nitrogen cycling in dense plantings of hybrid poplar and black alder. *Plant and Soil*, 87, 195–208.

Coulson, C.B., Davis, R.I. and Lewis, D.A. (1960), Polyphenols in plant, humus and soil. I Polyphenols of leaves, litter and superficial humus from mull and mor sites. II Reduction and transport by polyphenols of iron in model soil columns. *Journal of Soil Science*, 11, 20–29, 30–44.

Coupland, D. and Caseley, J.C. (1979), Presence of ^{14}C activity in root exudates and guttation fluid from *Agropyron repens* treated with ^{14}C-labelled glyphosate. *New Phytologist*, 83, 17–22.

Cowling, D.W. (1982), Biological nitrogen fixation and grassland production in the United Kingdom. *Philosophical Transactions of the Royal Society of London*, B296, 397–404.

Crank, J., McFarlane, N.R., Newby, J.C., Paterson, E.D. and Pedley, J.B. (1981), *Diffusion Processes in Environmental Systems*. The Macmillan Press Ltd, London.

Cress, W.A., Thronberry, Q.O. and Lindsey, D.L. (1979), Kinetics of phosphorus adsorption by mycorrhizal and non-mycorrhizal tomato roots. *Plant Physiology*, 63, 484–487.

Crosby, D.G. (1976), Nonbiological degradation of herbicides in the soil. In Audus, L.J. (ed.), *Herbicides, Physiology, Biochemistry, Ecology*. Academic Press, pp. 65–97.

Crosby, D.G. and Tang, C.S. (1969), Photodecomposition of 3-(p-chlorophenyl)-1, 1-dimethylurea (Monuron), *Journal of Food and Agricultural Chemistry*, 17, 1041–1044.

Crosby, D.G. and Wang, A.S. (1973), Photodecomposition of 2,4,5-T in water. *Journal of Agricultural and Food Chemistry*, 21, 1054–1058.

Cruse, R.M., Linden, D.R., Radke, J.K., Larson, W.E. and Larntz, K. (1980), A model to predict tillage effects on soil temperature. *Soil Science Society of America Journal*, 44, 378–383.

Cruse, R.M., Potter, K.N. and Allmaras, R.R. (1982), Modeling tillage effects on soil temperature. In Unger, P.W. and Van Dorev, D.M. (eds), *Predicting Tillage Effects on Soil Physical Properties and Processes*. ASA Special Pub. No 44, pp. 133–150.

Currie, J.A. (1961), Gaseous diffusion in the aeration of aggregated soils. *Soil Science*, 92, 40–45.

Curtis, C.D. (1975), Chemistry of rock weathering: fundamental reactions and controls. In Derbyshire, E. (ed.) *Geomorphology and Climate*, pp. 25–57. John Wiley, Chichester.

Curtis, C.D. (1976), Stability of minerals in surface weathering reactions: a general thermochemical approach. *Earth Surface Processes and Landforms*, 1, 63–70.

Cushman, J.H. (1982), Nutrient transport inside and outside the root rhizosphere: theory. *Soil Science Society of America Journal*, 46, 704–09.

Cushman, J.H. (1984), Nutrient transport inside and outside the root rhizosphere: generalised model. *Soil Science*, 138, 164–171.

Daft, M.J. and El-Giahmi, A.A. (1978), Effects of arbuscular mycorrhiza on plant growth: VIII: Effects of defoliation and light on selected hosts. *New Phytology*, 80, 365–372.

Daft, M.J. and Nicholson, T.H. (1966), Effect of *Endogone* mycorrhiza on plant growth. *New Phytology*, 65, 343–350.

Dalal, R.C. (1975), Effect of associated anions on ammonium adsorption by and desorption from soils. *Journal of Agricultural and Food Chemistry*, 23, 684–687.

Darrah, P.R., Nye, P.H. and White, R.E. (1983), Diffusion of NH_4^+ and NO_3^- mineralized from organic N in soil. *Journal of Soil Science*, 34, 693–707.

Davies, D.B., Finney, J.B. and Richardson, S.J. (1973), Relative effects of tractor weight and wheelslip in causing soil compaction. *Journal of Soil Science*, 24, 399–409.

Davies, R.I. (1971), Relation of polyphenols to decomposition of organic matter and to pedogenetic processes. *Soil Science*, 111, 80–85.

Dawson, H.J., Ugolini, F.C., Hrutfiord, B.F. and Zachara, J. (1978), Role of soluble organics in the soil processes of a podzol, Central Cascades, Washington. *Soil Science*, 126, 290–296.

Deb, B.C. (1949), The movement and precipitation of iron oxides in podzol soils. *Journal of Soil Science*, 1, 112–122.

De Coninck, F. (1980), Major mechanisms in formation of spodic horizons. *Geoderma*, 24, 101–128.

Dekker, G.C., Bruce, W.N. and Biggar, J.H. (1965), The accumulation and dissipation of residues resulting from use of aldrin in soils. *Journal of Economic Entomology*, 58, 266–271.

Devitt, D., Letey, J., Lund, L.J. and Blair, J.W. (1976), Nitrate-nitrogen movement through soil as affected by soil profile characteristics. *Journal of Environmental Quality*, 5, 283–288.

de Vries, D.A. (1963), Thermal properties of soils. In Van Wijk, W.R. (ed.). Physics of the Plant Environment, pp. 210–235. North Holland Publishing Co., Amsterdam.

Dexter, A.R. (1976), Internal structure of a tilled soil. *Journal of Soil Science*, 27, 267–278.

Dexter, A.R. (1978), A stochastic model for the growth of roots in tilled soil. *Journal of Soil Science*, 29, 102–126.

Dexter, A.R. (1979), Prediction of soil structures produced by tillage. *Journal of Terramechanics*, 16(3), 117–127.

Dick, W.A. (1983), Organic carbon, nitrogen and phosphorus concentrations and pH in soil profiles as affected by tillage intensity. *Soil Science Society of America Journal*, 47, 102–107.

Dickinson, C.H. (1974), Decomposition of litter in soil. In Dickinson, C.H. and Pugh, G.J.F. (eds), *Biology of Plant Litter Decomposition, Vol 2*, 633–658. Academic Press, London.

Dieguez-Carbonell, D. and Pascual, C.R. (1975), Adsorption of the herbicide '2,4-D' by montmorillonite. *Environmental Quality and Safety*, III 237–242.

Dighton, J. (1983), Phosphatase production by mycorrhizal fungi. *Plant and Soil*, 71, 455–462.

Dimbleby, G.W. (1962), *The Development of British Heathlands and their Soils*, Clarendon Press, Oxford.

Dixit, S.P., Gombeer, R. and D'Hoore, J. (1975), The electrophoretic mobility of natural clays and their potential mobility within the pedon. *Geoderma*, 13, 325–330.

Dixon, J.B., Moore, D.E., Agnihotri, N.P. and Lewis, D.E. Jr (1970), Exchange of diquat^{2+} in soil clays, vermiculite and smectite. *Soil Science Society of America Proceedings*, 34, 805–808.

Dobereiner, J. and Day, J.M. (1975), Nitrogen fixation in the rhizosphere of tropical grasses. In Stewart, W.D.P. (ed.) *Nitrogen Fixation By Free-Living Microorganisms. International Biological Programme*, Vol 6, pp. 39–56, Cambridge University Press, Cambridge.

Dobereiner, J., Day, J.M. and Dart, P.J. (1972), Nitrogenase activity and oxygen sensitivity of the *Paspalum notatum-Azotobacter pasopli* association. *Journal of . General Microbiology*, 71, 103–116.

Dommergues, Y.R. (1978), The plant–microorganism system. In Dommergues,

Y.R. and Krupa, S.V. (eds) Interactions between non-pathogenic soil microorganisms and plants. *Developments in Agriculture and Managed-Forest Ecology*, Vol 4, pp. 1–37. Elsevier Science, Amsterdam.

Domsch, K.H. (1984), Effects of pesticides and heavy metals on biological processes in soil. *Plant and Soil*, 76, 367–378.

Domsch, K.H., Jagnow, G. and Anderson, T.H. (1983), An ecological concept for the assessment of side-effects of agrochemicals on soil microorganisms. *Residue Review*, 86, 65–105.

Douglas, J.T. (1976), The effect of cultivation on the stability of aggregates from the soil surface. *Letcomb Laboratory Annual Report*, 1976, pp. 46–48, Agricultural Research Council.

Dowdell, R.J. and Cannell, R.Q. (1975), Effect of ploughing and direct drilling on soil nitrate content. *Journal of Soil Science*, 26, 53–61.

Dowdell, R.J. and Webster, C.P. (1980), A lysimeter study using nitrogen-15 on the uptake of fertilizer nitrogen by perennial ryegrass swards and losses by leaching. *Journal of Soil Science*, 31, 65–75.

Dowdell, R.J. and Webster, C.P. (1981), Fate of fertilizer N applied to permanent grassland: lysimeter studies. *Letcombe Laboratory Annual Report*, 1980, pp. 50–51, Agricultural Research Council.

Dowdell, R.J., Crees, R., Burford, J.R. and Cannell, R.Q. (1979), Oxygen concentrations in a clay soil after ploughing or direct drilling. *Journal of Soil Science*, 30, 239–295.

Dowdell, R.J., Smith, K.A., Crees, R. and Restall, S.W.F. (1972), Field studies of ethylene in the soil atmosphere – equipment and preliminary results. *Soil Biology and Biochemistry*, 4, 325–331.

Dowdell, R.J., Webster, C.P. and Mercer, E.R. (1984), A lysimeter study of the fate of fertilizer nitrogen in spring barley crops grown on shallow soil overlying chalk : crop uptake and leaching losses. *Journal of Soil Science*, 35, 169–181.

Dowding, P. (1974), Nutrient losses from litter on IBP tundra sites. In Holding, A.J. *et al.* (eds) *Soil Organisms and Decomposition in Tundra*. IBP, Tundra Biome, pp. 363–373.

Drever, J.I. (1982), *The Geochemistry of Natural Waters*. Prentice-Hall Inc., Englewood Cliffs, New Jersey.

Drew, M.C. and Nye, P.H. (1969), The supply of nutrient ions by diffusion to plant roots in soil. II: The effect of root hairs on the uptake of potassium by roots of ryegrass (*Lolium multiflorum*). *Plant and Soil*, 31, 407–424.

Drew, M.C. and Saker, L.R. (1978), Effects of direct drilling and ploughing on root distribution in spring barley, and on the concentrations of extractable phosphate and potassium in the upper horizons of a clay soil. *Journal of the Science of Food and Agriculture*, 29, 201–206.

Dubey, H.D. and Rodriguez, R.L. (1970), Effect of dyrene and maneb on nitrification and ammonification, and their degradation in tropical soils. *Soil Science Society of America Proceedings*, 34, 435–439.

Duddridge, J.A., Malibari, A. and Read, D.J. (1980), Structure and function of mycorrhizal rhizomorphs with special reference to their role in water transport. *Nature (London)*, 287, 834–836.

Dunham, R.J. and Nye, P.H. (1974), The influence of soil water content on the uptake of ions by roots. II: Chloride uptake and concentration gradients in soil. *Journal of Applied Ecology*, 11, 581–596.

Duvigneaud, P. and Denaeyer-de Smet, S. (1970), Biological cycling of minerals in temperature deciduous forests. In Reichle, D.E. (ed.), *Analysis of Temperate Forest Ecosystems, Ecological Studies No 1*, pp. 199–225.

Duxbury, J.M., Bouldin, D.R., Terry, R.E. and Tate, R.L. (III) (1982), Emissions of nitrous oxide from soils. *Nature (London)*, 298, 462–464.

Eck, H.V. and Davies, R.G. (1971), Profile modification and root yield, distribution and activity. *Agronomy Journal*, 63, 934–937.

Eck, H.V. and Unger, P.W. (1985), Soil profile modification for increasing crop production. *Advances in Soil Science*, 1, 65–100.

Edwards, C.A. (1969), Soil pollutants and soil animals. *Scientific American*, 220(4), 88–99.

Edwards, C.A. and Lofty, J.R. (1972), *Biology of Earthworms*. Chapman and Hall Ltd, London.

Edwards, C.A. and Lofty, J.R. (1980), Effects of earthworm inoculation upon the root growth of direct drilled cereals. *Journal of Applied Ecology*, 17, 533–543.

Eggington, G.M. and Smith, K.A. (1986), Nitrous oxide emission from a grassland soil fertilized with slurry and calcium nitrate. *Journal of Soil Science*, 37, 59–67.

Ehlers, W. (1975), Observations on earthworm channels and infiltration on tilled and untilled loess soil. *Soil Science*, 119, 242–249.

Ehlers, W. (1976), Rapid determination of unsaturated hydraulic conductivity in tilled and untilled loess soil. *Soil Science Society of America Journal*, 40, 837–840.

Ehlers, W. (1977), Measurement and calculation of hydraulic conductivity in horizons of tilled and untilled loess-derived soil, Germany. *Geoderma*, 19, 293–306.

Eijsackers, H. and van der Drift, J. (1976), Effects on soil fauna. In Audus, L.J. (ed.), *Herbicides. Physiology, Biochemistry, Ecology*, Vol. 2, pp. 149–174, Academic Press, London.

Ellis, F.B. and Barnes, B.T. (1980), Growth and development of root systems of winter cereals grown after different tillage methods including direct drilling. *Plant and Soil*, 55, 283–295.

Ellis, F.B., Elliott, J.G., Barnes, B.T. and Howse, K.R. (1977), Comparison of direct drilling, reduced cultivation and ploughing on the growth of cereals. *Journal of Agricultural Science, Cambridge*, 89, 631–642.

Ellis, J.R., Larsen, H.J., Boosalis, M.G. (1985), Drought resistance of wheat plants inoculated with vesicular-arbuscular mycorrhizae. *Plant and Soil*, 86, 369–378.

Emerson, W.W. (1959), The structure of soil crumbs. *Journal of Soil Science*, 10, 235–244.

Eno, C.F. and Everett, P.H. (1958), Effects of soil applications of 10 chlorinated hydrocarbon insecticides on soil microorganisms and the growth of stringless black valentine beans. *Soil Science Society of America Proceedings*, 22, 235–238.

European Economic Community (1980), Council directive on the quality of water for human consumption. *Official Journal*, No 80/778, EEC L 229, p. 11.

Evans, A.C. (1947), Method for studying the burrowing activity of earthworms. *Annals and Magazine of Natural History*, 11, 643–650.

Evans, H.J., Euerich, D.W., Maier, R.J., Hanus, F.J. and Russell, S.A. (1979), Hydrogen cycling within the nodules of legumes and non-legumes and its role in nitrogen fixation. In Gordon, J.C., Wheeler, C.T. and Perry, D.A. (eds) *Symbiotic Nitrogen Fixation in the Management of Temperate Forests*, pp. 196–206.

Farmer, V.C. (1982), Significance of the presence of allophane and imogolite in podzol B_s horizons for podzolisation mechanisms: a review. *Soil Science and Plant Nutrition*, 28(4), 571–578.

Farmer, V.C. and Fraser, A.R. (1982), Chemical and colloidal stability of sols in the Al_2O_3-Fe_2O_3-SiO_2-H_2O system: their role in podzolisation. *Journal of Soil Science*, 33, 737–742.

Farmer, V.C., McHardy, W.J., Robertson, L., Walker, A. and Wilson, M.J. (1985), Micromorphology and sub-microscopy of allophane and imogolite in a podzol B_s horizon: evidence for translocation and origin. *Journal of Soil Science*, 36, 87–95.

Farmer, V.C., Russell, J.D. and Berrow, M.L. (1980), Imogolite and proto-imogolite allophane in spodic horizons: evidence for a mobile aluminium silicate complex in podzol formation. *Journal of Soil Science*, 31, 673–684.

Farmer, V.C., Russell, J.D. and Smith, B.F.L. (1983), Extraction of inorganic forms of translocated Al, Fe and Si from a podzol B_s horizon. *Journal of Soil Science*, 34, 571–576.

Farmer, W.J., Igue, K., Spencer, W.F., and Martin, J.P. (1972), Volatility of organochlorine insecticides from soil: I: Effect of concentration, temperature, air flow rate, and vapour pressure. *Soil Science Society of America Proceedings*, 36, 443–447.

Felsot, A. and Dahm, P.A. (1979), Sorption of organophosphorus and carbomate insecticides by soil. *Journal of Agricultural Food Chemistry*, 27, 557–563.

Fenn, L.B. and Hossner, L.R. (1985), Ammonia volatilization from ammonium or ammonia-forming nitrogen fertilizers. *Advances in Soil Science*, 1, 123–169.

Fenn, L.B. and Kissel, D.E. (1973), Ammonia volatilization from surface applications of ammonium compounds on calcareous soils. I: General theory. *Soil Science Society of America Proceedings*, 37, 855–859.

Fenn, L.B. and Kissel, D.E. (1975), Ammonium volatilisation from surface applications of ammonium compounds on calcareous soils. IV: Effect of calcium carbonate content. *Soil Science Society of America Proceedings*, 39, 631–633.

Fenn, L.B., Matocha, J.E. and Wu, E. (1981), Ammonia losses from surface-applied urea and ammonium fertilizers as influenced by rate of soluble Ca. *Soil Science Society of America Journal*, 45, 883–886.

Ferrier, R.C. and Alexander, I.J. (1985), Persistence under field conditions of excised fine roots and mycorrhizas of spruce. In Fitter, A.H. (ed.), *Ecological Interactions in Soil, Plants, Microbes and Animals*, 175–179. Special Publication No 4, British Ecology Society. Blackwell Scientific, Oxford.

Fink, A. (1982), *Fertilizers and Fertilization. Introduction and Practical Guide to Crop Fertilization*. Verlag Chemie.

Firestone, M.K. (1982), Biological denitrification. In Stevenson, F.J. (ed.), *Nitrogen in Agricultural Soils*, 289–326. No 22 in the Agronomy Series, Soil Society of America.

Fisher, G. and Yan, O. (1984), Iron mobilisation by heathland plant extracts. *Geoderma*, 32, 339–345.

Floate, M.J.S. (1970), Decomposition of organic materials from hill soils and pastures. II: Comparative studies on the mineralisation of carbon, nitrogen and phosphorus from plant materials and sheep faeces. *Soil Biology and Biochemistry*, 2, 173–185.

Focht, D.D., Stolzy, L.H. and Meek, B.D. (1979), Sequential reduction of nitrate and nitrous oxide under field conditions as brought about by organic amendments and irrigation management. *Soil Biology and Biochemistry*, 11, 37–46.

Fogel, R. (1983a), Root turnover and productivity of coniferous forests. *Plant and Soil*, 71, 75–85.

Fogel, R. (1983b), Roots as primary producers in below-ground ecosystems. In

Fitter, A.H. (ed.) *Ecological Interactions in Soil*. Special Publication, British Ecology Society No. 4, 23–36.

Fogel, R. and Cromack, K. (1977), Effect of habitat and substrate quality on Douglas fir litter decomposition in Western Oregon. *Canadian Journal of Botany*, 55, 1632–1640.

Fogel, R. and Hunt, G. (1983), Contribution of mycorrhizae and soil fungi to nutrient cycling in a Douglas fir ecosystem. *Canadian Journal of Forest Research*, 13, 219–232.

Foster, R.C., Rovira, A.D. and Cock, T.W. (1983), *Ultrastructure of the Root-Soil Interface*. The American Phytopathological Society, St Paul, Minnesota.

Foster, S.S.D., Cripps, A.C., and Smith-Carington, A. (1982), Nitrate leaching to groundwater. *Philosophical Transactions of the Royal Society of London*, 296, 477–489.

Frame, J. and Newbould, P. (1986), Agronomy of white clover. *Advances in Agronomy*, 40, 1–88.

Francis, R. and Read, D.J. (1984), Direct transfer of carbon between plants connected by vesicular-arbuscular mycorrhizal mycelium. *Nature (London)*, 307, 53–56.

Franklin, J.F., Dyrness, C.T., Moore, D.G. and Tarrant, R.F. (1968), Chemical soil properties under coastal Oregon stands of alder and conifers. In Trappe, J.M., Franklin, J.F., Tarrant, R.F., and Hansen, G.M. (eds), *Biology of Alder*, pp. 157–172. Pacific Northwest Forest and Range Experiment Station, Forest Service, USDA.

Freney, J.R. (1965), Increased growth and uptake of nutrients by cornplants treated with low levels of simazine. *Australian Journal of Agricultural Research*, 16, 257–263.

Frenkel, H., Goertzen, J.O. and Rhoades, J.D. (1978), Effects of clay type and content, exchangeable sodium percentage, and electrolyte concentration on clay dispersion and soil hydraulic conductivity. *Soil Science Society of America Journal*, 42, 32–39.

Fried, M. (1964), E, L and A values, *Transactions of the Eighth International Congress of Soil Science (Bucharest)*, IV, 29–39.

Frissel, M.J. and van Veen, J.A. (1981) (eds) *Simulation of Nitrogen Behaviour of Soil–plant Systems*. Centre for Agricultural Publishing & Documentation, Wageningen.

Frissel, M.J., van Veen, J.A. and Kolenbrander, G.J. (1980), The use of submodels in the simulation of nitrogen transformations in soils. In Banin, A. and Kafkafi, U. (eds), *Agrochemicals in Soils*, pp. 253–265.

Furley, P.A. (1968), Soil formation and slope development: 2: The relationship between soil formation and gradient angle in the Oxford area. *Zeitschrift für Geomorphologie, NF*, 12, 25–42.

Furley, P.A. (1971), Relationships between slope form and soil properties developed over chalk parent materials. In Brunsden, D. (ed.) *Slopes: Form and Processes*, pp. 141–163. Institute of British Geographers Special Publication, No 3.

Gaillardon, P. (1975), Etude des phénomènes desorption entre deux triazines herbicides et des acides humique. *Weed Research*, 15, 393–399.

Gaines, G.L. and Thomas, H.C. (1953). Adsorption studies on clay minerals. II: A formulation of the thermodynamics of exchange adsorption. *Journal of Chemical Physics*, 21, 714–718.

Gamar, Y. and Mustafa, M.A. (1975), Adsorption and desorption of diquat^{2+} and paraquat^{2+} on arid-zone soils. *Soil Science*, 119, 290–295.

Gamble, D.S. and Khan, S.U. (1985), Atrazine hydrolysis in soils: catalysis by the acidic functional groups of fulvic acid. *Canadian Journal of Soil Science*, 65, 435–443.

Gantzer, C.J. and Blake, G.R. (1978), Physical characteristics of le Sueur clay loam soil following no-till and conventional tillage. *Agronomy Journal*, 70, 853–857.

Gapon, Ye. N. (1933), On the theory of exchange adsorption in soils. *Journal of General Chemistry, USSR* (English translation), 3, 144–160.

Garrels, R.M. (1976), Genesis of some ground waters from igneous rock. In Abelson, P.H. (ed.) *Researches in Geochemistry Vol 2*, pp. 405–520. John Wiley, Chichester.

Garrels, R.M. and Christ, C.L. (1965), *Solutions, Minerals and Equilibria*. Harper & Row/John Weatherhill Inc., New York.

Garretson, A.L. and San Clemente, C.K. (1968), Inhibition of nitrifying chemolithotrophic bacteria by several insecticides. *Journal of Economic Entomology*, 61, 285–288.

Gasser, J.K.R. (1958), Use of deep-freezing in the presentation and preparation of fresh soil samples. *Nature (London)*, 181, 1334–1335.

Gasser, J.K.R. (1969), Some processes affecting nitrogen in the soil. In *Nitrogen and Soil Organic Matter*. MAFF Technical Bulletin, 15, 15–19.

Gasser, J.K.R. and Hamlyn, F.G. (1968), The effects of winter wheat of ammonium sulphate, with and without a nitrification inhibitor, and of calcium nitrate. *Journal of Agricultural Science, Cambridge*, 243–249.

Gatterdam, P.E., Casida, J.E. and Stoutamire, D.W. (1959), Relation of structure to stability, antiesterase activity and toxicity with substituted-vinyl phosphate insecticides. *Journal of Economic Entomology*, 52, 270–276.

Gauer, A.C. and Misra, K.C. (1977), Effects of simazine, lindane and ceresan on soil respiration and nitrification rates. *Plant and Soil*, 46, 5–15.

Gausman, H.W., Leamer, R.W., Noriega, J.R., Rodriguez, R.R. and Wiegand, C.L. (1977), Field-measured spectroradiometric reflectances of disked and non-disked soil with and without wheat straw. *Soil Science Society of America Journal*, 41, 793–796.

Gburek, W.J. and Heald, W.R. (1974), Soluble phosphate output of an agricultural watershed in Pennsylvania. *Water Resource Research*, 10, 113–118.

Gemmell, R.P. (1977), *Colonisation of Industrial Wasteland, Studies in Ecology 80*. Edward Arnold, London.

Gerstl, Z. and Yaron, B. (1983), Behaviour of bromacil and napropamide in soils: II: Distribution after application from a point source. *Soil Science Society of America Journal*, 47, 478–483.

Ghilarov, M.S. (1970), Soil biocoenoses. In Phillipson, J. (ed.) *Methods of Study in Soil Ecology*. pp. 67–77, Unesco.

Gibson, A.H. (1976), Recovery and compensation by nodulated legumes to environmental stress. In Nutman, P.S. (ed.) *Symbiotic Nitrogen Fixation in Plants*. International Biology Programme, 7, 385–403. Cambridge University Press, Cambridge.

Giddens, J.E. (1957), Rate of loss of carbon from Georgia soils. *Soil Science Society of America Proceedings*, 21, 513–515.

Gildon, A. and Tinker, P.B. (1983), Interactions of vesicular-arbuscular mycorrhizal infection and heavy metals in plants. II: The effects of infection on uptake of copper. *New Phytology*, 95, 263–268.

Giles, H., MacEwan, T.H., Nakhwa, S.N. and Smith, D. (1960), Studies in adsorption. Part XI: A system of classification of solution adsorption isotherms and its use in diagnosis of adsorption mechanisms and in measurement of specific surface areas of solids. *Journal of the Chemical Society*, 111, 3973–3993.

Gish, C.D. (1970), Pesticides in soil, organochlorine insecticide residues in soils and soil invertebrates from agricultural lands. *Pesticide Monitoring Journal*, 3, 241–252.

Glentworth, R. and Muir, J.W. (1963), The soils of the county round Aberdeen, Inverurie and Fraserburgh. (Sheets 77, 76 and 87/97) Dept. of Ag., Fish for Scotland. *Memoirs of the Soil Survey of Great Britain*. HMSO, Edinburgh.

Goldberg, S. and Sposito, G. (1984), A chemical model of phosphate adsorption by soils. I: Reference oxide minerals. *Soil Science Society of America Journal*, 48, 772–778.

Golovleva, L.A. and Skryabin, G.K. (1978), Co-metabolism of foreign compounds. In *Symposium on Environmental Transport and Transformation of Pesticides*. EPA-600/9-78-003, pp. 73–85.

Gooderham, P.T. (1976), The effect on soil conditions of mechanised cultivation at high moisture content and of loosening by hand digging. *Journal of Agricultural Science*, 86, 567–571.

Gore, A.J.P. and Olson, J.S. (1967), Preliminary models for accumulation of organic matter in an *Eriophorum/Calluna* ecosystem. *Aquilo, Serie Botanica*, 6, 297–313.

Goreham, E. and Sanger, J. (1967), Caloric values of organic matter in woodland swamp and lake soils. *Ecology*, 48, 492–494.

Goss, M.J., Howse, K.R. and Harris, W. (1978), Effects of cultivation on soil water retention and water use by cereals in clay soils. *Journal of Soil Science*, 29, 475–488.

Gould, W.D., Cooke, F.D. and Bulat, J.A. (1978), Inhibition of urease activity by heterocyclic sulphur compounds. *Soil Science Society of America Journal*, 42, 66–72.

Goulding, K.W.T. (1983a), Thermodynamics and potassium exchange in soils and clay minerals. *Advances in Agronomy*, 36, 215–264.

Goulding, K.W.T. (1983b), Adsorbed ion activities and other thermodynamic parameters of ion exchange defined by mole or equivalent fractions. *Journal of Soil Science*, 34, 69–74.

Goulding, K.W.T. and Talibudeen, O. (1984a), Thermodynamics of K-Ca exchange in soils. I: Effects of potassium and organic matter residues in soils from the Broadbalk and Saxmundham Rotation I Experiments. *Journal of Soil Science*, 35, 397–408.

Goulding, K.W.T and Talibudeen, O. (1984b), Thermodynamics of K-Ca exchange in soils. II: Effect of minerology, residual K and pH in soils from long-term ADAS experiments. *Journal of Soil Science*, 35, 409–420.

Granhall, U. and Lid-Torsvik, V. (1974), Nitrogen fixation by bacteria and freeliving blue-green algae in tundra areas. In Wielgolaski, F.E. (ed.) *Plants and Microorganisms, Vol 1*. Springer Verlag, Berlin.

Granhall, U. and Lindberg, T. (1978), Nitrogen fixation in some coniferous forest ecosystems. In Granhall, U. (ed.), *Environmental Role of Nitrogen-Fixing Blue-Green Algae and Asymbiotic Bacteria. Ecology Bulletin*, 26, 172–177.

Gray, T.R.G. and Williams, S.T. (1971), *Soil Micro-organisms*, Oliver and Boyd, Edinburgh.

Greaves, M.P. and Malkomes, H.P. (1980), Effects on soil microflora. In Hance, R.J. (ed.), *Interactions Between Herbicides and the Soil*, pp. 223–253. Academic Press, London.

Green, L.A. (1978), Nitrates in water supply abstractions in the Anglian region: current trends and remedies under investigation. *Water Pollution Control*, 77, 478–491.

Green, R.E. (1974), Pesticide–Clay–Water interactions. In Guenzi, W.D. (ed) *Pesticides in Soil and Water*, 3–37. Soil Science Society of America Inc.

Greenland, D.J. (1977). Soil damage by intensive arable cultivation: temporary or permanent? *Philosophical Transactions of the Royal Society of London, B*, 281, 193–208.

Greenland, D.J. (1979), The physics and chemistry of the soil–root interface: some comments. In Harley, J.L. and Russell, R.S. (eds) *The Soil–Root Interface*, pp. 81–98. Academic Press, London.

Greenland, D.J. and Nye, P.H. (1959), Increases in the carbon and nitrogen contents of tropical soils under natural fallows. *Journal of Soil Science*, 10, 284–299.

Greenwood, D.J. (1961), The effect of oxygen concentration on decomposition of organic materials in soil. *Plant and Soil*, 14, 360–376.

Grieve, I.C. (1984), Relations among dissolved organic matter, iron and discharge in a small moorland stream. *Earth Surface Processes and Land Forms*, 9, 35–41.

Grieve, I.C. (1985), Annual losses of iron from moorland soils and their relation to free iron contents. *Journal of Soil Science*, 36, 307–312.

Griffin, R.A. and Jurinak, J.J. (1973), The interaction of phosphate and calcite. *Soil Science Society of America Proceedings*, 37, 847–850.

Grossbard, E. (1976), Effects on the soil microflora. In Audus, L.H. (ed.), *Herbicides, Physiology, Biochemistry, Ecology, Vol. 2*, p. 99–147. Academic Press, London.

Grossbard, E. and Marsh, J.A.P. (1974) The effect of several substituted urea herbicides on the soil microflora. *Pesticide Science*, 5, 609–623.

Grover, R., Maybank, J., Yoshida, K. and Plimmer, J.R. (1973), Droplet and volatility drift hazards for pesticide application. *Paper 73–106, presented at the 66th meeting of the Air Pollution Control Association*, Pittsburgh, USA.

Guenzi, W.D. and Beard, W.E. (1970), Volatilisation of lindane and DDT from soils. *Soil Science Society of America Proceedings*, 34, 443–447.

Guillet, B., Rouiller, J. and Souchier, B. (1975), Podzolisation and clay migration in spodosols of eastern France. *Geoderma*, 14, 223–245.

Gunary, D. (1970), A new adsorption isotherm for phosphate in soil. *Journal of Soil Science*, 21, 72–77.

Gunner, H.B., Zuckerman, B.M., Walker, R.W., Miller, C.W., Deubert, K.H. and Longley, R.E. (1966), The distribution and persistence of diazinon applied to plant and soil and its influence on rhizosphere and soil microflora. *Plant and Soil*, 25, 249–264.

Haigh, R.A. and White, R.E. (1986), Nitrate leaching from a small, underdrained, grassland, clay catchment. *Soil Use and Management*, 2, 65–70.

Haines, B.L. and Best, G.R. (1976), *Glomus mosseae*, endomycorrhizal with *Liquidambar styraciflua* L. seedlings retards NO_3, NO_2 and NH_4 nitrogen loss from a temperate forest soil. *Plant and Soil*, 45, 257–261.

Haith, D.A. (1986), Simulated regional variations in pesticide runoff. *Journal of Environmental Quality*, 15, 5–8.

Hale, M.G. and Moore, L.D. (1979), Factors affecting root exudation II: 1970–1978, *Advances in Agronomy*, 31, 93–124.

Hall, J.K., Hartwig, N.L. and Hoffman, L.D. (1983), Application mode and alternate cropping effects on atrazine losses from a hillside. *Journal of Environmental Quality*, 12, 336–340.

Halvorson, A.D. and Black, A.L. (1985), Fertilizer phosphorus recovery after seventeen years of dryland cropping. *Soil Science Society of America Journal*, 49, 933–937.

Hamaker, J.M. and Thomson, J.M. (1972), Adsorption. In Goring, C.A.I. and Hamaker, J.W. (eds), *Organic Chemicals in the Soil Environment*, pp. 49–143. Marcel Dekker, New York.

Hamaker, J.W. (1972), Decomposition: quantitative aspects. In *Organic Chemicals in the Soil Environment*, Goring, C.A.I. and Hamaker, J.W. (eds), pp. 253–340, Marcel Dekker, New York.

Hamaker, J.W., Goring, C.A.I. and Youngson, C.R. (1966), Sorption and leaching of 4-amino,-3-4-6-trichloro-picolinic acid in soils. In Gould, F. (ed.), *Organic Pesticides in the Environment*, pp. 23–37. *Advances in Chemistry, Series 60*. American Chemistry Society, Washington, DC.

Hamblin, A.P. and Tennant, D. (1981), The influence of tillage on soil water behaviour. *Soil Science*, 132, 233–239.

Hance, R.J. (1969), Influence of pH, exchangeable cation and the presence of organic matter on the adsorption of some herbicides by montmorillonite. *Canadian Journal of Soil Science*, 49, 357–364.

Hance, R.J. (1972), Complex formation as an adsorption mechanism for linuron and atrazine. *Weed Research*, 17, 197–201.

Hance, R.J. (1976), Adsorption of glyphosate by soils. *Pesticide Science*, 7, 363–369.

Hance, R.J. (1979), Effect of pH on the degradation of atrazine, dichlorprop, linuron and propyzamide in soil. *Pesticide Science*, 10, 83–86.

Hance, R.J. (1980), Transport in the vapour phase. In Hance, R.J. (ed.), *Interactions Between Herbicides and the Soil*, pp. 59–81. Academic Press, London.

Haque, A. and Ebing, W. (1983), Toxicity determination of pesticides to earthworms in the soil substrate. *Zeitschrift für Pflanzenkrankheiten und Pflanzenschutz*, 90(4), 395–408.

Haque, I. and Walmsley, D. (1972), Incubation studies on mineralisation of organic sulphur and organic nitrogen. *Plant and Soil*, 37, 255–264.

Hardie, K. and Layton, L. (1981), The influence of vesicular-arbuscular mycorrhiza on growth and water relations of red clover. I: In phosphate deficient soil. *New Phytologist*, 89, 599–608.

Hardy, R.W.F., Criswell, J.G. and Havelka, U.D. (1977), Investigations of possible limitations of nitrogen fixation by legumes: (1) methodology, (2) identification, and (3) assessment of significance. In Newton, W.E. (ed.), *Nitrogen Fixation*, pp. 451–467. Academic Press, London.

Hardy, R.W.F. and Havelka, U.D. (1976), photosynthate as a major factor limiting nitrogen fixation by field-grown legumes with emphasis on soybeans. In Nutman, P.S. (ed.), *Symbiotic Nitrogen Fixation in Plants, International Biology Programme*, 7, 421–439, Cambridge University Press, Cambridge.

Hargrove, W.L. and Kissel, D.E. (1979), Ammonia volatilisation from surface applications of urea in the field and laboratory. *Soil Science Society of America Journal*, 43, 359–363.

Harley, J.L. and McCready, C.C. (1959), Uptake of phosphate by excised mycorrhizas of beech. *New Phytologist*, 49, 388–397.

Harley, J.L. and Smith, S.E. (1983), *Mycorrhizal Symbiosis*. Academic Press, London.

Harmsen, G.W. and van Schreven, D.A. (1955), Mineralisation of organic nitrogen in soil. *Advances in Agronomy*, 7, 300–398.

Harris, C.R., Chapman, R.A., Harris, C. and Tu, C.M. (1983), Biodegradation of pesticides in soil: rapid induction of carbamate degrading factors after carbofuran treatment. *Journal of Environmental Science and Health*, B19, 1–11.

Harrison, A.G. (1982), Labile organic phosphorus mineralization in relationship to soil properties. *Soil Biology and Biochemistry*, 14, 343–352.

Harter, R.D. and Barek, D.E. (1977), Applications and misapplications of the Langmuir equation to soil adsorption phenomena. *Soil Science Society of America Journal*, 41, 1077–1080.

Harvey, J. Jr. and Han, J.C.Y. (1978), Decomposition of oxamyl in soil and water. *Journal of Agricultural and Food Chemistry*, 26, 536–541.

Havelka, U.D., Boyle, M.G. and Hardy, R.W.F. (1982), Biological nitrogen fixation. In Stevenson, F.J. (ed.), *Nitrogen in Agricultural Soils*. No. 22 in the Series Agronomy, pp. 365–422, American Society of Agronomy/Soil Science Society of America.

Haworth, F. and Cleaver, T.J. (1963), Soil potassium and the growth of vegetable seedlings. *Journal of Science and Food Agriculture*, 14, 264–233.

Hay, R.K.M., Holmes, J.C. and Hunter, E.A. (1978), The effects of tillage, direct drilling and nitrogen fertilizer on soil temperature under a barley crop. *Journal of Soil Science*, 29, 174–183.

Hayes, M.H.B. and Swift, R.S. (1978), *The Chemistry of Soil Constituents*, pp. 179–320. John Wiley, Chichester.

Hayman, D.S. (1970), Endogone spore numbers in soil and vesicular-arbuscular mycorrhiza in wheat as influenced by season and soil treatment. *Transactions of the British Mycology Society*, 54, 53–63.

Hayman, D.S. and Mosse, B. (1972), The role of vesicular-arbuscular mycorrhiza in the removal of phosphorus from soil by plant roots. *Review of Ecology, Biology and Soil*, 9, 463–470.

Haynes, R.J. (1982), Effects of liming on phosphate availability in acid soils. *Plant and Soil*, 68, 289–308.

Heal, O.W. (1979), Decomposition and nutrient release in even-aged plantations. In Ford, D.C., Atterson, J. and Malcolm, D.C. (eds), *The Ecology of Even-Aged Plantations*. Proceedings of Meeting, Division I, IUFRO, Edinburgh 1978.

Heal, O.W. and French, D.D. (1974), Decomposition of organic matter in tundra. In Holding, A.J., Heal, O.W., MacLean, S.F. and Flanagan, P.W. (eds), *Soil Organisms and Decomposition in Tundra*, IBP Tundra Biome, pp. 279–309.

Heal, O.W., Latter, P.M. and Howson, G. (1978), A study of the rates of decomposition of organic matter. In Heal, O.W. and Perkins, D.F. (eds), *Production Ecology of British Moors and Montane Grasslands*. Ecology Studies 27, pp. 136–159. Springer Verlag, New York.

Hedly, M.J., Stewart, J.W.B. and Chauhan, B.S. (1982), Changes in inorganic and organic soil, phosphorus fractions induced by cultivation practices and by laboratory incubations. *Soil Science Society of America Journal*, 46, 970–976.

Helling, C.S., Chesters, G. and Corey, R.B. (1964), Contribution of organic matter and clay to soil cation exchange capacity as affected by the pH of the saturating solutions. *Soil Science Society of America Proceedings*, 28, 517–520.

Helling, C.S., Kearney, P.C. and Alexander, M. (1971), Behaviour of pesticides in soils. *Advances in Agronomy*, 23, 147–240.

Hendrix, P.F. and Parmelee, R.W. (1985), Decomposition, nutrient loss and microarthropod densities in herbicide-treated grass litter in a Georgia piedmont agroecosystem. *Soil Biology and Biochemistry*, 17, 421–428.

Henriksson, E. (1971), Algal nitrogen fixation in temperate regions. In Lie, J.A. and Mulder, E.G. (eds) *Plant & Soil Special Volume, Biological Nitrogen Fixation in Natural and Agricultural Habitats*, pp. 415–419.

Hera, C. (1976), Effect of inoculation and fertilizer application on the growth of soybeans in Rumania. In Nutman, P.S. (ed.), *Symbiotic Nitrogen Fixation in Plants, International Biology Programme*, 7, pp. 269–279. Cambridge University Press, Cambridge.

Hergert, G.W., Klausner, S.D., Bouldin, D.R. and Zwerman, P.J. (1981), Effects of dairy manure on phosphorus concentrations and losses in tile effluent. *Journal of Environmental Quality*, 10, 345–349.

Hill, D. (1984), Diffusion coefficients of nitrate, chloride, sulphate and water in cracked and uncracked Chalk, *Journal of Soil Science*, 35, 27–33.

Hill, I.R. (1978), Microbial transformation of pesticides. In Hill, I.E. and Wright, S.J.L. (eds), *Pesticide Microbiology*, pp. 137–202. Academic Press, London.

Hill, R.L. and Cruse, R.M. (1985), Tillage effects on bulk density and strength of two mollisols. *Soil Science Society of America Journal*, 49, 1270–1273.

Hill, R.L., Horton, R. and Cruse, R.M. (1985), Tillage effects on soil water retention and pore size distribution of two mollisols. *Soil Science Society of America Journal*, 49, 1264–1270.

Hiltbold, A.E. and Buchanan, G.A. (1977), Influence of soil pH on persistence of atrazine in the field. *Weed Science*, 25, 515–520.

Hingston, F.J. (1963), Activity of polyphenolic constituents of leaves of *Eucalyptus* and other species in complexing and dissolving iron oxide. *Australian Journal of Soil Research*, 1, 63–73.

Hingston, F.J., Atkinson, R.J., Posner, A.M. and Quirk, J.P. (1968), Specific adsorption of anions by goethite. *Transactions of the Ninth International Congress of Soil Science, Vol. 1*, pp. 669–678. Adelaide, Australia.

Hingston, F.J., Posner, A.M. and Quirk, J.P. (1972), Anion adsorption by goethite and gibbsite. I: The role of the proton in determining adsorption envelopes. *Journal of Soil Science*, 25, 16–26.

Hingston, F.J., Posner, A.M. and Quirk, J.P. (1974), Anion adsorption by goethite and gibbsite. II: Desorption of anions from hydrous oxide surfaces. *Journal of Soil Science*, 25, 16–26.

Hinman, W.C. (1970), Effects of freezing and thawing on some chemical properties of three soils. *Canadian Journal of Soil Science*, 50, 179–182.

Hodder, A.P.W. (1984), Thermodynamic interpretation of weathering indices and its application to engineering properties of rocks. *Engineering Geology*, 20, 241–251.

Holford, I.C.R. (1982), The comparative significance and utility of the Freundlich and Langmuir parameters for characterising sorption and plant availability of phosphate in soils. *Australian Journal of Soil Research*, 20, 233–242.

Holford, I.C.R. and Mattingly, G.E.C. (1975), The high and low energy phosphate adsorbing surfaces in calcareous soils. *Journal of Soil Science*, 26, 407–417.

Holford, I.C.R., Wedderburn, R.W.M. and Mattingly, G.E.G. (1974), A Langmuir two-surface equation as a model for phosphate absorption by soils. *Journal of Soil Science*, 25, 242–255.

Holmgren, G.S. (1967), A rapid citrate-dithionite extractable iron procedure. *Soil Science Society of America Proceedings*, 31, 210–211.

Hood, A.E.M. (1976), Nitrogen, grassland and water quality in the United Kingdom. *Outlook on Agriculture*, 8, 320–327.

Houseworth, L.D. and Tweedy, B.G. (1973), Effect of atrazine in combination with captan or thiram upon fungal and bacterial populations in the soil. *Plant and Soil*, 38,(3), 493–500.

Huang, P.M., Grover, R. and McKercher, R.B. (1984), Components and particle size fractions involved in atrazine adsorption by soils. *Soil Science*, 138, 20–24.

Hunt, H.W. (1978), A simulation model for decomposition in grasslands. In Innis, G.S. (ed.), *Grassland Simulation Model, Ecological Studies 26*, pp. 155–183. Springer-Verlag, New York.

Hunt, H.W. and Adamsen, F.J. (1985), Empirical representation of ammonium adsorption in two soils. *Soil Science*, 139, 205–210.

Hurle, K. and Walker, A. (1980), Persistence and its prediction. In Hance, R.J. (1980) (ed.), *Interactions Between Herbicides and the Soil*, pp. 83–122. Academic Press, London.

Ivey, M.J. and Andrews, H. (1965), Leaching simazine, atrazine, diuron and DCPA in soil columns. *Proceedings of the Southern Weed Conference*, 18, 678–684.

Jaakkola, A. (1984), Leaching losses of nitrogen from a clay soil under grass and cereal crops in Finland. *Plant and Soil* 76, 59–66.

Jackson, M.L. and Sherman, G.D. (1953), Chemical weathering of minerals in soils. *Advances in Agronomy*, 5, 219–318.

Jackson, R.M. and Mason, P.A. (1984), *Mycorrhiza*. Studies in Biology No. 159, 59. Edward Arnold.

Jacobs, T.C. and Gilliam, J.W. (1985), Riparian losses of nitrate from agricultural drainage waters. *Journal of Environmental Quality*, 14, 472–478.

Jager, G. (1968), The influence of drying and freezing of soil on its organic matter decomposition. *Stikstof*, 12, 75–88.

Jalali, B.L. (1976), Biochemical nature of root exudates in relation to root rot of wheat. III: Carbohydrate shifts in response to foliar treatments. *Soil Biology and Biochemistry*, 8, 127–129.

Jeffries, R.A., Bradshaw, A.D. and Putwain, P.D. (1981), Growth nitrogen accumulation and nitrogen transfer by legume species established on mine spoils. *Journal of Applied Ecology*, 18, 945–956.

Jenkinson, D.S. (1969), Radiocarbon dating of soil organic matter. *Annual Report, Rothamsted Experimental Station for 1968*, Part 1, p. 73.

Jenkinson, D.S. (1977), The nitrogen economy of the Broadbalk Experiments: II: Nitrogen balance in the experiments. *Annual Report Rothamsted Experimental Station for 1976*, Part 2, pp. 103–109.

Jenkinson, D.S. (1981), The fate of plant and annual residues in soil. In Greenland, D.J. and Hayes, M.H.B. (eds), *The Chemistry of Soil Processes*, pp. 505–561.

Jenkinson, D.S. and Rayner, J.H. (1977), The turnover of soil organic matter in some of the Rothamsted Classical Experiments. *Soil Science*, 123, 298–305.

Jenny, H. (1941), *Factors of Soil Formation*. McGraw-Hill, New York.

Jenny, H. (1980), *The Soil Resource: Origin and Behaviour, Ecological Studies 37*, Springer-Verlag, New York.

Jenny, H., Gessel, S.P. and Bingham, F.T. (1949), Comparative study of decomposition rates of organic matter in temperate and tropical regions. *Soil Science*, 68, 419–432.

Johnson, C.B. and Moldenhauer, W.C. (1979), Effect of chisel versus mouldboard plowing on soil erosion by water. *Soil Science Society of America Journal*, 43, 177–179.

Jones, H.E. and Gore, A.J.P. (1978), A simulation of production and decay in blanket bog. In Heal, O.W. and Perkins, D.F. (eds), *Production Ecology of British Moors and Montane Grasslands. Ecological Studies, 27*, pp. 160–186. Springer Verlag, New York.

Jones, R.L. and Hinesly, T.D. (1986), Potassium losses in runoff and drainage waters from cropped, large-scale lysimeters. *Journal of Environmental Quality*, 15, 137–140.

Juma, N.G., Paul, E.A. and Mary, B. (1984), Kinetic analysis of net nitrogen mineralisation in soil. *Soil Science Society of America Journal*, 48, 753–757.

Jurgensen, M.F., Arno, S.F., Harvey, A.E., Larsen, M.J., and Pfister, R.D. (1979), Symbiotic and non-symbiotic nitrogen fixation in Northern Rocky Mountain forest ecosystems. In Gordon, J.C., Wheeler, C.T. and Perry, D.A. (eds), *Symbiotic Nitrogen Fixation in the Management of Temperate Forests, Proceedings of a Workshop, 1979, Corvallis, Oregon*, pp. 294–308. National Science Foundation.

Jury, W.A. (1982), Simulation of solute transport using a transfer function model. *Water Resources Research*, 18, 363–368.

Jury, W.A., Gardner, W.R., Saffiena, P.G. and Tanner, C.B. (1976), Model for predicting simultaneous movement of nitrate and water through a loamy sand. *Soil Science*, 122, 36–43.

Jury, W.A., Spencer, W.F. and Farmer, W.J. (1983), Behaviour assessment model for trace organics in soil. I: Model desorption. *Journal of Environmental Quality*, 12, 558–564.

Jury, W.A., Stolzy, L.H. and Shouse, P. (1982), A field test of the transfer function model for predicting solute transport. *Water Resources Research*, 18, 369–375.

Kapoor, B.S. (1972), Weathering of micaceous clays in some Norwegian podzols. *Clay Minerals*, 9, 383–394.

Katznelson, H., Rouatt, J.W. and Payne, T.M.B. (1955), The liberation of amino acids and reducing compounds by plant roots. *Plant and Soil*, 7, 35–48.

Kaufman, D.D. (1974), Degradation of pesticides by soil microorganisms. In Guenzi, W.D. (ed.), *Pesticides in Soil and Water*, pp. 133–202. Soil Science Society of America Inc.

Kaufman, D.D. and Kearney, P.C. (1976), Microbial transformations in the soil. In Audus, L.J. (ed.) *Herbicides: Physiology, Biochemistry, Ecology*, pp. 29–64.

Kavanagh, B.V., Posner, A.M. and Quirk, J.P. (1980), Effect of adsorption of phenoxy acetic acid herbicides on the surface charge of goethite. *Journal of Soil Science*, 31, 33–39.

Kearney, P.C., Kaufman, D.D., von Endt, D.W. and Guardia, F.S. (1969), TCA metabolism by soil micro-organisms. *Journal of Agriculture and Food Chemistry*, 17, 581–584.

Keeney, D.R. and Bremner, J.M. (1964), Effect of cultivation on the nitrogen distribution in soils. *Soil Science Society of America Proceedings*, 28, 653–656.

Keeney, D.R., Fillery, I.R. and Marx, G.P. (1979), Effect of temperature on the gaseous nitrogen products of denitrification in a silt loam soil. *Soil Science Society of America Journal*, 43, 1124–1128.

Kellar, P.E. (1979), *Annual Revue Meeting of American Society of Limnology and Oceanography. Corpus Christi, Texas*.

Kepert, D.G., Robson, A.D. and Posner, A.M. (1979), The effect of organic root products on the availability of phosphorus to plants. In Harley, J.L. and

Scott Russell, R. (eds), *The Soil–Root Interface*, pp. 115–124. Academic Press, London.

Kerr, H.W. (1928), The nature of base exchange and soil acidity. *Journal of the American Society of Agronomy*, 20, 309–335.

Khan, S.U. (1973a), Interaction of bipiridylium herbicides with organo-clay complex. *Journal of Soil Science*, 24, 244–248.

Khan, S.U. (1973b), Interaction of humic substances with bipyridylium herbicides. *Canadian Journal of Soil Science*, 53, 199–204.

Khan, S.U. (1973c), Equilibrium and kinetic studies of the adsorption of 2,4-D and picloram on humic acid. *Canadian Journal of Soil Science*, 53, 429–434.

Khan, S.U. (1978), Kinetics of hydrolysis of atrazine in aqueous fulvic acid solution. *Pesticide Science*, 9, 39–43.

Khan, S.U. (1980), *Pesticides in the Soil Environment. Fundamental Aspects of Pollution Control and Environmental Science*, Vol. 5. Elsevier Science.

Khan, S.U. and Schnitzer, M. (1978), UV irradiation of atrazine in aqueous fulvic acid solution. *Journal of Environmental Science and Health*, B13, 299–310.

Khasawneh, F.E. and Doll, E.C. (1978), The use of phosphate rock for direct applications to soils. *Advances in Agronomy*, 30, 159–206.

Kilham, O.W. and Alexander, M. (1984), A basis for organic matter accumulation in soils under anaerobiosis. *Soil Science*, 137(6), 419–427.

Kirda, C. Nielsen, D.R. and Biggar, J.W. (1973), Simultaneous transport of chloride and water during infiltration. *Soil Science Society of America Proceedings*, 37, 339–345.

Kirkby, M.J. (1977), Soil development models as a component of slope models, *Earth Surface Processes*, 2, 203–230.

Kirkby, M.J. (1985), A basis for soil profile modelling in a geomorphic context, *Journal of Soil Science*, 36, 97–121.

Klyuchnikov, L. Yu., Petrova A.N., Polesko and Yu. A. (1964), Influence of simazine and atrazine on the microflora of sandy soil. *Microbiology*, 33(6), 879–882.

Knight, B.A.G. and Denny, P.J. (1970), The interaction of paraquat with soil: adsorption by an expanding lattice clay mineral. *Weed Research*, 10, 40–48.

Knight, B.A.G. and Tomlinson, T.E. (1967), The interaction of paraquat with mineral soils. *Journal of Soil Science*, 18, 233–243.

Knisel, W.G. (ed.) (1980), CREAMS: a field-scale model for chemicals, runoff, and erosion, from agricultural management systems. *USDA Conservation Research Report*, 26, USDA Washington.

Kolenbrander, G.J. (1969), Nitrate content and nitrogen loss in drain water. *Netherlands Journal of Agricultural Science*, 17, 246–255.

Konrad, J.G., Armstrong, D.E. and Chesters, G. (1967), Soil degradation of diazinon, a phosphorothioate insecticide. *Agronomy Journal*, 59, 591–594.

Konrad, J.G., Chesters, G. and Armstrong, D.E. (1969), Soil degradation of malathion, a phosphorodithioate insecticide. *Soil Science Society of America Proceedings*, 33, 259–262.

Koskinen, W.C. and Keeney, D.R. (1982), Effect of pH on the rate of gaseous products of denitrificaton in a silt loam soil. *Soil Science Society of America Journal*, 46, 1165–1167.

Koskinen, W.C. and Cheng, H.H. (1983), Effects of experimental variables on 2,4,5-T adsorption-desorption in soil. *Journal of Environmental Quality*, 12, 325–330.

Kowalenko, C.G. and Cameron, D.R. (1976), Nitrogen transformations in an incubated soil as affected by combinations of moisture content and temperature and adsorption-fixation of ammonium. *Canadian Journal of Soil Science*, 56, 63–70.

Kowalenko, C.G. and Cameron, D.R. (1978), Nitrogen transformations in soil–plant systems in three years of field experiments using tracer and non-tracer methods on an ammonium fixing soil. *Canadian Journal of Soil Science*, 58, 195–208.

Kozak, J., Weber, J.B. and Sheets, T.J. (1983), Adsorption of prometryn and metolachlor by selected soil organic matter fractions. *Soil Science*, 136, 94–101.

Krishnoppa, A.M. and Shinde, J.E. (1978), Working Paper No 41, presented at the 4th Research Coordination Meeting of the Joint FAO/IAEA/GSF Coordination Program on N Residues held at Piracicciloa, Brazil. In Savant, N.K. and De Datta, S.K. (1982) Nitrogen transformations in wetland soils. *Advances in Agronomy*, 35, 241–302.

Kronberg, B.I. and Nesbitt, H.W. (1981), Quantification of weathering, soil geochemistry and soil fertility. *Journal of Soil Science*, 32, 453–459.

Kunc, F., Tichy, P. and Vancura, V. (1985), 2,4-D in the soil: mineralization and changes in the counts of bacterial decomposers. In *Comportement et effets secondaires des pesticides dans le soil. Versailles, France INRA, 4–8 June, 1984*, pp. 175–181.

Lamb, D. (1975), Patterns of nitrogen mineralisation in the forest floor of stands of *Pinus radiata* on different soils. *Journal of Ecology*, 63, 615–625.

Lanowska, J. (1966), Influence of different sources of nitrogen on the development of mycorrhiza in *Pisum satirum*. *Pamietnik Pulowski*, 21, 365–386.

Lapidus, L. and Amundson, N.R. (1952), Mathematics of adsorption in beds. VI: the effect of longitudinal diffusion in ion exchange and chromatographic columns. *Journal of Physical Chemistry*, 56, 984–988.

Larsen, S. (1952), The use of ^{32}P in studies of the uptake of phosphorus by plants. *Plant and Soil*, 4, 1–10.

Larsen, S. (1967), Soil phosphorus. *Advances in Agronomy*, 19, 151–210.

LaRue, T.A. and Patterson, T.G. (1981), How much nitrogen do legumes fix? *Advances in Agronomy*, 34, 15–38.

Latter, P.M. and Howson, G. (1977), The use of cotton strips to indicate cellulose decomposition in the field. *Pedobiologia*, 17, 145–155.

Lavy, T.L., Roeth, F.W. and Fenster, C.R. (1973), Degradation of 2,4-D and atrazine at three soil depths in the field. *Journal of Environmental Quality*, 2, 132–137.

Lee, M. and Lockwood, J.L. (1977), Enhanced severity of *Thielaviopis basicola* root rot induced in soybean by the herbicide chloramben. *Phytopathology*, 67, 1360–1367.

Leonard, R.A., Langdale, G.W. and Fleming, W.G. (1979), Herbicide runoff from upland piedmont watersheds – data and implications for modeling pesticide transport. *Journal of Environmental Quality*, 8, 223–229.

Lethbridge, G. and Burns, R.G. (1976), Inhibition of soil urease by organophosphorus insecticides. *Soil Biology and Biochemistry*, 8, 99–102.

Li, G.C. and Felbeck, G.T. Jr. (1972), Atrazine hydrolysis as catalyzed by humic acids. *Soil Science*, 114, 201–209.

Lindsay, W.L. (1979), *Chemical Equilibria in Soils*. John Wiley, Chichester.

Lindsay, W.L. and Stephenson, H.F. (1959), Nature of the reactions of monocalcium phosphate monohydrate in soils. I: The solution that reacts with the soil. *Soil Science Society of America Proceedings*, 23, 12–18.

Lindstrom, F.T. and Boersma, L. (1970), Theory of chemical transport with simultaneous sorption in a water saturated porous medium. *Soil Science*, 110, 1–9.

Lippold, H., Forster, I., Hagemann, O. and Matzel, W. (1981), Messung der Denitrifizierung auf Grunland mit Hilfe der Gaschromatographic und der ^{15}N-Technik. *Archiv vor Ackerund Pflanzenbau und Bodenkunde, Berlin*, 25, 79–86.

Loughnan, F.C. (1969), *Chemical Weathering of the Silicate Minerals*. American Elsevier Publishing Co. Inc., New York.

Loveland, P.J. and Digby, P. (1984), The extraction of Fe and Al by 0.1M pyrophosphate solutions: a comparison of some techniques. *Journal of Soil Science*, 35, 243–250.

Low, A.J. (1972), The effect of cultivation on the structure and other physical characteristics of grassland and arable soils (1945–1970). *Journal of Soil Science*, 3, 363–380.

Luce, R.W., Bartlett, R.W. and Parks, G.A. (1972), Dissolution kinetics of magnesium silicate. *Geochemica Cosmochimica Acta*, 36, 35–50.

Lynch, J.M. (1984), Interactions between biological processes, cultivation and soil structure. *Plant and Soil*, 76, 307–318.

Lynch, J.M. and Bragg, E. (1985), Microorganisms and soil aggregate stability. *Advances in Soil Science*, 2, 133–171.

Lynd, J.Q., Hanlon, E.A. and Odell, G.V. (1984), Nodulation and nitrogen fixation by arrowleaf clover: effects of phosphorus and potassium. *Soil Biology and Biochemistry*, 16, 589–594.

MacFadyen, A. (1967), Thermal energy as a factor in the biology of soils. In Rose, A.H. (ed.), *Thermobiology*, pp. 535–553.

Mackney, D. (1961), A podzol development sequence in oakwoods and heath in Central England. *Journal of Soil Science*, 12, 23–40.

MacKowen, C.T. and Tucker, T.C. (1985), Ammonium nitrogen movement in a coarse-textured soil amended with Zeolite. *Soil Science Society of America Journal*, 49, 235–238.

Madhun, Y.A., Young, J.L. and Freed, V.H.J. (1986), Binding of herbicides by water soluble organic materials from soil. *Journal of Environmental Quality*, 15, 64–68.

Mahler, R.L., Bezdicek, D.F. and Witters,, R.E. (1979), Influence of slope position on nitrogen fixation and yield of dry peas. *Agronomy Journal*, 71, 348–351.

Malcolm, D.C. and Cuttle, S.P. (1983), The application of fertilizers to drained peat. I: Nutrient losses in drainage. *Forestry*, 56, 155–174.

Malcolm, R.L. and McCracken, R.J. (1968), Canopy drip: a source of mobile soil organic matter for mobilisation of iron and aluminium. *Soil Science Society of America Proceedings*, 32, 834–838.

Malhi, S.S. and Nyborg, M. (1983), Field study of the fate of fall-applied ^{15}N labelled fertilizers in three Alberta soils. *Agronomy Journal*, 75, 71–74.

Manguiat, I.J. and Broadbent, F.E. (1977), Recoveries of tagged N (^{15}N-labelled) fertilizer under some management practices for lowland rice, *Phillipine Agriculturalist*, 60, 367–377.

Marshall, C.E. (1975), *The Physical Chemistry and Mineralogy of Soils. Vol. I: Soil Materials*, Robert E. Krieger Publishing Co., New York.

Marshall, C.E. (1977), *The Physical Chemistry and Mineralogy of Soils. Vol. II: Soils in Place*, Wiley-Interscience, New York.

Marshall, T.J. (1958), A relation between permeability and size distribution of pores. *Journal of Soil Science*, 9, 1–8.

Martel, Y.A., De Kimpe, C.R. and Laverdièr, M.R. (1978), Cation exchange capacity of clay-rich soils in relation to organic matter, mineral composition and surface area. *Soil Science Society of America Journal*, 42, 764–767.

Martin, J.P. (1971), Decomposition and binding action of polysaccharides in soil. *Soil Biology Biochemistry*, 3, 33–41.

Martin, N.J. and Holding, A.J. (1978), Nutrient availability and other factors limiting microbial activity in the blanket peat. In Heal, O.W. and Perkins, D.F. (eds), *Production Ecology of British Moors and Montane Grasslands, Studies in Ecology 27*, 115–138.

Marx, D.H., Hatch, A.B. and Mendicino, J.F. (1977), High fertility decreases sucrose content and susceptibility of Loblolly pine roots to ectomycorrhizal infection by *Pisolithus tinctorius*. *Canadian Journal of Botany*, 55, 1569–1574.

Mary, B. and Remy, J.C. (1979), Essai d'appréciation de la capacité de minéralisation de l'azote des sols de grande culture: I: Signification des cinétiques de minéralisation de la matière organique humifiée. *Annales Agronomique*, 30, 513–527.

Mattingly, G.E.G. (1970), Residual value of basic slag, Gafsa rock phosphate and superphosphate in a sandy podzol. *Journal of Agricultural Science*, 75, 413–418.

Mattson, S. (1929), The laws of soil colloidal behaviour. I. *Soil Science*, 28, 179–220.

Mattson, S. and Karlsson, N. (1944), The pedography of hydrologic soil series: II: The composition and base status of the vegetation in relation to the soil. *Annals of the Royal Agricultural College, Sweden*, 12, 186.

McClennan, G.H. and Gremillion, L.R. (1980), Evaluation of phosphatic raw materials. In Khasawneh, F.E., Sample, E.C. and Kamprath, E.J. (eds), *The Role of Phosphorus in Agriculture*, pp. 43–80. American Society of Agronomy, Madison, Wisconsin.

McElhannon, W.S., Mills, H.A., and Bush, P.B. (1984), Simazine and atrazine – suppression of denitrification. *Horticultural Science*, 19, 218–219.

McGlamery, M.D. and Slife, F.W. (1966), The adsorption and desorption of atrazine as affected by pH, temperature and concentration. *Weeds*, 14, 237–239.

McHardy, W.J., Thomson, A.P. and Goodman, B.A. (1974), Formation of iron oxides by decomposition of iron-phenolic chelates. *Journal of Soil Science*, 25, 471–483.

McKeague, J.A. (1965), Properties and genesis of three members of the uplands catena. *Canadian Journal of Soil Science*, 45, 63–77.

McKeague, J.A. and Brydon, J.E. (1970), Mineralogical properties of the reddish brown soils from the Atlantic Provinces in relation to parent materials and pedogenesis. *Canadian Journal of Soil Science*, 50, 47–55.

McKeague, J.A., Wang, C., Coen, G.M., DeKimpe, C.R., Laverdièr, M.R., Evans, L.J., Kloosterman, B., and Green, A.J. (1983), Testing chemical criteria for spodic horizons on podzolic soils in Canada. *Soil Science Society of America Journal*, 47, 1052–1054.

McNeal, B.L. and Coleman, N.T. (1966), Effect of solution composition on soil hydraulic conductivity. *Soil Science Society of America Journal*, 30, 308–312.

McNeal, B.L., Layfield, D.A. Norvell, W.A. and Rhoades, J.D. (1968), Factors influencing hydraulic conductivity of oats in the presence of mixed-salt solutions. *Soil Science Society of America Journal*, 32, 187–190.

Mehra, O.P. and Jackson, M.L. (1960), Iron oxide removal from soils and clays by dithionite-citrate systems buffered with sodium bicarbonate. *Clays and Clay Minerals*, 7, 317–325.

Meikle, R.W., Youngson, C.R., Hedlung, R.T., Goring, C.A.I., Hamaker, J.W. and Addington, W.W. (1973), Measurement and prediction of picloram disappearance rates from soil. *Weed Science*, 21, 549–555.

Mekaru, T. and Uehara, G. (1972), Anion adsorption in ferruginous tropical soils. *Soil Science Society of America Proceedings*, 36, 296–300.

Mellanby, K. (1967), *Pesticides and Pollution*, Fontana New Naturalist Series, Collins.

Mercado, A. and Billings, G.K. (1975), The kinetics of mineral dissolution in carbonate aquifers as a tool for hydrological investigations: I: Concentration-time relationships. *Journal of Hydrology*, 24, 303–331.

Mersie, W. and Foy, C.L. (1986), Adsorption, desorption and mobility of chlorsulfuron in soils. *Journal of Agricultural and Food Chemistry*, 34, 89–92.

Meyer, R.D., Olson, R.A. and Roades, H.F. (1961), Ammonia loss from fertilized Nebraska soils. *Agronomy Journal*, 53, 241–244.

Micev, N. (1970), Effect of lumeton herbicide on soil edaphosphere and rhizosphere microflora of wheat. *Contemporary Agriculture*, 18(3), 245–249.

Miller, H.G. (1969), Nitrogen nutrition of pines on the sands of Culbin Forest, Morayshire. *Journal of Science and Food Agriculture*, 20, 417–419.

Miller, M.H. (1979), Contribution of nitrogen and phosphorus to subsurface drainage water from intensively cropped mineral and organic soils in Ontario. *Journal of Environmental Quality*, 8, 42–48.

Miller, R.D. and Johnson, D.D. (1964), The effect of soil moisture tension on CO_2 evolution, nitrification and nitrogen mineralisation. *Soil Science Society of America Proceedings*, 28, 644–647.

Miller, R.E. and Murray, M.D. (1979), Fertilizer versus red alder for adding nitrogen to Douglas fir forests of the Pacific Northwest. In Gordon, J.C., Wheeler, C.T. and Perry, D.A. (eds), *Symbiotic Nitrogen Fixation in the Management of Temperate Forests*. Proceedings of Workshop, Oregon, 1979, 356–373. National Science Foundation.

Mills, A.C. and Biggar, J.W. (1969), Solubility-temperature effect on the adsorption of gamma and beta-BHC from aqueous and hexane solutions by soil materials. *Soil Science Society of America Proceedings*, 33, 210–216.

Minchin, F.R. and Pate, J.S. (1973), The carbon balance of a legume and the functional economy of its root nodules. *Journal of Experimental Botany*, 24, 259–271.

Minderman, G. (1968), Addition, decompositon and accumulation of organic matter in forests. *Journal of Ecology*, 56, 355–362.

Mingelgrin, U., Saltzman, S. and Yaron, B. (1977), A possible model for the surface-induced hydrolysis of organophosphorus pesticides on kaolinite clays, *Soil Science Society of America Journal*, 41, 519–523.

Mingelgrin, U., Yariv, S. and Saltzman, S. (1978), Differential infrared spectroscopy in the study of parathion-dentonite complexes. *Soil Science Society of America Journal*, 42, 664–665.

Mokma, D.L. (1983), New chemical criteria for defining the spodic horizon. *Soil Science Society of American Journal*, 47, 972–976.

Mokma, D.L. and Buurman, P. (1982), *Podzols and Podzolisation in Temperate Regions. ISM Monograph No 1*, pp. 1–131, International Soil Museum, Wageningen.

Moore, I.C. and Madison, F.W. (1985), Description and application of an annual waste phosphorus loading model. *Journal of Environmental Quality*, 14, 364–369.

Mortland, M.M. (1970), Clay–organic complexes and interactions. *Advances in Agronomy*, 22, 75–117.

Mortland, M.M. and Raman, K.V. (1967), Catalytic hydrolysis of some organic phosphate pesticides by copper (II). *Journal of Agricultural and Food Chemistry*, 15, 163–167.

Mosse, B. (1986), Mycorrhiza in a sustainable agriculture. In Lopez-Real, J.M. and Hodges, R.D. (eds), *The Role of Microorganisms in a Sustainable Agriculture*, pp. 105–123. AB Academic Publishers.

Mosse, B., Stribley, D.P. and Le Tacon, F. (1981), Ecology of mycorrhizae and mycorrhizal fungi. *Advances in Microbiology and Ecology*, 5, 137–210.

Moustafa, E., Bell, R. and Field, T.R.O. (1969), The use of acetylene reduction to study the effect of nitrogen fertilizer and defoliation on nitrogen fixation by field-grown white clover. *New Zealand Journal of Agricultural Research*, 12, 691–696.

Muir, J.W. and Logan, J. (1982), Eluvial/illuvial coefficients of major elements and the corresponding losses and gains in three soil profiles. *Journal of Soil Science*, 33, 295–308.

Murdoch, C.L., Jakobs, J.A. and Gerdemann, J.W. (1967), Utilization of phosphorus sources of different availability by mycorrhizal and non-mycorrhizal maize. *Plant and Soil*, 27, 329–334.

Murray, K. and Linder, P.W. (1983), Fulvic acids: structure and metal binding. I: A random molecular model. *Journal of Soil Science*, 34, 511–523.

Murray, K. and Linder, P.W. (1984), Fulvic acids: structure and metal binding. II: Predominant metal binding sites. *Journal of Soil Science*, 35, 217–222.

Murrmann, R.P. and Peech, M. (1969), Effects of pH on labile and soluble phosphate in soils. *Soil Science Society of America Proceedings*, 33, 205–210.

Nakos, G. (1980), Effects of herbicides used in forestry on soil nitrification. *Soil Biology and Biochemistry*, 12, 517–519.

Narine, D.R. and Guy, R.D. (1982), Binding of diquat and paraquat to humic acid in aquatic environments. *Soil Science*, 133, 356–363.

Nash, R.G. (1983), Comparative volatilisation and dissipation rates of several pesticides from soil. *Journal of Agricultural Food Chemistry*, 31, 210–217.

Nearpass, D.C. (1965), Effects of soil acidity in the adsorption, penetration and persistence of simazine. *Weeds*, 13, 341–346.

Nearpass, D.C. (1967), Effect of predominating cation on the adsorption of simazine and atrazine by Bayboro clay soil. *Soil Science*, 103, 177–182.

Nearpass, D.C. (1976), Adsorption of picloram by humic acids and humin. *Soil Science*, 121, 272–277.

Negi, S.C., Raghavan, G.S.V. and Taylor, F. (1981), Hydraulic characteristics of conventionally and zero-tilled field plots. *Soil Tillage Research*, 2, 281–292.

Nelson, D.W. (1982), Gaseous losses of nitrogen other than through denitrification. In Stevenson, F.J. (ed.), *Nitrogen in Agricultural Soils, Agronomy Monograph No 22*, pp. 327–363. ASA-CSSA-SSSA.

Nelson, D.W., Huber, D.M. and Warren, H.L. (1980), Nitrification inhibitors – powerful tools to conserve fertilizer nitrogen. In Banin, A. and Kafkafi, U. (eds), *Agrochemicals in Soils*, 57–64.

Nelson, L.M., Yaron, B. and Nye, P.H. (1982), Biologically induced hydrolysis of parathion in soil: kinetics and modelling. *Soil Biology and Biochemistry*, 14, 223–227.

Neuman, S.P., Feddes, R.A. and Bresler, E. (1975), Finite element analysis of two-dimensional flow in soils considering water uptake by roots. I Theory. II Field applications. *Soil Science Society of America Proceedings*, 39, 224–237.

Newbould, P. (1982), Biological nitrogen fixation in upland and marginal areas of the UK. *Philosophical Transactions of the Royal Society of London*, B296, 405–417.

Newman, A.C.D. (1969). Cation exchange properties of micas. I The relation between mica composition and potassium exchange in solutions of different pH. *Journal of Soil Science*, 20, 357–373.

Nicholaichuk, W. and Grover, R. (1983). Loss of fall-applied 2,4-D in spring runoff from a small agricultural watershed. *Journal of Environmental Quality*, 12, 412–414.

Nielsen, D.R., Biggar, J.W. and Erh, K.J. (1973), Spatial variability of field-measured soil water properties. *Hilgardia*, 42, 215–259.

Nielsen, D.R., Jackson, R.D., Cary, J.W. and Evans, D.D. (1972) (eds), Chapter 3: Theoretical analysis. In *Soil Water*. American Society of Agronomy and Soil Science Society of America.

Nihlgard, B. (1972), Plant biomass, primary production and distribution of chemical elements in a beech and a planted spruce forest in southern Sweden. *Oikos*, 23, 69–81.

Nkedi-Kizza, P., Rao, P.S.C. and Johnson, J.W. (1983), Adsorption of diuron and 2,4,5-T on soil particle size separates. *Journal of Environmental Quality*, 12, 195–197.

Nommik, H. (1956), Investigations on denitrification in soil. *Acta Agricultura Scandinavica*, 6, 195–228.

Nommik, H. (1957), Fixation and defixation of ammonium in soils. *Acta Agricultura Scandinavica*, 7, 395–436.

Nommik, H. (1968), Nitrogen mineralisation and turnover in Norway Spruce (*Picea abies* (L.) Karst.) raw humus as influenced by liming. *Transactions of the Ninth International Congress of Soil Science*, 2, 533–545.

Nommik, H. and Vahtras, K. (1982), Retention and fixation of ammonium and ammonia in soils. In Stevenson, F.J. (ed.) *Nitrogen in Agricultural Soils*, pp. 123–171, No 22 in the series Agronomy, ASA, SSSA, CSSA.

Nye, P.H. (1979), Diffusion of ions and uncharged solutes in soils and soil clays. *Advances in Agronomy*, 31, 225–272.

Nye, P.H. and Tinker, P.B. (1977), *Solute Movement in the Soil–Root System. Studies in Ecology*, Vol. 4, Blackwell Scientific, Oxford.

Oades, J.M. (1978), Mucilages at the root surface. *Journal of Soil Science*, 29, 1–16.

O'Connor, G.A. and Anderson, J.U. (1974), Soil factors affecting the adsorption of 2,4,5-T. *Soil Science Society of America Proceedings*, 38, 433–436.

Oghoghorie, C.G.O. and Pate, J.S. (1971), The nitrate stress syndrome of the nodulated field pear (*Pisum arvense* L.). *Plant and Soil (Special Volume)*, 185–202.

Ojeniyi, S.O. and Dexter, A.R. (1979a), Soil factors affecting the macro-structures produced by tillage. *Transactions of the American Society of American Engineers*, ASAE 22(2), 339–343.

Ojeniyi, S.O. and Dexter A.R. (1979b), Soil structural changes during multiple pass tillage. *Transactions of the American Society of Agricultural Engineers*, 22(5), 1068–1072.

Oke, T.R. (1978), *Boundary Layer Climates*. Methuen and Co Ltd., London.

Ollier, C. (1969), *Weathering*. Oliver and Boyd, Edinburgh.

Olness, A. and Clapp, C.E. (1975), Influence of polysaccharide structure on dextran adsorption by montmorillonite. *Soil Biology and Biochemistry*, 7, 113–118.

Olsen, S.R. and Khasawneh, F.E. (1980), Use and limitations of physical-chemical criteria for assessing the status of phosphorus in soils. In Khasawneh, F.E., Sample, E.C. and Kamprath, E.J. (eds), *The Role of Phosphorus in Agriculture*, pp. 471–514. American Society of Agronomy.

Olsen, S.R. and Watanabe, F.S. (1957), A method to determine a phosphorus adsorption maximum of soils as measured by the Langmuir isotherm. *Soil Science Society of America Proceedings*, 21, 144–149.

Omoti, U. and Wild, A. (1979), Use of fluorescent dyes to mark the pathways of solute movement through soils under leaching conditions: 2 Field Experiments. *Soil Science*, 128, 98–104.

Ottow, J.C.G. (1971), Iron reduction and gley formation by nitrogen-fixing *Clostridia*. *Oecologia (Berlin)*, 6, 164–175.

Ou, L-T. (1985), Methyl parathion and metabolism in soil: influence of high soil water contents. *Soil Biology Biochemistry*, 17, 241–243.

Ou, L-T, Gancarz, D.H., Wheeler, W.B., Rao, P.S.C. and Davidson, J.M. (1982), Influence of soil temperature and soil moisture on degradation and metabolism of carbofuran in soils. *Journal of Environmental Quality*, 11, 293–298.

Overrein, L.N. (1968a), Lysimeter studies on tracer ^{15}N in forest soil. I: N losses by leaching and volatilization after addition of urea-^{15}N. *Soil Science*, 106, 281–290.

Overrein, L.N. (1968b), Lysimeter studies on tracer N in forest soil. II: Comparative losses of N through leaching and volatilisation after the addition of urea-, NH_4- and NO_3-^{15}N. *Soil Science*, 107, 149–159.

Overrein, L.N. (1971), Isotope studies on N in forest soil. I: Relative losses of N through leaching during a period of 40 months. *Meddelelser fra Det Norske Skogforsøksvesen*, 29, 261–280.

Overrein, L.N. (1972), Isotope studies on nitrogen in forest soil. II: Distribution and recovery of ^{15}N enriched fertilizer nitrogen in a 40-month lysimeter investigation. *Meddelelser fra Det Norske Skogforsyksvesen*, 30, 306–324.

Ovington, J.D. (1962), Quantitative ecology and the woodland ecosystem concept. *Advances in Ecology Research*, 103–192.

Ovington, J.D. and Heitkamp, D. (1961), The accumulation of energy in forest plantations in Britain. *Journal of Ecology*, 48, 639–646.

Ozanne, P.G. and Shaw, T.C. (1961), The loss of phosphorus from sandy soils. *Australian Journal of Agricultural Research* 12, 409–423.

Paces, T. (1985), Sources of acidification in Central Europe estimated from elemental budgets in small basins. *Nature (London)*, 315, 31–36.

Page, A.L., Burge, W.D. and Ganje, T.J. (1967), Potassium and ammonium fixation by vermiculite soils. *Soil Science Society of America Proceedings*, 31, 337–341.

Pairunan, A.K., Robson, A.D. and Abbott, L.K. (1980), The effectiveness of vesicular-arbuscular mycorrhizas in increasing growth and phosphorus uptake of subterranean clover from phosphorus sources of different solubilities. *New Phytologist*, 84, 327–334.

Pang, P.C. and Paul, E.A. (1980), Effects of vesicular-arbuscular mycorrhiza on ^{14}C and ^{15}N distribution in nodulated faba beans. *Canadian Journal of Soil Science*, 60, 241–250.

Parfitt, R.L. (1978), Anion adsorption by soils and soil materials, *Advances in Agronomy*, 30, 1–50.

Parfitt, R.L. and Atkinson, R.J. (1976), Phosphate adsorption on goethite (α Fe OOH). *Nature (London)*, 264, 740–742.

Parfitt, R.L. Fraser, A.R., Russell, J.D. and Farmer, V.C. (1977a), Adsorption on hydrous oxides: II Oxalate, benzoate and phosphate on gibbsite. *Journal of Soil Science*, 28, 40–47.

Parfitt, R.L., Fraser, A.R. and Farmer, V.C. (1977b), Adsorption on hydrous oxides: II: Fulvic and humic acid on goethite, gibbsite and imogolite. *Journal of Soil Science*, 28, 289–296.

Parker, L.W. and Doxtader, K.G. (1983), Kinetics of the microbial degradation of 2,4-D in soil: effects of temperature and moisture. *Journal of Environmental Quality*, 12, 553–558.

Patrick, W.H. Jr. (1978), Critique of 'Measurement and Prediction of Anaerobics in Soils'. In Neilson, D.R. and MacDonald, J.G. (eds), *Nitrogen in the Environment, Vol. 1. Nitrogen Behaviour in Field Soil*, pp. 449–457. Academic Press, London.

Patrick, W.H. Jr. and Delaune, R.D. (1972), Characterisation of the oxidised and reduced zones in flooded soil. *Soil Science Society of America Proceedings*, 36(4), 573–576.

Patrick, W.H. Jr. and Gotoh, S. (1974), The role of oxygen in nitrogen loss from flooded soils. *Soil Science*, 118, 78–81.

Patrick, W.H. Jr. and Reddy, K.R. (1976a), Nitrification-denitrification reactions in flooded soils and water bottoms: dependence on oxygen supply and ammonium diffusion. *Journal of Environmental Quality*, 5(4), 469–472.

Patrick, W.H. Jr. and Reddy, K.R. (1976b), Fate of fertilizer nitrogen in a flooded rice soil. *Soil Science Society of America Journal*, 40, 678–681.

Patrick, W.H. Jr. and Wyatt, R. (1964), Soil nitrogen loss as a result of alternate submergence and drying. *Soil Science Society of America*, 28, 647–653.

Pawluk, S. (1972), Measurement of crystalline and amorphous iron removed in soils. *Canadian Journal of Soil Science*, 53, 119–123.

Peck, A.J., Luxmoore, R.J. and Stolzy, J.L. (1977), Effects of spatial variability of soil hydrologic properties on water budget modeling. *Water Resources Research*, 3, 348–354.

Peck, D.E., Corwin, D.L. and Farmer, W.J. (1980), Adsorption-desorption of diuron by freshwater sediments. *Journal of Environmental Quality*, 9, 101–106.

Persson, H. (1982), Changes in the tree and dwarf shrub fine-roots after clear cutting in a mature Scots Pine stand. *Swedish Coniferous Forest Project, Technical Report, 31*, 19pp.

Peterson, J.R., Adams, R.S. Jr. and Cutkomp, L.K. (1971), Soil properties influencing DDT bioactivity. *Soil Science Society of America Proceedings*, 35, 72–78.

Pettijohn, F.J. (1941), Persistence of heavy minerals and geologic age. *Journal of Geology*, 49, 610–625.

Philen, O.D. Jr, Weed, S.B. and Weber, J.B. (1970), Estimation of surface charge density of mica and vermiculate by competitive adsorption of diquat vs. paraquat. *Soil Science Society of America Proceedings*, 34, 527–531.

Phillips, D.A. (1980), Efficiency of symbiotic nitrogen fixation in legumes. *Annual Review of Plant Physiology*, 31, 29–49.

Phillips, D.A. and Bennett, J.P. (1978), Measuring symbiotic nitrogen fixation in rangeland plots of *Trifolium subterraneum* L. and *Bromus mollis* L. *Agronomy Journal*, 70, 671–674.

Phillipson, J., Putman, R.J., Steel, J. and Woodell, S.R.J. (1975), Litter input, litter decomposition and the evolution of CO_2 in a beech woodland – Wytham

Woods, Oxford. *Oecologia (Berlin)*, 20, 203–217.

Piene, H. and Van Cleve, K. (1976), A low-cost unit for measuring CO_2 evolution from organic matter under field conditions. *Canadian Journal of Forestry Research*, 6, 33–39.

Pierce, R.H., Jr., Olney, C.E. and Felbeck, G.T. Jr. (1971), Pesticide adsorption in soils and sediments. *Environmental Letters*, 1, 157–172.

Pigeon, J.D. and Soane, B.D. (1977), Effects of tillage and direct drilling on soil properties during the growing season in a long-term barley mono-culture system. *Journal of Agricultural Science*, 88, 431–442.

Pionnke, H.B. and Chesters, G. (1973), Pesticide–sediment–water interactions. *Journal of Environmental Quality*, 2, 29–45.

Plummer, L.N. and Wigley, T.M.L. (1976), The dissolution of calcite in CO_2-saturated solutions at 25 °C and 1 atmosphere total pressure. *Geochemica et Cosmochemica Acta*, 40, 191–202.

Ponnamperuma, F.N. (1972), The chemistry of submerged soils. *Advances in Agronomy*, 24, 29–96.

Popović B. (1977), Kvävenmineralisering i ett äldre och ett yngre tallbestånd. *Swedish Coniferous Forest Project. Internal Report No 60*, 21pp.

Posner, A.M. and Bowden, J.W. (1980), Adsorption isotherms: should they be split? *Journal of Soil Sciences*, 31, 1–10.

Potter, K.N., Cruse, R.M. and Horton, R. (1985), Tillage effects on soil thermal properties. *Soil Science Society of America Journal*, 49, 968–973.

Powers, D.H. and Skidmore, E.L. (1984), Soil structure as influenced by simulated tillage. *Soil Science Society of America Journal*, 48, 879–884.

Powlson, D.S. and Jenkinson, D.S. (1980), The effect of mechanical disturbance on organic matter changes in soil. *Rothamsted Annual Report for 1979 Part I*, 233–234.

Prasad, R., Rajale, G.B. and Lakhdiue, B.A. (1971), Nitrification retarders and slow-release nitrogen fertilizers. *Advances in Agronomy*, 23, 337–383.

Prasad Reddy, B.V., Dhanaraj, P.S. and Narayana Rao, V.V.S. (1984), Effects of insecticides on soil microorganisms. In Lal, R. (ed.) *Insecticide Microbiology*, pp. 169–201. Springer Verlag, New York.

Pratt, P.F., Cannell, G.H., Garber, M.J. and Bair, F.L. (1967), The effect of three nitrogen fertilizers on gains, losses and distribution of various elements in irrigated lysimeters. *Hilgardia*, 38, 265–283.

Pratt, P.F., Lund, L.J. and Warneke, J.E. (1980), Nitrogen losses in relation to soil profile characteristics. In Banin, A. and Kafkafi, U. (eds), *Agrochemicals in Soils*, pp. 33–45. International Irrigation Information Centre & Israel Society of Soil Science (Pergamon Press).

Probst, G.W., Golab, T. and Wright, W.L. (1975), Degradation of herbicides. In Kearney, P.C. and Kaufman, D.D. (eds), *The Degradation of Herbicides*. Marcel Dekker, New York.

Pupisky, H. and Shainberg, I. (1979), Salt effects on the hydraulic conductivity of a sandy soil. *Soil Science Society of America Journal*, 43, 429–433.

Quirk, J.P. and Schofield, R.K. (1955), The effect of electrolyte concentration on soil permeability. *Journal of Soil Science*, 6, 163–178.

Raats, P.A.C. (1984), Tracing parcels of water and solutes in unsaturated zones. In Yaron, B., Dagan, G. and Goldshmid, J. (eds), *Pollutants in Porous Media. The Unsaturated Zone Between Soil Surface and Groundwater. Ecological Studies 47*, 4–16. Springer Verlag, New York.

Rachhpal-Singh, and Nye, P.H. (1986), Model of ammonia volatilisation from applied urea. I: Development of the model. *Journal of Soil Science*, 37, 9–20.

Rajagopal, B.S., Chendrayan, K., Reddy, B.R. and Sethunathan, N. (1983), Persistence of carbaryl in flooded soils and its degradation by soil enrichment cultures. *Plant and Soil*, 73, 35–45.

Rajatam, K.P. and Sethunathan, N. (1975), Effect of organic sources on the degradation of parathion in flooded alluvial soil. *Soil Science*, 119, 296–300.

Ranem, W.A. (1965), Physical factors of the soil as they affect soil micro-organisms. In Baker, K.F. and Snyder, W.C. (eds), *Ecology of Soil-Borne Plant Pathogens*, pp. 115–118.

Rao, P.S.C., Hornsby, A.G. and Jessup, R.E. (1985), Indices for ranking the potential for pesticide contamination of groundwater. *Soil and Crop Science Society of Florida Proceedings*, 44, 1–8.

Rao, P.S.C., Ralston, D.E., Jessup, R.E. and Davidson, J.M. (1980), Solute transport in aggregated porous media: theoretical and experimental evaluation. *Soil Science Society of America Journal*, 44, 1139–1146.

Rapoport, E.H. and Cangioli, G. (1963), Herbicides and the soil fauna. *Pedobiologia*, 2, 235–238.

Ratkowsky, D.A. (1986), A statistical study of seven curves for describing the sorption of phosphate by soil, *Journal of Soil Science*. 37, 183–189.

Raven, J.A. and Smith, F.A. (1976), Nitrogen assimilation and transport in vascular land plants in relation to intracellular pH regulation. *New Phytologist*, 76, 415–431.

Read, D.J., Francis, R. and Finlay, R.D. (1985), Mycorrhizal mycelia and nutrient cycling in plant communities. In Fitter, A.H. (ed.), *Ecological Interactions in Soil. Plants, Microbes and Animals*, pp. 193–217. Special Publication No. 4, British Ecology Society.

Reaves, R.C. and Cooper, A.W. (1960), Stress distribution in soils under tractor loads. *Agricultural Engineering*, 41, 20–21.

Reddy, K.R. and Patrick, W.H. Jr. (1976), Effect of frequent changes in aerobic and anaerobic conditions on redox potential and nitrogen loss in a flooded soil. *Soil Biology and Biochemistry*, 8, 491–495.

Reddy, K.R. and Patrick, W.H. Jr. (1977), Effect of placement and concentration of applied NH_4-N on nitrogen loss from flooded soil. *Soil Science*, 123, 142–148.

Reddy, K.R. and Patrick, W.H. Jr. (1986), Fate of fertilizer nitrogen in the rice root zone. *Soil Science Society of American Journal*, 50, 649–651.

Reddy, K.R., Rao, P.S.C., Patrick, W.H. Jr. (1980), Factors influencing oxygen consumption rates in flooded soils. *Soil Science Society of America Journal*, 44, 741–744.

Reddy, N.V., Saxena, M.C. and Srinivasulu, R. (1982), E- L- and A-values for estimation of plant available soil phosphorus. *Plant and Soil*, 69, 3–11.

Redman, F.H. and Patrick, W.H. Jr. (1965), *Effect of Submergence on Several Biological and Chemical Soil Properties*, Louisiana State University and Agricultural and Mechanical College, Bulletin No. 592, 28pp.

Rehfuess, K.E. (1979), Underplanting of pines with legumes in Germany. In Gordon, J.C., Wheeler, C.T. and Perry, D.A. (eds), *Symbiotic Nitrogen Fixation in the Management of Temperate Forests*. Proceedings of the Workshop. Oregon, 1979, pp. 374–387. National Science Foundation.

Reiche, P. (1943), Graphic representation of chemical weathering. *Journal of Sedimentation and Petrology*, 13, 58–68.

Reiche, P. (1950), A survey of weathering processes and products. *University New Mexico, Publications in Geology*, 3, 95pp.

Reicosky, D.C. (1983), Soil management for efficient water use: soil profile modification effects on plant growth and yield in the southeastern United States. In Taylor, H.M., Jordan, W.R. and Sinclair, T.R. (eds) *Limitations to Efficient Water Use in Crop Production.* pp. 471–477. American Society of Agronomy Inc. (SSSA/CSSA).

Reid, C.R.P. and Hurtt, W. (1970), Root exudation of herbicides by woody plants: allelopathic implications. *Nature (London)*, 225, 291.

Reid, J.M., MacLeod, D.A. and Cresser, M.S. (1981), The assessment of chemical weathering rates within an upland catchment in north east Scotland. *Earth Surface Proceedings and Landforms*, 6, 447–457.

Remacle, J. (1977), Microbial transformation of nitrogen in forests. *Ecology of Plants*, 12, 33–43.

Remacle, J. and Vanderhoven, Ch. (1972), Evolution of carbon and nitrogen contents in incubated litters. *Plant and Soil*, 39, 201–202.

Rhodes, L.H. and Gerdemann, J.W. (1978), Hyphal translocation and uptake of sulphur by versicular-arbuscular mycorrhizae of onions. *Soil Biology and Biochemistry*, 10, 355–360.

Ricci, E.D., Hubert, W.A. and Richard, J.J. (1983), Organochlorine residues in sediment cores of a midwestern reservoir. *Journal of Environmental Quality*, 12, 418–421.

Richards, B.N. (1974), *Introduction to the Soil Ecosystem.* Longman Group, London.

Richards, K.S., Arnett, R.R. and Ellis, S. (1985), Introduction. In Richards, K.S., Arnett, R.R. and Ellis, S. (eds) *Geomorphology and Soils*, pp. 1–9. George Allen and Unwin.

Riley, D., Wilkinson, W. and Tucker, B.V. (1976), Biological unavailability of bound paraquat residues in soil. In *ACS Symposium Series No. 29. Bound and Conjugated Pesticide Residues.* American Chemical Society, Washington DC, pp. 301–353.

Rinaudo, G., Balandreau, J. and Dommergues, Y. (1971), Algal and bacterial non-symbiotic nitrogen fixation in paddy soils. In Lie, T.A. and Mulder, E.G. (eds), *Biological Nitrogen Fixation in Natural and Agricultural Habitats. Plant and Soil (Special Volume)*, pp. 471–479.

Ritchie, G.S.P. and Posner, A.M. (1982), The effect of pH and metal binding on the transport properties of humic acids. *Journal of Soil Sciences*, 33, 233–247.

Robbins, C.W., Jurinak, J.J. and Wagenet, R.J. (1980), Calculating cation exchange in a salt transport model. *Soil Science Society of America Journal*, 44, 1195–1200.

Robinson, R.A. and Stokes, R.H. (1959), *Electrolyte Solutions.* Butterworths, London.

Rode, A.A. (1935), To the problems of the degree of podzolisation. Dokuchaev Institute, *Studies in the Genesis and Geography of Soils*, pp. 55–70. Academy of Sciences Press, Moscow.

Rolston, D.E., Hoffman, D.L. and Toy, D.W. (1978), Field measurement of denitrification. I: Flux of N_2 and N_2O. *Soil Science Society of America Journal*, 42, 863–869.

Rolston, D.E., Sharpley, A.N., Toy, D.W. and Broadbent, F.E. (1982), Field measurement of denitrification. II: Rates during irrigation cycles. *Soil Science Society of America Journal*, 46, 289–296.

Rose, D.A. (1973), Some aspects of the hydrodynamic dispersion of solutes in porous materials. *Journal of Soil Science*, 24, 284–295.

Rose, D.A. and Passioura, J.B. (1971), Gravity segregation during miscible displacement experiments. *Soil Science*, 111, 258–265.

Rose, C.W., Chichester, F.W., Williams, J.R. and Ritchie, J.T. (1982), A contribution to simplified models of field solute transport. *Journal of Environment Quality*, 11, 146–150.

Roslycky, E.B. (1982), Glyphosate and the response of the soil microbiota. *Soil Biology and Biochemistry*, 14, 87–92.

Ross, G.J. (1980a), Mineralogical, physical and chemical characteristics of amorphous constituents in some podzolic soils from British Columbia. *Canadian Journal of Soil Science*, 60, 31–43.

Ross, G.J. (1980b), The mineralogy of spodosols. In Theng, B.K.G. (ed.), *Soils With Variable Charge*, New Zealand Society of Soil Science, pp. 127–143.

Ross, S.M. (1979), *Nutrient mobilisation in a cultivated heathland soil*, Unpublished PhD thesis, University of Edinburgh.

Ross, S.M. (1984), Site preparation and soil modification with particular reference to the afforestation of upland heaths. *Proceedings SW England Soils Discussion Group*, 2, 32–43.

Ross, S.M. (1985), Field and incubation temperature effects on mobilisation of nitrogen, phosphorus and potassium in peat. *Soil Biology and Biochemistry*, 17, 479–482.

Ross, S.M. (1987), Energetics of soil processes. In Gregory, K.J. (ed.), *Energetics of Physical Environment*, pp. 119–143. John Wiley & Sons Ltd, Chichester.

Ross, S.M. and Malcolm, D.C. (1982), Effects of intensive forestry ploughing practices on an upland heath soil in South East Scotland. *Forestry*, 55(2), 155–171.

Ross, S.M. and Malcolm, D.C. (1988), Effects of intensive soil cultivation on nutrient mobilisation in a peaty ferric stagnopodzol. *Plant and Soil*, 107, 113–121.

Ross, S.M. and Smith, R.E. (1987), Use of catechol solutions as extractants for mobilisable iron in forest soils. *Communications in Soil Science and Plant Analysis*, 18, 147–160.

Rovira, A.D. and Campbell, R. (1974), Scanning electron microscopy of microorganisms on the roots of wheat. *Microbiological Ecology*, 1, 15–30.

Rovira, A.D. and Davey, C.B. (1974), Biology of the rhizosphere. In Carson, E.W. (ed.), *The Plant Root and its Environment*, pp. 153–204. University Press of Virginia, Charlottesville.

Rovira, A.D. and Graecen, E.L. (1957), The effect of aggregate disruption on the activity of microorganisms in the soil. *Australian Journal of Agricultural Research*, 8, 659–673.

Rovira, A.D., Foster, R.C. and Martin, J.K. (1979), Note on terminology: origins, nature and nomenclature of the organic materials in the rhizosphere. In Harley, J.L. and Scott Russell, R. (eds), *The Soil–Root Interface*, pp. 1–14. Academic Press, London.

Rowell, D.L. (1981), Oxidation and reduction. In Greenland, D.J. and Hayes, M.H.B. (eds), *The Chemistry of Soil Processes*, pp. 401–461. John Wiley & Sons Ltd, Chichester.

Rowell, D.L., Martin, M.W. and Nye, P.H. (1967), The measurement of ion diffusion in soils. III: The effect of moisture content and soil solution concentration on the self-diffusion of ions in soils. *Journal of Soil Science*, 18, 204–222.

Royal Society of Chemistry (1983), *The Agrochemicals Handbook*. The Royal Society of Chemistry, The University, Nottingham NG7 2RD, England.

Runge, E.C.A. (1973), Soil development sequences and energy models. *Soil Science*, 115, 183–193.

Ruschel, A.P., Victoria, R.L., Salati, E. and Henis, Y. (1978), Nitrogen fixation in sugarcane (*Saccharum officinarum* L.). In Granhall, U. (ed.), *Environmental Role of Nitrogen-fixing Blue-Green Algae and Asymbiotic Bacteria. Ecology Bulletin*, 26, 297–303.

Russell, E.W. (1973), *Soil Conditions and Plant Growth*, 10th edn. Longman, London.

Russell, J.D., Guy, M., White, J.L., Bailey, G.W., Payne, W.R., Pope, J.O. and Teasley, J.I. (1968), Model of chemical degradation of triazines by montmorillonite. *Science*. 160, 1340–1342.

Russell, R.S. (1977), *Plant Root Systems: Their Function and Interaction with the Soil*. McGraw-Hill, London.

Russell, R.S., Erickson, J.B. and Adams, S.N. (1954), Isotopic equilibria between phosphates in soils and their significance in the assessment of fertility by tracer methods. *Journal of Soil Science*, 5, 85–105.

Ryden, J.C. (1981), N_2O exchange between a grassland soil and the atmosphere. *Nature (London)*, 292, 235–237.

Ryden, J.C. (1983), Denitrification loss from a grassland soil in the field receiving different rates of nitrogen as ammonium nitrate. *Journal of Soil Science*, 34, 355–365.

Ryden, J.C. and Lund, L.J. (1980a), Nature and extent of directly measured denitrification losses from some irrigated vegetable crop production units. *Soil Science Society of America Journal*, 44, 505–511.

Ryden, J.C. and Lund, L.J. (1980b), Nitrous oxide evolution from irrigated land. *Journal of Environmental Quality*, 9, 387–393.

Ryden, J.C., McLaughlin, J.R. and Syers, J.K. (1977), Mechanisms of phosphate sorption by soils and hydrous ferric oxide gel. *Journal of Soil Science*, 28, 72–79.

Ryle, G.J.A., Powell, C.E. and Gordon, A.J. (1979a), The respiratory costs of nitrogen fixation in soyabean, cowpea and white clover. I: Nitrogen fixation and the respiration of the nodulated root. *Journal of Experimental Botany*, 30, 135–144.

Ryle, G.J.A., Powell, C.E. and Gordon, A.J. (1979b), The respiratory costs of nitrogen fixation in soyabean, cowpea and white clover. II: Comparisons of the cost of nitrogen fixation and the utilisation of combined nitrogen. *Journal of Experimental Botany*, 30, 145–153.

Sadler, J.M. and Stewart, J.W.B. (1977), Labile residual fertilizer phosphorus in Chernozemic soils. I: Solubility and quantity/intensity studies. *Canadian Journal of Soil Science*, 57, 65–73.

Salonius, P.O. (1983), Effects of organic-mineral soil mixtures and increasing temperature on the respiration of coniferous raw humus material. *Canadian Journal of Forestry Research*, 13, 102–107.

Saltzman, S., Kliger, L. and Yaron, B. (1972), Adsorption-desorption of parathion as affected by soil organic matter. *Journal of Agricultural and Food Chemistry*, 20, 1224–1226.

Saltzman, S., Mingelgrin, U. and Yaron, B. (1976), Role of water in the hydrolysis of parathion and methyl parathion on kaolinite. *Journal of Agricultural Food Chemistry*, 24, 739–743.

Sample, E.C., and Taylor, A.W. (1964), Rapid, nondestructive method for estimating rate and extent of movement of phosphorus from fertilizer granules in soil. *Soil Science Society of America Proceedings*, 28, 296–297.

Sample, E.C., Soper, R.J., Racz, G.J. (1980), Reactions of phosphate fertilizers in soils. In Khasawneh, F.E., Sample, E.C. and Kamprath, E.J. (eds). *The*

Role of Phosphorus in Agriculture. American Society of Agronomy (CSSA/SSSA), pp. 263–310.

Sanchez, P.A. and Uehara, G. (1980), Management considerations for acid soils with high phosphorus fixation capacity. In Khasawneh, F.E., Sample, E.C. and Kamprath, E.J. (eds) *The Role of Phosphorus in Agriculture*, pp. 471–514. American Society of Agronomy, Madison, Wisconsin.

Sánchez Camazano, M. and Sánchez Martin, M.J. (1983), Factors influencing interactions of organophosphorus pesticides with montmorillonite. *Geoderma*, 29, 107–118.

Sanders, F.E. and Sheikh, N.A. (1983), The development of vesicular-arbuscular mycorrhizal infection in plant root systems. *Plant and Soil*, 71, 223–246.

Sanders, F.E. and Tinker, P.B. (1973), Phosphate flow into mycorrhizal roots. *Pesticide Science*, 4, 385–395.

Savant, N.K. and DeDatta, S.K. (1982), Nitrogen transformations in wetland rice soils. *Advances in Agronomy*, 35, 241–302.

Saxton, K.E., Schuman, G.E. and Burwell, R.E. (1977), Modeling nitrate movement and dissipation in fertilized soils. *Soil Science Society of America Journal*, 41, 265–271.

Schachtschabel, P. (1940), Untersuchungen über die Sorption der Tonmineralien und organischen Boden-kolloide. *Kolloid Beihefte*, 51, 199–243.

Schaff, B.E. and Skogley, E.O. (1982), Diffusion of potassium, calcium and magnesium in Bozeman silt loam as influenced by temperature and moisture. *Soil Science Society of America Journal*, 46, 521–524.

Schmidt, E.L. (1978), Legume symbiosis. A. Ecology of the legume root nodule bacteria. In Domergues, Y.R. and Krupa, S.V. (eds), *Interactions Between Non-pathogenic Soil Microorganisms and Plants*, 269–303. Elsevier Science, Amsterdam.

Schnitzer, M. and Desjardins, J.G. (1969), Chemical characteristics of a natural soil leachate from a humic podzol. *Canadian Journal of Soil Science*, 49, 151–158.

Schnitzer, M. and Hansen, E.H. (1970), Organo-metallic interactions in soils: 8. An evaluation of methods for the determination of stability constants of metal-fulvic acid complexes. *Soil Science*, 109, 333–340.

Schnitzer, M. and Khan, S.U. (1972), *Humic Substances in the Environment*. Marcel Dekker, New York.

Schnitzer, M. and Skinner, S.I.M. (1963), Organo-metallic interactions in soils. 1. Reactions between a number of metal ions and the organic matter of a podzol B_h horizon. *Soil Science*, 96, 86–93.

Schnitzer, M. and Skinner, S.I.M. (1965), Organo-metallic interactions in soils. 4. Carboxyl and hydroxyl groups in organic matter and metal retention. *Soil Science*, 99, 278–284.

Schofield, R.K. (1947), Calculation of surface areas from measurements of negative adsorption. *Nature (London)*, 160, 408–410.

Schubert, K.R. and Evans, H.J. (1977), The relation of hydrogen reactions to nitrogen fixation in nodulated symbionts. In Newton, W., Postgate J.R. and Rodriguez-Barrueco, C. (eds), *Recent Developments in Nitrogen Fixation*, pp. 469–485. Academic Press, London.

Schuman, G.E., Spomer, R.G. and Piest, R.F. (1973), Phosphorus losses from four agricultural watersheds on Missouri Valley loess. *Soil Science Society of America Proceedings*, 37, 424–427.

Schwertman, U. (1985), The effect of pedogenic environments on iron oxide minerals. *Advances in Soil Science*, 1, 171–200.

Schwintzer, C.R., Berry, A.M. and Disney, L.D. (1982), Seasonal patterns of root nodule growth, endophyte morphology, nitrogenase activity and shoot development in *Myrica gale*. *Canadian Journal of Botany*, 60, 746–757.

Scotter, D.R. (1978), Preferential solute movement through larger soil voids. I Some computations using simple theory. *Australian Journal of Soil Research*, 16, 257–267.

Selby, M.J. (1982), *Hillslope Materials and Processes*. Oxford University Press, Oxford.

Selim, H.M. (1978), Transport of reactive solutes during transient, unsaturated water flow in multilayered soils. *Soil Science*, 126, 127–135.

Senesi, N. and Testini, C. (1980), Adsorption of some nitrogenated herbicides by soil humic acids. *Soil Science*, 130, 314–320.

Sethunathan, N. (1973), Organic matter and parathion degradation in flooded soil. *Soil Biology and Biochemistry*, 5, 641–644.

Shaler, T.A. and Klécka, G.M. (1986), Effets of dissolved oxygen concentration on biodegradation of 2,4-D. *Applied and Environmental Microbiology*, 51, 950–955.

Sharpe, R.R., Boswell, F.C. and Hargrove, W.L. (1986), Phosphorus fertilization and tillage effect on dinitrogen fixation in soybeans. *Plant and Soil*, 96, 31–44.

Sharpley, A.N. (1980), Phosphorus cycling in unfertilized and fertilized agricultural soils. *Soil Science Society of America Journal*, 49, 905–911.

Sharpley, A.N. (1985), Phosphorus cycling in unfertilised and fertilised agricultural soils. *Soil Science Society of America Journal*, 49, 905–911.

Sharpley, A.N. and Syers, J.K. (1979), Phosphorus inputs into a stream draining an agricultural watershed. II Amounts and relative significance of runoff types. *Water, Air, Soil Pollution*, 9, 417–428.

Sharpley, A.N., Menzel, R.G., Smith, S.J., Rhoades, E.D. and Olness, A.E. (1981), The sorption of soluble phosphorus by soil material during transport in runoff from cropped and grassed watersheds. *Journal Environmental Quality*, 10, 211–215.

Shin, Y.O., Chodan, J.J. and Wolcott, A.R. (1970), Adsorption of DDT by soils, soil fractions and biological materials. *Journal of Agricultural and Food Chemistry*, 18, 1129–1133.

Sibbesen, E. (1981), Some new equations to describe phosphate sorption by soils. *Journal of Soil Science*, 32, 67–74.

Sidiras, N., Henklain, J.C. and Derpsch, R. (1982), Comparison of three different tillage systems with respect to aggregate stability; the soil water conservation and the yields of soybean and wheat on an oxisol. In *Ninth ISTRO Conference, Yugoslavia*, pp. 537–544. International Soil Tillage Research Organisation.

Sillen, L.G. (1967), Mater variables and activity scales. In *Equilibrium Concepts in Natural Water Systems, Advances in Chemistry Series No 67*, pp. 45–69, American Chemical Society, Washington DC.

Silvester, W.B. and Bennett, K.H. (1973), Acetylene reduction by roots and associated soil by New Zealand conifers. *Soil Biology Biochemistry*, 5, 171–179.

Silvester, W.B. and Smith, D.R. (1969), Nitrogen fixation by *Gunnera–Nostoc* symbiosis. *Nature (London)*, 224, 1321–

Silvester, W.B., Carter, D.A. and Sprent, J.I. (1979), Nitrogen input by *Lupinus* and *Coriaria* in *Pinus radiata* forest in New Zealand. In Gordon, J.C., Wheeler, C.T. and Perry, D.A. (eds), *Symbiotic Nitrogen Fixation in the Management of Temperate Forests. Proceedings of the Workshop, 1979*,

Corvallis, Oregon, pp. 253–265. National Science Foundation.

Simon-Sylvestre, G. (1974), Complexité des effets des pesticides sur l'equilibre biologique des sols et repercussion sur les cultures. *Agrochemica*, 18, 334–343.

Simon-Sylvestre, G., and Fournier, J.-C. (1979), Effects of pesticides on the soil microflora. *Advances in Agronomy*, 31, 1–92.

Sipos, S., Sipos, E., Dékány, I., Deér, A., Meisel, J. and Lakatos, B. (1978), Biopolymer–metal complex systems. II: Physical properties of humic substances and their metal complexes. *Acta Agronomica Academiae Sciéntiamuna Hungarica*, 27, 31–42.

Skeffington, R.A. and Bradshaw, A.D. (1980), Nitrogen fixation by plants grown on reclaimed china clay waste. *Journal of Applied Ecology*, 17, 469–477.

Skipper, H.D., Volk, V.V., Mortland, M.M. and Rainan, K.V. (1978), Hydrolysis of arazine on soil colloids. *Weed Science*, 26, 46–50.

Slade, P. (1966), The fate of paraquat applied to plants. *Weeds*, 6, 158–167.

Sleight, D.M., Sander, D.H. and Peterson, G.A. (1984), Effect of fertilizer phosphorus placement on the availability of phosphorus. *Soil Science Society of America Journal*, 48, 336–340.

Sloger, C., Bezdicek, D., Milberg, R. and Boonkerd, N. (1975), Seasonal and diurnal variations in $N_2(C_2H_2)$-fixing activity in field soybeans. In Stewart, W.D.P. (ed.), *Nitrogen Fixation by Free-Living Microorganisms. International Biology Programme No 6*, pp. 271–284. Cambridge University Press, Cambridge.

Small, J.G. and Leonard, O.A. (1969), Translocation of ^{14}C-labelled photosynthate in nodulated legumes as influenced by nitrate-nitrogen. *American Journal of Botany*, 56, 187–194.

Smettem, K.R.J. (1986), Solute movement in soils. In Trudgill, S.T. (ed.), *Solute Processes*, pp. 141–165. John Wiley and Sons Ltd, Chichester.

Smettem, K.R.J., Trudgill, S.T. and Pickles, A.M. (1983), Nitrate loss in soil drainage waters in relation to by passing flow and discharge on an arable site. *Journal of Soil Science*, 34, 499–509.

Smith, A.E. (1982), Herbicides and the soil environment in Canada. *Canadian Journal of Soil Science*, 62, 433–460.

Smith, A.E. and Walker, A. (1977), A quantitative study of asulam persistence in soil. *Pesticide Science*, 8, 449–456.

Smith, C.A.S., Coen, G.M. and Pluth, D.J. (1981), Podzolic soils with Luvisolic-like morphologies in the upper subalpine subzone of the Canadian Rockies: I: Stratigraphy and mineralogy. *Canadian Journal of Soil Science*, 61, 325–335.

Smith, C.J. and Patrick, W.H. Jr. (1983), Nitrous oxide emission as affected by alternate anaerobic and aerobic conditions from soil suspensions enriched with ammonium sulphate. *Soil Biology and Biochemistry*, 15, 693–697.

Smith, C.V. (1977), Work days from weather data. *Journal of the Proceedings of the Institute of Agricultural Engineers*, 32, 96–97.

Smith, D.G. and Lorimer J.W. (1964), An examination of the humic acids of *Sphagnum* peat. *Canadian Journal of Soil Science*, 44, 76–87.

Smith, J.L., Schnabel, R.R., McNeal, B.L. and Campbell, G.S. (1980), Potential errors in the first-order model for estimating soil nitrogen mineralisation potentials. *Soil Science Society of America Journal*, 44, 996–1000.

Smith, K.A. (1977), Soil aeration. *Soil Science*, 123, 284–291.

Smith, K.A. and Dowdell, R.J. (1974), Field studies of the soil atmosphere I. Relationship between ethylene, oxygen, soil moisture content and temperature. *Journal of Soil Science*, 25, 219–230.

Smith, K.A. and Restall, S.W.F. (1971), The occurrence of ethylene in anaerobic soil. *Journal of Soil Science 22*, 430–443.

Smith, K.A., Restall, S.W.F. and Robertson, P.D. (1969), Further studies on the occurrence of ethylene in soil and its effects on plant growth, *ARC Annual Report Letcombe Laboratory*, pp. 54–58.

Smith, M.S. and Weeraratna, C.S. (1974), The influence of some biologically active compounds on microbial activity of plant nutrients in soils. *Pesticide Science*, 5, 721–729.

Smith, O.L. (1979a), An analytical model of the decomposition of soil organic matter. *Soil Biology and Biochemistry*, 11, 585–606.

Smith, O.L. (1979b), Application of a model of the decomposition of soil organic matter. *Soil Biology and Biochemistry*, 11, 607–618.

Smith, O.L. (1982), Soil microbiology: a model of decomposition and nutrient cycling, *CRC Series: Mathematical Models in Microbiology*. CRC Press Inc., Florida.

Smith, S.C. and Rice, C.W. (1986), The role of microorganisms in the soil nitrogen cycle. In Mitchell, M.J. and Nakas, J.P. (eds), *Microfloral and Faunal Interactions in Natural and Agro-Ecosystems*, pp. 245–284. Martinus Nijhoff/ Dr. W. Junk, The Hague.

Smith, S.E. and Bowen, G.D. (1979), Soil temperature, mycorrhizal infection and nodulation of *Medicago trunculata* and *Trifolium subterraneum*. *Soil Biology and Biochemistry*, 11, 469–473.

Smith, S.J., Young, L.B. and Miller, G.E. (1977), Evaluation of soil nitrogen mineralisation potentials under modified field conditions. *Soil Science Society of America, Journal*, 41, 74–76.

Smith, W.H. (1971), *Influence of Artificial Defoliation on Exudates of Sugar Maple*, Pergamon Press, Oxford

Smith, W.H. (1976), Character and significance of forest tree root exudates. *Ecology*, 57, 324–331.

Smith, W.H. (1977), Tree root exudates and the forest soil ecosystem: exudate chemistry, biological significance and alteration by stress. In Marshall, J.K. (ed.), *The Below Ground Ecosystem: A Synthesis of Plant-Associated Processes. Range Science Dept Science Series No 26*, pp. 289–301, Colorado State University, Fort Collins.

Smith, W.W. (1962), Weathering of some Scottish basic igneous rocks with reference to soil formation. *Journal Soil Science*, 13, 202–215.

Snellgrove, R.C., Splittstoesser, W.E., Stribley, D.P. and Tinker, P.B. (1982), The distribution of carbon and the demand of the fungal symbiont in leek plants with vesicular-arbuscular mycorrhizas. *New Phytologist*, 92, 75–87.

Soane, B.D. and Pidgeon, J.D. (1975), Tillage requirements in relation to soil physical properties. *Soil Science*, 119(5) 376–384.

Soil Survey Staff (1975), *Soil Taxonomy. A Basic System of Soil Classification for Making and Interpreting Soil Surveys*, Agricultural Handbook No. 436, USDA. US Government Printing Office, Washington, DC.

Sorensen, L.H. (1974), Rate of decomposition of organic matter in soil as influenced by repeated air-drying-rewetting and repeated addition of organic material. *Soil Biology and Biochemistry*, 6, 287–292.

Soulas, G. (1982), Mathematical model for microbial degradation of pesticides in the soil. *Soil Biology and Biochemistry*, 14, 107–115.

Soulas, G., Codaccioni, P. and Fournier, J.C. (1983), Effect of cross-treatment on the subsequent breakdown of 2,4-D, MCPA and 2,4,5-T in the soil.

Behaviour of the degrading microbial populations. *Chemosphere*, 12, 1101–1106.

Sowden, F.J., MacLean, A.A. and Ross, E.J. (1978), Native clay-fixed ammonium content and the fixation of added ammonium of some soils of Eastern Canada. *Canadian Journal of Soil Science*, 58, 27–38.

Spoor, G. (1979), Soil type and workability. In Jarvis, M.G. and Mackney, D. (eds), *Soil Survey Applications. Soil Survey Technical Monograph 13*.

Spoor, G. and Godwin, R.J. (1979), Soil deformation and shear strength characteristics of some clay soils at different moisture contents. *Journal of Soil Science*, 30, 483–498.

Sposito, G. (1977), The Gapon and the Vanselow selectivity coefficients. *Soil Science Society of America Journal*, 41, 1025–1026.

Sposito, G. (1981a), *The Thermodynamics of Soil Solutions*. Clarendon Press, Oxford.

Sposito, G. (1981b), Cation exchange in soils: an historical and theoretical perspective. In Stelly, M. (ed.), *Chemistry in the Soil Environment*. ASA Special Publication No 40, p. 13–30, ASA/SSSA.

Sposito, G. (1982), On the use of the Langmuir equation in the interpretation of 'adsorption' phenomena. II: The 'two surface' Langmuir equation. *Soil Science Society of America Journal*, 46, 1147–1152.

Sprent, J.I. (1972), The effects of water stress on nitrogen-fixing root nodules. II: Effects on the fine structure of detached soybean nodules. *New Phytologist*, 71, 443–450.

Sprent, J.I. (1976), Nitrogen fixation by legumes subjected to water and light stresses. In Nutman, P.S. (ed.), *Symbiotic Nitrogen Fixation in Plants. International Biology Programme*. Cambridge University Press, Cambridge.

Sprent, J.I. and Silvester, W.B. (1973), Nitrogen fixation by *Lupinus arboreus* grown in the open and under different aged stands of *Pinus radiata*. *New Phytologist*, 72, 991–1003.

Springett, J.A. (1983), The effect of five species of earthworm on some soil properties. *Journal of Applied Ecology*, 20, 865–872.

Stanford, G. and Epstein, E. (1974), Nitrogen mineralisation–water relations in soil. *Soil Science Society of America Proceedings*, 38, 103–106.

Stanford, G., Frere, M.H. and Schwaninger, D.H. (1973), Temperature coefficient of soil nitrogen mineralisation. *Soil Science*, 115, 321–323.

Stanford, G., Frere, M.H. and van der Pol, R.A. (1975), Effect of fluctuating temperatures on soil nitrogen mineralisation. *Soil Science*, 119, 222–226.

Stanford, G. and Pierre, W.H. (1947), The relation of potassium fixation to ammonium fixation. *Soil Science Society of America Proceedings*, 11, 155–160.

Stanford, G. and Smith, S.J. (1972), Nitrogen mineralisation potentials of soils. *Soil Science Society of America Proceedings*, 36, 465–472.

Steichen, J.M. (1984), Infiltration and random roughness of a tilled and untilled claypan soil. *Soil and Tillage Research*, 4, 251–262.

Steubing, L. (1977), Soil microbial activity under beech and spruce stands. *Naturaliste Canadien*, 104, 143–150.

Stevenson, F.J. (1967), Organic acids in soil. In McLaren, A.D. and Peterson, G.H. (eds). *Soil Biochemistry*, 119–146.

Stevenson, F.J. (1982) (ed.), *Nitrogen in Agricultural Soils*. No 22 in the Agronomy Series. American Society of Agronomy (CSSA/SSSA).

Stewart, W.D.P. (1966), *Nitrogen Fixation in Plants*, Athlone Press.

Stewart, W.D.P. and Rosswall, T. (1982) (eds), The nitrogen cycle. *Philosophical*

Transactions of the Royal Society of London, B, 296, 299–576.

Stojanovic, B.J., Kennedy, M.V. and Shuman, Jr., F.L. (1972), Edaphic aspects of the disposal of unused pesticides, pesticide wastes and pesticide containers. *Journal of Environmental Quality*, 1, 54–62.

Stribley, D.P., Tinker, P.B. and Rayner, J.H. (1980), Relations of internal phosphorus concentration and plant weight in plants infected by vesicular-arbuscular mycorrhizas. *New Phytologist*, 86, 261–266.

Stribley, D.P., Tinker, P.B. and Snellgrove, R.C. (1980), Effect of vesicular-arbuscular mycorrhizal fungi on the relations of plant growth, internal phosporus concentration and soil phosphate analyses. *Journal of Soil Science*, 31, 655–672.

Strutt Report (1970), *Modern Farming and the Soil. Report of the Agricultural Advisory Council on Soil Structure and Soil Fertility*. Agricultural Advisory Council/HMSO, London.

Stumm, W. and Morgan, J.J. (1970), *Aquatic Chemistry. An Introduction Emphasizing Chemical Equilibria in Natural Waters*. Wiley-Interscience, New York.

Stumpe, J.N., Vlek, P.L.G. and Lindsay, W.L. (1984), Ammonia volatalization from urea and urea phosphates in calcareous soils. *Soil Science Society of America Journal*, 48, 921–927.

Summers, L.A. (1980), *The Bipyridinium Herbicides*. Academic Press, New York.

Swaby, R.J. (1949), The relationship between micro-organisms and soil aggregation. *Journal of General Microbiology*, 3, 236–254.

Swift, M.J. (1977), The roles of fungi and animals in the immobilisation and release of nutrient elements from decomposing branch wood. In Lohm, U. and Persson, T. (eds), *Soil Organisms as Components of Ecosystems. Ecology Bulletin (Stockholm)*, 25, 193–202.

Swift, M.J., Heal, O.W. and Anderson, J.M. (1979), *Decomposition in Terrestrial Ecosystems. Studies in Ecology*, Vol. 5, Blackwell Scientific, Oxford.

Swift, R.S. (1980), The effect of adsorbed organic materials on the cation exchange of clay minerals. In Banin, A. and Kafkafi, U. (eds), *Agrochemicals in Soils*, pp. 123–129.

Szegi, J. (1972), Effect of a few herbicides on the decomposition of cellulose. *Symposia Biology Hungary*, 11, 349–354.

Tabatabai, M.A. (1977), Effects of trace elements on urease activity in soils. *Soil Biology and Biochemistry*, 9, 9–13.

Tabatabai, M.A. and Hanway, J.J. (1969), Potassium supplying power of Iowa soils at their 'minimal' levels of exchangeable potassium. *Soil Science Society of America Proceedings*, 33, 105–109.

Talibudeen, O. (1981), Cation exchange in soils. In Greenland, D.J. and Hayes, M.H.B. (eds), *The Chemistry of Soil Processes*, pp. 115–177, John Wiley & Sons Ltd, Chichester.

Talley, S.N., Lim, E. and Rains, D.W. (1977), Application of *Azolla* in crop production. In Lyons, J.M., Valentine, R.C., Phillips, D.A., Rains, D.W. and Huffaker, R.C. (eds), *Genetic Engineering for Nitrogen Fixation*. Plenum Press, New York.

Talley, S.N. and Rains, D.W. (1980), *Azolla filiculoides* LAM as a fallow season green manure for rice in a temperate climate. *Agronomy Journal*, 72, 11–18.

Tan, K.H. (1977), High and low molecular weight fractions of humic and fulvic acids. *Plant and Soil*, 48, 89–101.

Tan, K.H. and Dowling, P.S. (1984), Effect of organic matter on CEC due to permanent and variable charges in selected temperate region soils. *Geoderma*, 32, 89–101.

Tardy, Y., Bocquier, G., Paquet, H. and Millot, G. (1973), Formation of clay from granite and its distribution in relation to climate and topography, *Geoderma* 10, 271–284.

Taslim, H. and Verstraeten, L.M.J. (1977), Mineralisation of slow-release N-compounds in waterlogged conditions. *Zeitschrift für Pflanzenernahrung und Bodenkunde*, 140, 183–191.

Terman, G.L. (1979), Volatilisation losses of nitrogen as ammonia from surface-applied fertilizers, organic amendments and crop residues. *Advances in Agronomy*, 31, 189–223.

Terman, G.L., Parr, J.F. and Allen, S.E. (1968), Recovery of nitrogen by corn from solid fertilizers and solutions. *Journal of Agricultural and Food Chemistry*, 16, 685–690.

Thomas, B. (1967), *The effect of leaf leachate on the degradation of some minerals and on the movement of ions in soil*. PhD thesis, University of Wales.

Thomlinson, T.E. (1970), Nutrient losses from agricultural land. *Outlook on Agriculture*, 3, 272–278.

Thomlinson, T.E. (1974), Soil structural aspects of direct drilling. *Transactions of the Tenth International Congress of Soil Science*, Vol. I, 203–213.

Thompson, A.R. and Edwards, C.A. (1974), Effects of pesticides on nontarget invertebrates in freshwater and soil. In Guenzi, W.D. (ed.), *Pesticides in Soil and Water*. pp. 341–386. Soil Science Society of America Inc., Madison.

Thornthwaite, C.W. (1931), The climates of North America according to a new classification. *Geographical Review*, 21, 633–656.

Tiessen, H. and Stewart, J.W.B. (1983), Particle-size fractions and their use in studies of soil organic matter. II: Cultivation effects on organic matter composition in size fractions. *Soil Science Society of America Journal*, 47, 509–514.

Tinker, P.B. (1975). The soil chemistry of phosphorus and mycorrhizal effects on plant growth. In Sanders, F.E., Mosse, B. and Tinker, P.B. (eds), *Endomycorrhizae*, pp. 353–372. Academic Press, London.

Tinker, P.B. (1978), Effects of vesicular-arbuscular on plant nutrition and growth. *Physiologic Vegetable*, 16, 743–751.

Tinker. P.B. (1980), Root–soil interactions in crop plants. In Tinker, P.B. (ed.), *Soils and Agriculture*, 1–34. *Critical Reports on Applied Chemistry*, vol. 2. Blackwell Scientific, Oxford.

Tinker, P.B. (1984), The role of microorganisms in mediating and facilitating the uptake of plant nutrients from soil. *Plant and Soil*, 76, 77–91.

Tinker, P.H.B. (1976) Transport of water to plant roots in soil. *Philosophical Transactions of the Royal Society, London*, B, 273, 445.

Tisdall, J.M. and Oades, J.M. (1979), Stabilisation of soil aggregates by the root system of ryegrass. *Australian Journal of Soil Research*, 17, 429–441.

Tisdall, J.M. and Oades, J.M. (1982), Organic matter and water-stable aggregates in soil. *Journal of Soil Science*, 32, 335–350.

Todd, R.L., Meyer, R.D. and Waide, J.B. (1978), Nitrogen fixation in a deciduous forest in the South Eastern United States. In Granhall, U. (ed.), *Environmental Role of Nitrogen-Fixing Blue-Green Algae and Asymbiotic Bacteria*. Ecology Bulletin, No 26, pp. 172–177.

Tollner, E.W., Hargrove, W.L. and Langdale, G.W. (1984), Influence of conventional and no-till practices on soil physical properties in the southern Piedmont. *Journal of Soil and Water Conservation*, 39(1), 73–76.

Torstensson, L. (1985), Behaviour of glyphosate in soils and its degradation. In Grossbard, E. and Atkinson, D. (eds), *The Herbicide Glyphosate*, pp. 137–150. Butterworths, London.

Torstensson, N.T.L. (1975), Degradation of 2,4-D and MCPA in soils of low pH. In Coulston, F. and Korte, F. (eds), *Environmental Quality and Safety*. Supple. Vol. III. pp. 262–265. Georg Thieme, Stuttgart.

Torstensson, N.T.L. and Rosswall, T. (1977), The effect of 20 year's application of 2,4-D and MCPA on the soil-flora. *International Symposium on Interaction of Soil Microflora and Environmental Pollution*, Vol. 1, 170–176.

Tsukano, Y. (1986), Transformations of selective pesticides in flooded rice-field soils. *Journal of Contaminant Hydrology*, 1, 47–63.

Tu, C.M. (1970), Effects of four organochlorine insecticides on microbial activities in soil. *Applied Microbiology*, 19, 479–484.

Tu, C.M. and Bollen, W.B. (1968), Effect of paraquat on microbial activities in soils. *Weed Research*, 8, 28–37.

Turchenek, L.W. and Oades, J.M. (1978), Organo-clay particles in soils. In Emerson, W.W. and Dexter, A.R. (eds), *Modification of Soil Structure*, 137–144, John Wiley, Chichester.

Turner, F.T. and Patrick, W.H. Jr. (1968), Chemical changes in waterlogged soils as a result of oxygen depletion. *Transactions of the Ninth International Congress of Soil Science*, 4, 53–65.

Turner, J. (1977), Effect of nitrogen availability on nitrogen cycling in a Douglas fir stand. *Forest Science*, 23, 307–316.

Tusneem, M.E. and Patrick, W.H. Jr. (1971), *Nitrogen Transformations in Waterlogged Soil. Bulletin No 657*, Louisana State University, Department of Agronomy and Agriculture Experimental Station, 75pp.

Usoroh, N.J. and Hance, R.J. (1974), The effect of temperature and water content on the rate of degradation of the herbicide linuron in soil. *Weed Research*, 14, 19–21.

van Breemen, N. and Brinkman, R. (1978), Chemical equilibria and soil formation. In Bolt, G.H. and Bruggenwert, M.G.M. (eds), *Soil Chemistry: Vol A: Basic elements*, pp. 141–170.

van Breemen, N., Driscoll, C.T. and Mulder, J. (1984), Acidic deposition and internal proton sources in acidification of soils and waters. *Nature (London)*, 307, 599–604.

van Cleve, K. (1974), Organic matter quality in relation to decomposition. In Holding, A.J. *et al.* (eds), *Soil Organisms and Decomposition in Tundra*. IBP Tundra Biome, pp. 311–324.

Vancura, V. and Garcia, J.L. (1969), Root exudates of reversibly wilted millet plants (*Panicum miliaceum* L.). *Oecologia Plantarum*, 4, 93–98.

Vancura, V., Prikryl, Z., Kalachova, L. and Wurst, M. (1977), Some quantitative aspects of root exudation. In Lhom, U. and Persson, T. (eds), *Soil Organisms as Components of Ecosystems. Ecology Bulletin (Stockholm)*, 25, 381–386.

van Dijk, C. (1979), Endophyte distribution in the soil. In Gordon, J.C., Wheeler, C.T. and Perry, D.A. (eds), *Symbiotic Nitrogen Fixation in the Management of Temperate Forests*, 84–94. *Proceedings of the Workshop, Corvallis, Oregon, April 1979*, pp. 84–94. National Science Foundation.

van Duin, R.H.A. (1956), On the influence of tillage on conduction of heat, diffusion of air and infiltration of water in soil. *Versl. Landbouwkunde Onderz.*, No. 62.7.

van Genuchten, M.Th. and Cleary, R.W. (1979), Movement of solutes in soil: computer-simulated and laboratory results. In Bolt, G.H. (ed.) *Soil Chemistry*.

B: *Physico-Chemical Models*, pp. 349–386. Elsevier, Amsterdam.

van Genuchten, M.Th. and Wierenga, P.J. (1974), An evaluation of kinetic and equilibrium equations for the prediction of pesticide movement through porous media. *Soil Science Society of America Journal*, 38, 29–35.

van Genuchten, M.T. and Wierenga, P.J. (1976), Mass transfer in sorbing porous media. I: Analytical solutions. *Soil Science Society of America Proceedings*, 40, 473–480.

van Olphen, H. (1977), *An Introduction to Clay Colloid Chemistry For Clay Technologists, Geologists and Soil Scientists*, 2nd edn. John Wiley and Sons, Chichester.

van Ouwerkerk, C. and Boone, F.R. (1970), Soil physical aspects of zero-tillage experiments. *Netherlands Journal of Agricultural Science*, 18, 247–261.

van der Pol, R.M., Wierenga, P.J. and Nielsen, D.R. (1977), Solute movement in a field soil. *Soil Science Society of America Journal*, 41, 10–13.

van Rhee, J.A. (1972), *Landbouw en Plantenziekten*. Publ. Lanbd. Voorl Dienst Verlag, pp. 93–102.

van Schreven, D.A. and Sieben, W.H. (1972), The effect of storage of soils under waterlogged conditions upon subsequent mineralisation of nitrogen, nitrification and fixation of ammonia. *Plant and Soil*, 37, 245–253.

van Schuylenborgh, J. (1965), The formation of sesquioxides in soils. In Hallsworth, E.G. and Crawford, D.V. (eds), *Experimental Pedology*, pp. 113–125.

Vanselow, A.P. (1932), Equilibria of the base-exchange reactions of dentomites, permutites, soil colloids and zeolites. *Soil Science*, 33, 95–113.

van Wijk, W.R. and de Vries, D.A. (1963), Periodic temperature variations in a homogeneous soil. In van Wijk, W.R. (ed.), *Physics of Plant Environment*, pp. 102–143. North-Holland Publishing Co., Amsterdam.

Veneman, P.L.M., Murray, J.R. and Baker, J.H. (1983), Spatial distribution of pesticide residues in a former apple orchard. *Journal of Environmental Quality*, 12, 101–104.

Ventura, W.B. and Yoshida, T. (1977), Ammonia volatilisation from a flooded tropical soil. *Plant and Soil*, 46, 521–523.

Vlassak, K., Paul, E.A. and Harris, R.E. (1973), Assessment of biological nitrogen fixation in grassland and associated sites. *Plant and Soil*, 38, 637–649.

Vogt, K.A., Grier, C.C., Meier, C.E. and Keyes, M.R. (1983), Organic matter and nutrient dynamics in forest floors of young and mature *Abies amabilis* ecosystems in western Washington, as affected by fine-root input. *Ecology Monographs*, 53, 139–157.

Vogt, K.A., Grier, C.C. and Vogt, D.J. (1986), Production, turnover and nutrient dynamics of above and below ground detritus of world forests. *Advances in Ecology Research*, 15, 303–377.

Volk, G.M. (1966), Efficiency of urea as affected by method of application, soil moisture and lime. *Agronomy Journal*, 58, 249–252.

Voorhees, W.B. (1983), Relative effectiveness of tillage and natural forces in alleviating wheel-induced soil compaction. *Soil Science Society of America Journal*, 47, 129–133.

Vyn, T.J., Daynard, T.B. and Ketcheson, J.W. (1982), Effect of reduced tillage systems on soil physical properties and maize grain yield in Ontario. *Proceedings of the Ninth International Conference on Soil Tillage Research Organization*, pp. 151–161.

Wadisirisuk, P. and Weaver, R.W. (1985), Importance of bacteroid number in nodules and effective nodule mass to dinitrogen fixation by cowpeas. *Plant and Soil*, 87, 223–231.

Wagar, B.I., Stewart, J.W.B. and Moir, J.O. (1986), Changes with lime in the form and availability of residual fertilizer phosphorus on Chernozemic soils. *Canadian Journal of Soil Science*, 66, 105–119.

Wagenet, R.J. (1983), Principles of salt movement in soils. In Nelson, D.W., Elrick, D.E. and Tanji, K.K. (eds), *Chemical Mobility and Reactivity in Soil Systems*, pp. 123–140. *SSSA Special Publication No 11*. SSSA and American Society Agronomy.

Waksman, S.A. (1936), *Humus*. Baillière, Tindall & Cox, London.

Walker, A. (1974), A simulation model for prediction of herbicide persistence. *Journal of Environmental Quality*, 3, 396–401.

Walker, A. (1976), Simulation of herbicide persistence in soil. I: Simazine and prometryne; II: Simazine and linuron in long-term experiments; III: Propyzamide in different soil types. *Pesticide Science*, 7, 41–64.

Walker, A. and Barnes, A. (1981), Simulation of herbicide persistence in soil; a revised computer model. *Pesticide Science*, 12, 123–132.

Walker, A. and Crawford, D.V. (1968), The role of organic matter in adsorption of the triazine herbicides by soils. In *Isotopes and Radiation in Adsorption of the Triazine Herbicides by Soils. Proceedings of the Second Symposium on International Atomic Energy Agency, Vienna*, pp. 91–108.

Walker, A. and Smith, A.E. (1979), Persistence of 2,4,5-T in a heavy clay soil. *Pesticide Science*, 10, 151–157.

Walling, D.E. and Webb, B.W. (1981), Water quality. In Lewin, J. (ed.), *British Rivers*, pp. 126–169. George Allen & Unwin.

Wang, C. and McKeague, J.A. (1982), Illuviated clay in sandy podzolic soils of New Brunswick. *Canadian Journal of Soil Science*, 62, 79–89.

Wang, C., Ross, G.J. and Rees, H.W. (1980), The landform and the characteristics of a residual podzol developed from granite in Northern Central New Brunswick. *Geoderma*.

Ward, T.M. and Weber, J.B. (1968), Aqueous solubility of alkylamino-s-triazines as a function of pH and molecular structure. *Journal of Agricultural and Food Chemistry*, 16, 959–961.

Warrick, A.W., Mullen, G.J. and Nielsen, D.E. (1977), Scaling field-measured soil hydraulic properties using a similar media concept. *Water Resources Research*, 13, 355–361.

Watanabe, I. and Roger, P.A. (1984), Nitrogen fixation in wetland rice field. In Subba Rao, N.S. (ed.), *Current Developments in Biological Nitrogen Fixation*, pp. 237–276. Edward Arnold, London.

Wauchope, R.D. (1978), The pesticide content of surface water draining from agricultural fields – a review. *Journal of Environmental Quality*, 7, 459–472.

Wauchope, R.D. and Leonard, R.A. (1980), Maximum pesticide concentrations in agricultural runoff: a semiempirical prediction formula. *Journal of Environmental Quality*, 9, 665–672.

Waylen, M.J. (1979), Chemical weathering in a drainage basin underlain by old red sandstone. *Earth Surface Processes*, 4, 167–178.

Weber, J.B. (1970), Adsorption of s-triazines by montmorillonite as a function of pH and molecular structure. *Soil Science Society of America Proceedings*, 34, 401–404.

Weber, J.B. (1972), Interaction of organic pesticides with particulate matter in aquatic and soil systems. In Gould, R.F. (ed.), *Fate of Organic Pesticides in the Aquatic Environment*, pp. 55–120. *Advances in Chemistry Series, No. 111*, American Chemical Society, Washington, DC.

Weber, J.B., Perry, P.W. and Upchurch, R.P. (1965), The influence of temperature on the adsorption of paraquat, diquat, 2,4,-D and prometone by clays, charcoal and an anion-exchange resin. *Soil Science Society of America Proceedings*, 29, 678–688.

Webster, C.P. and Dowdell, R.J. (1982), Nitrous oxide emission from permanent grass swards. *Journal of Science and Food Agriculture*, 33, 227–230.

Webster, C.P. and Dowdell, R.J. (1984), Effect of different rainfall regimes on the fate of nitrogen fertilizer applied to permanent grassland: summary of results and final balance sheet. *ARC Letcombe Laboratory Annual Report, 1983*, pp. 63–65.

Webster, C.P. and Dowdell, R.J. (1985), A lysimeter study of the fate of nitrogen applied to perennial ryegrass swards: soil analyses and the final balance sheet. *Journal of Soil Science*, 36, 605–611.

Webster, J.R. (1962), The composition of wet heath vegetation in relation to aeration of the groundwater and soil. I Field studies of groundwater and soil aeration in several communities. *Journal of Ecology*, 50, 619–638.

Webster, R. (1985), Quantitative spatial analysis of soil in the field. *Advances in Soil Science*, 3, 1–70.

Weed, S.B. and Weber, J.B. (1974), Pesticide–organic matter interactions. In Guenzi, W.D. (ed.), *Pesticides in Soil and Water*, pp. 39–66. Soil Science Society of America Inc., Madison, Wisconsin.

Weed, S.M. and Weber, J.B. (1969), The effect of cation exchange capacity on the retention of diquat^{2+} and paraquat^{2+} by three layer clay minerals: 1: Adsorption and release. *Soil Science Society of America Proceedings*, 33, 379–382.

Wehtje, G., Miekle, L.N., Leavitt, J.R.C. and Schepers, J.S. (1984), Leaching of atrazine in the root zone of an alluvial soil in Nebraska. *Journal of Environmental Quality*, 13, 507–513.

Welch, L.F. and Scott, A.D. (1960), Nitrification of fixed ammonium in clay minerals as affected by added potassium. *Soil Science*, 90, 79–85.

Wendt, R.C. and Corey, R.B. (1980), Phosphorus variations in surface runoff from agricultural lands as a function of land use. *Journal of Environmental Quality*, 9, 130–136.

Wheelan, A.M. and Alexander, M. (1986), Effects of low pH and high Al, Mn and Fe levels on the survival of *Rhizobium trifolii* and nodulation of subterranean clover. *Plant and Soil*, 92, 363–371.

Wheeler, C.T. and McLaughlin, M.E. (1979), Environmental modulation of niutrogen fixation in actinomycete nodulated plants. In Gordon, J.C., Wheeler, C.T. and Perry, D.A. (eds), *Symbiotic Nitrogen Fixation in the Management of Temperate Forests*, pp. 124–139. *Proceedings of the Workshop, Corvallis, Oregon, April 1979*. National Science Foundation.

White, D.C. (1967), Absence of nodule formation on *Ceanothus cuneatus* in serpentine soils. *Nature (London)*, 215, 875–877.

White, R.E. (1979), *Introduction to the Principles and Practice of Soil Science*. Blackwell Scientific, Oxford.

White, R.E. (1980), Retention and release of phosphate by soil and soil constituents. In Tinker, P.B. (ed.), *Soils and Agriculture, Critical Reports on Applied Chemistry*, Vol 2, pp. 71–114. Blackwell Scientific, Oxford.

White, R.E. (1985), The influence of macropores on the transport of dissolved and suspended matter through soil. *Advances in Soil Science*, 3, 95–120.

White, R.E., Dyson, J.S., Gerstl, Z. and Yaron, B. (1986), Leaching of herbicides through undisturbed cores of a structured clay soil. *Soil Science*

Society of America Journal, 50, 277–283.

White, R.E. and Taylor, A.W. (1977), Effect of pH on phosphate adsorption and isotopic exchange in acid soils at low and high additions of soluble phosphate. *Journal of Soil Science,* 28, 48–61.

White, R.E., Thomas, G.W. and Smith, M.S. (1984), Modelling water flow through undisturbed soil cores using a transfer function model devised from ^{3}HOH and Cl transport. *Journal of Soil Science,* 35, 159–168.

Whittaker, R.H. and Marks, P.L. (1975), Methods of assessing terrestrial productivity. In Leith, H. and Whittaker, R.H. (eds), *Primary Productivity of the Biosphere. Ecological Studies No. 14,* pp. 55–118.

Whittingham, J. and Read, D.J. (1982), Vesicular-arbuscular mycorrhizas in natural vegetation systems. III: Nutrient transfers between plants with mycorrhizal interconnections. *New Phytologist,* 90, 277–284.

Widdowson, F.V. and Penny, A. (1973), Yields and N, P and K contents of the crops grown in the Rothamsted Reference Experiment, 1956–70. *Report of the Rothamsted Experimental Station for 1972, Part 2,* pp. 111–130.

Wierenga, P.J. and de Wit, C.T. (1970), Simulation of heat flow in soils. *Soil Science Society of America Proceedings,* 34, 845–854.

Wierenga, P.J., Nielson, D.R. and Hagan, R.M. (1969), Thermal properties of a soil based upon field and laboratory measurements. *Soil Science Society of America Proceedings,* 33, 354–360.

Wiklander, G. (1974), Leaching of plant nutrients in soils. I: General principles. *Acta Agriculturae Scandinavica,* 24, 349–356.

Wiklander, L. (1955), Cation and anion exchange phenomena. In Bear, F.E. (ed.), *Chemistry of the Soil,* 163–205. Reinhold, New York.

Wild, A. (1972), Nitrate leaching under bare fallow at a site in northern Nigeria. *Journal of Soil Science,* 23, 315–324.

Wild, A. (1981), Mass flow and diffusion. In Greenland, D.J. and Hayes, M.H.B. (eds), *The Chemistry of Soil Processes,* pp. 37–80.

Wild, A. and Cameron, K.C. (1980), Soil nitrogen and nitrate leaching. In Tinker, P.B. (ed.), *Soils and Agriculture,* pp. 35–70. *Critical Reports on Applied Chemistry Vol 2.* Blackwell Scientific, Oxford.

Wilding, L.P. and Rutledge, E.M. (1966), Cation exchange capacity as a function of organic matter, total clay and various clay fractions in a soil toposequence. *Soil Science Society of America Proceedings,* 30, 782–785.

Wilkinson, B. (1975), Soil types and direct drilling – a provisional assessment. *Outlook on Agriculture,* 8, 233–235.

Williams, A.G., Ternan, L. and Kent, M. (1986), Some observations on the chemical weathering of the Dartmoor Granite. *Earth Surface Processes and Landforms,* 11, 557–574.

Williams, B.L. (1972), Nitrogen mineralisation and organic matter decomposition in Scots pine humus. *Forestry,* 45, 177–188.

Williams, R.J.B. (1971), The chemical composition of water from land drains at Saxmundham and Woburn, and the influence of rainfall upon nutrient losses. *Report of the Rothamsted Experimental Station for 1970 Part 2,* pp. 36–67.

Williams, W.A., Jones, M.B. and Delwiche, C.C. (1977), Clover N-fixation measurements by total-N difference and ^{15}N A-values in lysimeters. *Agronomy Journal,* 69, 1023–1024.

Willis, G.H., McDowell, L.L., Harper, L.A., Southwick, L.M. and Smith, S. (1983), Seasonal disappearance and volatilisation of toxaphene and DDT from a cotton field. *Journal of Environmental Quality,* 12, 80–85.

Wollast, R. (1967), Kinetics of the alteration of K-feldspar in buffered solutions at low temperature. *Geochemica Cosmochemica Acta*, 31, 635–648.

Woods, F.W. and Brock, K. (1964), Interspecific transfer of ^{45}Ca and ^{32}P by root systems. *Ecology*, 45, 886–889.

Wright, W.L. and Warren, G.F. (1965), Photochemical decomposition of trifluralin. *Weeds*, 13, 329–331.

Wright, W.R. and Foss, J.E. (1972), Contributions of clay and organic matter to the cation exchange capacity of Maryland soils. *Soil Science Society of America Proceedings*, 36, 115–118.

Yamanaka, T. (1983), Effect of paraquat on growth of *Nitrosomonas europaea* and *Nitrobacter agilis*. *Plant and Cell Physiology*, 24(7), 1349–1352.

Yaron, B., Gerstl, Z. and Spencer, W.F. (1985), Behaviour of herbicides in irrigated soils. *Advances in Soil Science*, 3, 121–211.

Yeomans, J.C. and Bremner, J.M. (1985a), Denitrification in soil: effects of herbicides. *Soil Biology and Biochemistry*, 17, 447–452.

Yeomans, J.C. and Bremner, J.M. (1985b), Denitrification in soil: effects of insecticides and fungicides. *Soil Biology and Biochemistry*, 17, 453–456.

Yoshida, T. (1975), Microbial metabolism of flooded soil. In Paul, E.A. and McLaren, A.D. (eds) *Soil Biochemistry*, Vol 3, pp. 83–122. Marcel Dekker, New York.

Young, B.R., Newhook, F.J. and Allen, R.N. (1977), Ethanol in the rhizosphere of seedlings of *Lupinus augustifolium* L., *New Zealand Journal of Botany*, 15, 189–191.

Young, C.P. and Gray, E.M. (1978), *Nitrate in Groundwater*, Water Resources Centres Technical Report No 69.

Youngberg, C.T. and Wollum, A.G. (1976), Nitrogen accretion in developing *Ceanothus velutinus* soils. *Soil Science Society of America Journal*, 40, 109–112.

Zavitkovski, J. and Newton, M. (1968), Effect of organic matter and combined nitrogen on nodulation and nitrogen fixation in red alder. In Trappe, J.M., Franklin, J.F., Tarrant, R.F. and Hansen, G.M. (eds), *Biology of Alder*, pp. 209–223. Pacific NW Forest and Range Experimental Station, USDA Forest Service, Portland.

Zöttl, H. (1960), Dynamik der Stichstoffmineralisation im organischen Waldbodenmaterial. III: pH-Wert und Mineralstickstoff-Nachlieferung. *Plant and Soil*, 8, 207–223.

Index